Multilevel and Longitudin

Using Stata

Volume I: Continuous Responses

Third Edition

M000310219

Multilevel and Longitudinal Modeling Using Stata

Volume I: Continuous Responses

Third Edition

SOPHIA RABE-HESKETH
University of California–Berkeley
Institute of Education, University of London

ANDERS SKRONDAL
Norwegian Institute of Public Health

A Stata Press Publication
StataCorp LP
College Station, Texas

Published by Stata Press, 4905 Lakeway Drive, College Station, Texas 77845
Typeset in LaTeX 2_ε
Printed in the United States of America

10 9 8 7 6 5 4 3 2 1

ISBN-10: 1-59718-108-0 (volumes I and II)
ISBN-10: 1-59718-103-X (volume I)
ISBN-10: 1-59718-104-8 (volume II)
ISBN-13: 978-1-59718-108-2 (volumes I and II)
ISBN-13: 978-1-59718-103-7 (volume I)
ISBN-13: 978-1-59718-104-4 (volume II)

To my children Astrid and Inge
Anders Skrondal

To Simon
Sophia Rabe-Hesketh

Contents

Tables

Figures

Preface

This book is about applied multilevel and longitudinal modeling. Other terms for multilevel models include hierarchical models, random-effects or random-coefficient models, mixed-effects models, or simply mixed models. Longitudinal data are also referred to as panel data, repeated measures, or cross-sectional time series. A popular type of multilevel model for longitudinal data is the growth-curve model.

The common theme of this book is regression modeling when data are clustered in some way. In cross-sectional settings, students may be nested in schools, people in neighborhoods, employees in firms, or twins in twin-pairs. Longitudinal data are by definition clustered because multiple observations over time are nested within units, typically subjects.

Such clustered designs often provide rich information on processes operating at different levels, for instance, people's characteristics interacting with institutional characteristics. Importantly, the standard assumption of independent observations is likely to be violated because of dependence among observations within the same cluster. The multilevel and longitudinal methods discussed in this book extend conventional regression to handle such dependence and exploit the richness of the data.

Volume 1 is on multilevel and longitudinal modeling of continuous responses using linear models. The volume consists of four parts: I. Preliminaries (a review of linear regression modeling, preparing the reader for the rest of the book), II. Two-level models, III. Models for longitudinal and panel data, and IV. Models with nested and crossed random effects. For readers who are new to multilevel and longitudinal modeling, the chapters in part II should be read sequentially and can form the basis of an introductory course on this topic. A one-semester course on multilevel and longitudinal modeling can be based on most of the chapters in volume 1 plus chapter 10 on binary or dichotomous responses from volume 2. For this purpose, we have made chapter 10 freely downloadable from http://www.stata-press.com/books/mlmus3_ch10.pdf.

Volume 2 is on multilevel and longitudinal modeling of categorical responses, counts, and survival data. This volume also consists of four parts: I. Categorical responses (binary or dichotomous responses, ordinal responses, and nominal responses or discrete choice), II. Counts, III. Survival (in both discrete and continuous time), and IV. Models with nested and crossed random effects. Chapter 10 on binary or dichotomous responses is a core chapter of this volume and should be read before embarking on the other chapters. It is also a good idea to read chapter 14 on discrete-time survival before reading chapter 15 on continuous-time survival.

Our emphasis is on explaining the models and their assumptions, applying the methods to real data, and interpreting results. Many of the issues are conceptually demanding but do not require that you understand complex mathematics. Wherever possible, we therefore introduce ideas through examples and graphical illustrations, keeping the technical descriptions as simple as possible, often confining formulas to subsections that can be skipped. Some sections that go beyond an introductory course on multilevel and longitudinal modeling are tagged with the symbol ❖ . Derivations that can be skipped by the reader are given in displays. For an advanced treatment, placing multilevel modeling within a general latent-variable framework, we refer the reader to Skrondal and Rabe-Hesketh (2004a), which uses the same notation as this book.

This book shows how all the analyses described can be performed using Stata. There are many advantages of using a general-purpose statistical package such as Stata. First, for those already familiar with Stata, it is convenient not having to learn a new stand-alone package. Second, conducting multilevel-analysis within a powerful package has the advantage that it allows complex data manipulation to be performed, alternative estimation methods to be used, and publication-quality graphics to be produced, all without having to switch packages. Finally, Stata is a natural choice for multilevel and longitudinal modeling because it has gradually become perhaps the most powerful general-purpose statistics package for such models.

Each chapter is based on one or more research problems and real datasets. After describing the models, we walk through the analysis using Stata, pausing when statistical issues arise that need further explanation. Stata can be used either via a graphical user interface (GUI) or through commands. We recommend using commands interactively— or preferably in do-files—for serious analysis in Stata. For this reason, and because the GUI is fairly self-explanatory, we use commands exclusively in this book. However, the GUI can be useful for learning the Stata syntax. Generally, we use the `typewriter font` to refer to Stata commands, syntax, and variables. A "dot" prompt followed by a command indicates that you can type verbatim what is displayed after the dot (in context) to replicate the results in the book. Some readers may find it useful to intersperse reading with running these commands. We encourage readers to write do-files for solving the data analysis exercises because this is standard practice for professional data analysis.

The commands used for data manipulation and graphics are explained to some extent, but the purpose of this book is not to teach Stata from scratch. For basic introductions to Stata, we refer the reader to Acock (2010), Kohler and Kreuter (2009), or Rabe-Hesketh and Everitt (2007). Other books and further resources for learning Stata are listed at the Stata website.

If you are new to Stata, we recommend running all the commands given in chapter 1 of volume 1. A list of commands that are particularly useful for manipulating, describing, and plotting multilevel and longitudinal data is given in the appendix of volume 1. Examples of the use of these and other commands can easily be found by referring to the "commands" entry in the subject index.

We have included applications from a wide range of disciplines, including medicine, economics, education, sociology, and psychology. The interdisciplinary nature of this book is also reflected in the choice of models and topics covered. If a chapter is primarily based on an application from one discipline, we try to balance this by including exercises with real data from other disciplines. The two volumes contain over 140 exercises based on over 100 different real datasets. Solutions to exercises that are available to readers are marked with $\boxed{\text{Solutions}}$ and can be downloaded from http://www.stata-press.com/books/mlmus3-answers.html. Instructors can obtain solutions to all exercises from Stata Press.

All datasets used in this book are freely available for download; for details, see http://www.stata-press.com/data/mlmus3.html. These datasets can be downloaded into a local directory on your computer. Alternatively, individual datasets can be loaded directly into net-aware Stata by specifying the complete URL. For example,

```
. use http://www.stata-press.com/data/mlmus3/pefr
```

If you have stored the datasets in a local directory, omit the path and just type

```
. use pefr
```

We will generally describe all Stata commands that can be used to fit a given model, discussing their advantages and disadvantages. An exception to this rule is that we do not discuss our own `gllamm` command in volume 1 (see the `gllamm` companion, downloadable from http://www.gllamm.org, for how to fit the models of volume 1 in `gllamm`). In volume 1, we extensively use the Stata commands `xtreg` and `xtmixed`, and we introduce several more specialized commands for longitudinal modeling, such as `xthtaylor`, `xtivreg`, and `xtabond`. The new `sem` command for structural equation modeling is used for growth-curve modeling.

In volume 2, we use Stata's `xt` commands for the different response types. For example, we use `xtlogit` and `xtmelogit` for binary responses, and `xtpoisson` and `xtmepoisson` for counts. We use `stcox` and `streg` for multilevel survival modeling with shared frailties. `gllamm` is used for all response types, including ordinal and nominal responses, for which corresponding official Stata commands do not yet exist. We also discuss commands for marginal models and fixed-effects models, such as `xtgee` and `clogit`. The *Stata Longitudinal-Data/Panel-Data Reference Manual* (StataCorp 2011) provides detailed information on all the official Stata commands for multilevel and longitudinal modeling.

The `nolog` option has been used to suppress the iteration logs showing the progress of the log likelihood. This option is not shown in the command line because we do not recommend it to users; we are using it only to save space.

We assume that readers have a good knowledge of linear regression modeling, in particular, the use and interpretation of dummy variables and interactions. However, the first chapter in volume 1 reviews linear regression and can serve as a refresher.

Errata for different editions and printings of the book can be downloaded from http://www.stata-press.com/books/errata/mlmus3.html, and answers to exercises can be downloaded from http://www.stata-press.com/books/mlmus3-answers.html.

In this third edition, we have split the book into two volumes and have added five new chapters, comprehensive updates for Stata 12, 49 new exercises, and 36 new datasets. All chapters of the previous edition have been substantially revised.

Berkeley and Oslo   Sophia Rabe-Hesketh
February 2012   Anders Skrondal

Acknowledgments

The following have given very helpful comments on drafts of the third edition: Ed Bein, Bianca de Stavola, David Drukker, Leonardo Grilli, Bobby Gutierrez, Yulia Marchenko, Jeff Pitblado, Carla Rampichini, and Sophia Rabe-Hesketh's research group at University of California–Berkeley. Special thanks are due to Leonardo Grilli and Carla Rampichini, who have given us detailed comments on major parts of the book. We are also grateful to Nina Breinegaard, Leonardo Grilli, Bobby Gutierrez, Joe Hilbe, Katrin Hohl, and Carla Rampichini for their extensive and helpful comments on earlier editions of the book, to Deirdre Skaggs for diligent copyediting, and to Lisa Gilmore and others for efficient publishing. Several cohorts of students at the University of California–Berkeley and the London School of Economics have provided feedback that helped us improve the second edition. We thank Germán Rodríguez for correcting Stata errors and Raymond Boston for carefully checking and correcting the do-files for the previous edition. The book reviews of previous editions by Daniel Hall, Charlie Hallahan, Nick Horton, Brian Leroux, Thomas Loughlin, Daniel Stahl, S. F. Heil, and Rory Wolfe have been very useful for revising the book. Readers are encouraged to provide feedback on the current edition so that we may improve future editions.

We are grateful to the many people who have contributed datasets to this book, either by making them publicly available themselves or by allowing us to do so. G. Dunn, J. D. Finn, J. Neuhaus, A. Pebley, G. Rodríguez, D. Stott, T. Toulopoulou, and M. Yang kindly contributed their datasets to this book. Some of the datasets we use accompany software packages and are freely downloadable. For instance, aML (Lillard and Panis 2003), BUGS (Spiegelhalter et al. 1996a,b), HLM (Raudenbush et al. 2004), Latent GOLD (Vermunt and Magidson 2005), MLwiN (Rasbash et al. 2009), and SuperMix (Hedeker et al. 2008) all provide exciting, real datasets.

Some journals, such as *Biometrics*, *Journal of Applied Econometrics*, *Journal of Business & Economic Statistics*, *Journal of the Royal Statistical Society (Series A)*, and *Statistical Modelling*, encourage authors to make their data available on the journal website. We are grateful to J. Abrevaya, P. K. Chintagunta, N. Goldman, D. C. Jain, E. Lesaffre, D. Moore, E. Ross, G. Rodríguez, B. Spiessens, F. Vella, M. Verbeek, N. J. Vilcassim, and R. Winkelmann for making their data available this way.

We also used datasets from textbooks with accompanying webpages; some of these datasets are available (together with worked examples using various software packages, including Stata) through UCLA Technology Services (see http://www.ats.ucla.edu/stat/examples/default.htm). The books by Allison (1995, 2005); Baltagi (2008); Bollen and Curran (2006); Brown and Prescott (2006); Cameron and Trivedi (2005); De Boeck and Wilson (2004); Davis (2002); DeMaris (2004); Dohoo, Martin, and Stryhn (2010); Fitzmaurice, Laird, and Ware (2011); Fox (1997); Frees (2004); Gelman and Hill (2007); Greene (2012); Johnson and Albert (1999); Hand et al. (1994); Hayes and Moulton (2009); Littell et al. (2006); O'Connell and McCoach (2008); Rabe-Hesketh and Everitt (2007); Singer and Willett (2003); Skrondal and Rabe-Hesketh (2004b); Therneau and Grambsch (2000); Train (2009); Vonesh and Chinchilli (1997); Weiss (2005); West, Welch, and Galecki (2007); and Wooldridge (2010) were particularly helpful in this regard.

Sometimes data are printed in papers or books, allowing patient people like us to type them. Data were provided in this form by D. Altman, W. S. Aranov, G. E. Battese, M. Bland, D. T. Burwell, D. T. Danahy, W. A. Fuller, R. M. Harter, G. Koch, A. J. Macnab, R. Mare, R. Prakash, and P. Sham.

We thank Chapman & Hall/CRC for permission to use figures and tables from our book *Generalized Latent Variable Modeling: Multilevel, Longitudinal and Structural Equation Models.*

Any omissions are purely accidental.

Multilevel and longitudinal models: When and why?

Just as in standard regression analysis, the purpose of multilevel modeling is to model the relationship between a response variable and a set of explanatory variables. The difference is that multilevel modeling involves units of observation at different "levels".

For example, when considering the salaries of professors at universities, the levels could be individual professors (level 1), departments (level 2), or universities (level 3). Explanatory variables or covariates often reside at a given level. For instance, professors are characterized by their productivity, departments by the marketability of the discipline, and universities by whether they are private or public. If university-level explanatory variables are of interest, it might appear necessary to treat university as the unit of analysis by regressing university-mean-salaries on characteristics of universities.

If professor-level explanatory variables are aggregated to the university level for a university-level analysis, the estimated relationships are generally different from those found if professor is treated as the unit of analysis. Interpreting associations at the higher level as pertaining to the lower level is known as the *ecological fallacy*, whereas the reverse is called the *atomistic fallacy*. Instead of having to make a decision regarding the unit of analysis, we can use multilevel modeling to avoid the fallacies by considering all levels simultaneously and including explanatory variables from the professor, department, and university levels.

There is a preponderance of multilevel data. Examples discussed in this book include patients in hospitals[1], children in classes or schools[2], soldiers in army companies[3], residents in neighborhoods[4] or countries[5], siblings in families[6], rat pups in litters[7], cows in herds[8], eyes on heads[9], twins in twin-pairs[10], and police stops in city precincts[11].

Level-1 units within the same level-2 unit or cluster tend to be more similar to each other than to units in other clusters. One reason for this is that units do not end up in the same cluster by chance but through some mechanism that may be related to their characteristics. For instance, which schools children go to is influenced by their family background through the place of residence or parental choice, so children within a school already have something in common from the first day of school. Perhaps more importantly, children within a school are subsequently affected by their peers, teachers, and school policies, making them even more similar. Such within-school similarity or dependence will be particularly apparent if there are large between-school differences in terms of children's backgrounds and school environment and policies. In the same vain, siblings are similar at birth because they have the same parents and therefore share genes, and they subsequently become even more similar by being raised in the same family, where they share a common environment and experiences.

Within-cluster dependence violates the assumption of ordinary regression models that responses are conditionally independent given the covariates (the residuals are independent). Consequently, ordinary regression produces incorrect standard errors, a problem that can be overcome by using multilevel models.

More importantly, multilevel modeling allows us to disentangle processes operating at different levels, both by including explanatory variables at the different levels and by attributing unexplained variability to the different levels. One important challenge in multilevel modeling is to distinguish within- and between-cluster effects of lower-level covariates. For instance, for students nested in schools, an important student-level explanatory variable for achievement is socioeconomic status (SES). In addition to the effect of own SES, the school-average SES can be strongly associated with achievement, both through peer effects and because low-SES children tend to end up in worse schools.

Policy interventions often occur at the level of institutions, and it is important to understand how such higher-level variables affect the response variable. In the school setting, typical examples would be new curricula or different kinds of teacher professional development. The effect of such interventions on achievement is usually the primary concern, but it is also important to investigate the differential effects for different student subpopulations. For instance, a new science curriculum may be particularly beneficial for girls and hence reduce the gender gap in science achievement. If it does, then there is a cross-level interaction between the level-1 variable gender and the level-2 variable curriculum.

It will generally not be possible to explain all within-school and between-school variability in achievement. Quantifying the amount of unexplained variability at the different levels can be of interest in its own right, and sometimes multilevel models without explanatory variables are used to see how much the higher-level membership matters. For instance, achievement may vary less between schools in Scandinavia than in the U.S., and this tells us something about the societies and schooling systems.

Repeated observations on the same units are also clustered data, for instance, longitudinal or panel data on children's weight[12], mothers' postnatal depression[13], taxpayers' tax liability[14], employees' wages and union membership[15], and the investments[16] or number of patents[17] of firms. Although longitudinal data are quite different from clustered cross-sectional data, the same kinds of questions arise and the same kinds of models can be used to address them. For instance, for data on children's weight, growth is about the relationship between weight and the level-1 variable time. Level-2 explanatory variables include gender and ethnicity. Cross-level interactions between these variables and time represent differences in growth rates between groups. Clearly, not all differences in initial weight and growth rate can be explained by gender and ethnicity, and multilevel models can quantify how much variability remains within groups.

Disentangling the within- and between-effects of level-1 (or time-varying) covariates can be important for causal inference. For instance, evaluation of policies, such as bans of smoking in public places, typically rely on observational data where policies were implemented for some of the clusters, such as states, at different times during the longitudinal study. When considering the effect of the legislation on cigarette consumption, it is important to estimate the within-state effects, to control for any state-level variables that affect legislation as well as cigarette consumption. Within-state effects could of course be due to general time trends, which can be controlled by considering the mean difference in within-state differences for states that did and did not implement the policy.

Sources of multilevel and longitudinal data

Multistage surveys

In surveys, units are randomly selected according to a careful survey design to produce results that are representative of a population. The simplest design is simple random sampling, where every unit has the same probability of selection. However, such designs require a list of all eligible units, which is rarely available. Furthermore, it is often cheaper to sample units in batches, and for this reason, samples are often drawn in stages. In the first stage, primary sampling units (PSUs) such as areas or schools are sampled (often within strata). Lists of units within these PSUs are then assembled and units sampled from these lists. The resulting data are highly clustered by design. In contrast, simple random sampling will often lead to a few instances of multiple units per cluster.

In this book, we consider several multistage surveys, including the British Social Attitudes Survey (BSA)[18] and the Program for International Student Assessment (PISA)[19]. In the BSA, PSUs were postcode sectors, from which addresses were sampled in stage 2 and one respondent per household was selected in stage 3. In the PISA, a large number of countries conducted surveys by first sampling regions and then schools, or by directly sampling schools. In the final stage, students were sampled from schools.

Often the design falls short of being a proper survey in the sense that the PSUs, or clusters, are selected by convenience. Another important source of clustered data is administrative data—such as national death registries, hospital patient databases, and social security claims databases—collected by government and other institutions that cover an entire well-defined population. Administrative data are often linked with other administrative data, survey data, and census data.

Examples are the family birthweight data[20] from the Medical Birth Registry of Norway, and the neighborhood-effects data[21] from an education authority in Scotland. In the latter data, pupils' national examination test scores from the Examination Board were linked to survey data on individual characteristics, family background, and census data on the neighborhoods. Aggregate economic data[22], such as unemployment rates for regions of a country, are typically provided by the national statistical agencies, whereas international data are provided by organizations such as the Organization for Economic Cooperation and Development (OECD)[23].

A good book on multistage surveys is Heeringa, West, and Berglund (2010).

Cluster-randomized studies

Health, educational, and other interventions are often administered to groups or clusters of individuals. For instance, sex education for teenagers typically takes place in school classes. Randomized experiments to study the effectiveness of such interventions therefore naturally rely on assigning entire clusters of subjects to interventions. Such studies are called cluster-randomized trials. Another reason for not assigning different units in the same cluster to different interventions is that there may be a risk of contamination in the sense that someone assigned to the control intervention may benefit from the treatment given to another subject. An example would be if the intervention group is given a cookbook for healthy eating and these subjects end up sharing meals and recipes with those in the control group.

Cluster-randomized studies considered in this book include Scottish schools assigned to sex education programs[24], Kenyan schools assigned to nutritional interventions[25], and U.S. schools assigned to a smoking prevention and cessation program[26].

In quasi-experiments, assignment to treatments or interventions is not random. An example considered in this book is a study by the World Health Organization (WHO) where hospitals in one region of China implemented a program of case management to discourage inappropriate prescription of antibiotics to young children with acute respiratory infection[27]. Another example is where some cities were declared as enterprise zones that provided tax credits to reduce unemployment whereas other cities served as controls[28]. Here deprived cities were more likely to be declared as enterprise zones because the intervention was more needed there.

A good book on cluster-randomized trials is Hayes and Moulton (2009).

Multisite studies and meta-analysis

Clinical trials and other randomized intervention studies are often conducted at several sites, either because individual sites do not have enough eligible patients to obtain reliable results or because data from several settings can provide evidence that the effects hold more generally (external validity). The important difference between a multisite and a cluster-randomized design is that the randomization is within clusters in a multisite study, with each site constituting its own, usually small, randomized study. Examples include a hypertension trial conducted at 29 centers[29] and a class-size experiment conducted in 79 schools[30].

Meta-analysis is used to summarize the evidence accumulated to date about a treatment from a range of published studies. The idea is similar to a multisite trial in that each study constitutes a (typically randomized) trial in a different setting. However, the differences between studies tend to be more fundamental, often with important variations in the treatments, protocols, and outcome measures. Another distinguishing feature of a meta-analysis is that the data from the different studies are typically available only in aggregated form, in terms of estimated effect sizes and standard errors[31]. Meta-analysis is used not only to estimate treatment effects but is useful also for pooling estimates of any kind across studies.

A good book on meta-analysis is Borenstein et al. (2009).

Family studies

Data on groups of more-or-less genetically related subjects are often collected to examine similarities between the subjects[32] and possibly to disentangle the sources of similarity into nature (genetics) and nurture (environment). In twin designs[33], comparison of identical twins (who share all genes) and fraternal twins (who share half their genes) allows the proportion of variability that is due to genes (heritability) to be estimated. Data on parents and their children can also be used to estimate heritability, as we do with birthweights[34] in this book.

A good book on analysis of family data is Sham (1998).

Longitudinal studies

In longitudinal or panel data, each unit is observed at several occasions over time. This makes it possible to study individual change, either due to the passage of time or due to explanatory variables.

Clinical trials are usually longitudinal because treatments take some time to have an effect. For instance, in a clinical trial for the treatment of postnatal depression, women were randomized to receive an estrogen patch or a placebo. They were assessed for depression before randomization and then monthly for six months after beginning treatment[35].

Panel surveys, where respondents are interviewed annually (in panel "waves") or at other regular time intervals, are popular in the social sciences. For example, the antisocial-behavior data[36] consist of three biennial waves of data from the U.S. National Longitudinal Survey of Youth on children and their mothers. A major advantage of longitudinal data is that it allows comparisons to be made within subject, hence controlling for (possibly unknown) subject characteristics that are constant over time. For instance, we can estimate the effect of child poverty on antisocial behavior by considering the change in antisocial behavior for children who move into or out of poverty.

Instead of passively observing explanatory variables, such as child poverty, in an observational study, we can sometimes apply a sequence of treatments or conditions to each subject over a set of occasions to allow within-subject comparisons to be made. In this case, time or order effects are a nuisance and are typically dealt with by counterbalancing the order of the treatments (typically by random assignment). An example considered in this book is a four-period two-treatment double-blind cross-over trial where patients are randomized to receive different sequences of artificial sweetener and placebo, each taken for a week, and the number of headaches is recorded[37]. In contrast to regular clinical trials, the treatment effect is estimated using a within-subject comparison, giving more reliable estimates. Longitudinal designs with time-varying treatments are also important in experimental psychology, where they are called repeated-measures designs. For instance, in the verbal aggression data[38], there is a within-subject factorial design consisting of four situations that may cause aggression, three aggressive behaviors, and two modes of response.

Good books on the analysis of longitudinal data include Fitzmaurice, Laird, and Ware (2011) in biostatistics and Wooldridge (2010) in econometrics.

Measurement studies

Variables are usually treated as if they represent the concept implied by their name. For instance, achievement scores are viewed as representing achievement, and peak-expiratory-flow measurements are viewed as representing peak expiratory flow (strength of breathing out). However, even if the measures are valid (measure what they are supposed to measure and not something else), they are invariably subject to measurement error (that is, they differ from the true score). The amount and sources of measurement error can be investigated by conducting a measurement study, or generalizability study, where measurements of the same truth are repeated under different conditions. For example, in the peak-expiratory-flow study, subjects were measured on two occasions (test–retest data) by two different methods[39]. In the essay-grading data, several graders graded each of 198 essays[40]. Multilevel models can be used to estimate the measurement error variance and partition the measurement error variance into components due to different sources, such as raters and methods.

Item response theory (IRT) models for binary responses to test or questionnaire items can also be viewed as multilevel models. A typical application is measurement of students' math achievement[41].

A good book on the design and analysis of measurement studies is Dunn (2004).

Spatial data

Units are often expected to be more similar if they are located close together in space. If it makes sense to partition space into regions, the data can be viewed as clustered within the regions, such as zip codes, neighborhoods, counties, states, or countries. For example, in the skin-cancer data[42], the subjects are nested in counties, the counties are nested in regions, and the regions are nested in countries.

In small-area estimation, sparse data on small areas are used to estimate area-level means, proportions, or incidence rates. Area-specific estimates tend to be imprecise because of the low numbers, and some amount of averaging or pooling of data across areas is necessary to obtain more reliable estimates. Multilevel models can be used to estimate the variability between areas, and this information can be used to determine the adequate degree of pooling (shrinkage estimation or empirical Bayes prediction). A good display of small-area estimates is a map where areas are shaded according to the magnitudes of the estimates. If the estimates are disease rates, producing the maps is called disease mapping[43].

A good book on small-area estimation and disease mapping, from a multilevel modeling perspective, is Lawson, Browne, and Vidal Rodeiro (2003).

Notes

[1] Exercises 8.4 and 16.3

[2] Exercise 8.6

[3] Exercise 4.5

[4] Exercises 2.4 and 3.1

[5] Exercise 13.7

[6] Exercise 14.5

[7] Exercise 3.4

[8] Exercises 10.5 and 8.8

[9] Exercise 14.6

[10] Exercises 2.3 and 8.10

[11] Exercise 13.3

[12] Section 7.2

[13] Exercise 6.2

[14] Exercise 5.1

[15] Chapter 5

[16] Section 9.2

[17] Exercise 13.4

[18] Exercise 11.5

[19] Exercise 10.8

[20] Exercise 4.7

[21] Exercises 2.4, 3.1, and 9.5

[22] Exercise 8.3

[23] Section 6.7.1

[24]Exercise 3.8

[25]Section 8.10

[26]Exercises 11.3 and 14.7

[27]Exercise 16.3

[28]Exercises 5.3, 5.4, and 7.8

[29]Exercise 8.4

[30]Exercises 8.6, 8.7, and 9.10

[31]Exercise 2.8

[32]Exercises 2.6, 14.5, and 16.2

[33]Exercises 2.3, 8.10, and 16.11

[34]Exercise 4.7

[35]Section 6.2

[36]Exercise 5.2

[37]Exercise 13.2

[38]Exercises 10.4 and 11.2

[39]Sections 2.2 and 8.2

[40]Exercise 2.5 and section 11.9

[41]Exercise 16.10

[42]Exercise 16.7

[43]Section 13.13

Part I

Preliminaries

1 Review of linear regression

1.1 Introduction

In this chapter, we review the statistical models underlying independent-samples t tests, analysis of variance (ANOVA), analysis of covariance (ANCOVA), simple regression, and multiple regression. We formulate all of these models as linear regression models.

We take a model-based approach in this chapter; that is, we state all aspects of the model, including a normal distribution of the residuals. However, as we will see, statistical inferences can be valid even if some model assumptions, such as normality, are violated. We focus on model specification and interpretation of regression coefficients, and we discuss in detail the use of dummy variables and interactions.

The regression models considered here are essential building blocks for multilevel models. Although linear multilevel or mixed models for continuous responses are sometimes viewed from an ANOVA perspective, the regression perspective is beneficial because it is easily generalizable to binary and other types of responses. Furthermore, the Stata commands for multilevel modeling follow a regression syntax.

This chapter is not intended as a first introduction to linear regression, but rather as a refresher for readers already familiar with most of the ideas. Even experienced regression modelers are likely to benefit from reading this chapter because it introduces our notation and terminology as well as Stata commands used in later chapters, including factor variables for specifying dummy variables and interactions within estimation commands instead of creating new variables. If you are able to answer the self-assessment exercise 1.6 at the end of this chapter, you should be well prepared for the rest of the book.

1.2 Is there gender discrimination in faculty salaries?

DeMaris (2004) analyzed data on the salaries of faculty (academic staff) at Bowling Green State University in Ohio, U.S.A., in the academic year 1993–1994. The data are provided with his book *Regression with Social Data: Modeling Continuous and Limited Response Variables* and have previously been analyzed by Balzer et al. (1996) and Boudreau et al. (1997). The primary purpose of these studies was to investigate whether evidence existed for gender inequity in faculty salaries at the university.

The data considered here are a subset of the data provided by DeMaris, comprising $n = 514$ faculty members, excluding faculty from the Fireland campus, nonprofessors (instructors/lecturers), those not on graduate faculty, and three professors hired as Ohio Board of Regents Eminent Scholars. We will use the following variables:

- `salary`: academic year (9-month) salary in U.S. dollars
- `male`: gender (1 = male; 0 = female)
- `market`: marketability of academic discipline, defined as the ratio of the national average salary paid in the discipline to the national average across all disciplines
- `yearsdg`: time since degree (in years)
- `rank`: academic rank (1 = assistant professor; 2 = associate professor; 3 = full professor)

We start by reading the data into Stata:

```
. use http://www.stata-press.com/data/mlmus3/faculty
```

If you already have a dataset open in Stata that you do not need to save, add the `clear` option:

```
. use http://www.stata-press.com/data/mlmus3/faculty, clear
```

1.3   Independent-samples t test

If we have an interest in gender inequity, an obvious first step is to compare mean salaries between male and female professors at the university. We can use the `tabstat` command to produce a table of means, standard deviations, and sample sizes by gender:

```
. tabstat salary, by(male) statistics(mean sd n)
Summary for variables: salary
      by categories of: male

  male |      mean         sd          N
-------+--------------------------------
 Women |   42916.6    9161.61        128
   Men |  53499.24   12583.48        386
-------+--------------------------------
 Total |  50863.87   12672.77        514
```

We see that the male faculty at the university earn, on average, over \$10,000 more than the female faculty. The standard deviation is also considerably larger for the men than for the women.

Due to chance or sampling variation, the large difference between the mean salary $\bar{y}_1$ of the n_1 men and the mean salary $\bar{y}_0$ of the n_0 women in the sample does not necessarily imply that the corresponding population means or *expectations* μ_1 and μ_0 for male and female faculty differ (the Greek letter μ is pronounced "mew"). Here *population* refers either to an imagined infinite population from which the data can be viewed as sampled or to the statistical model that is viewed as the data-generating mechanism for the observed data.

To define a statistical model, let y_i and x_i denote the salary and gender of the ith professor, respectively, where $x_i = 1$ for men and $x_i = 0$ for women. A standard model for the current problem can then be specified as

$$y_i | x_i \sim N(\mu_{x_i}, \sigma_{x_i}^2)$$

Here "$y_i | x_i \sim$" means "y_i, for a given value of x_i, is distributed as", and $N(\mu_{x_i}, \sigma_{x_i}^2)$ stands for a normal distribution with conditional mean parameter μ_{x_i} and conditional variance parameter $\sigma_{x_i}^2$ (the Greek letter σ is pronounced "sigma"). The term *conditional* simply means that we are considering only the subset of the population for which some condition is satisfied—in this case, that x_i takes on a particular value. In other words, we are considering the distribution of salaries for men ($x_i = 1$) separately from the distribution for women ($x_i = 0$).

When conditioning on a categorical variable like gender, the expression "conditional on gender" can be replaced by "within gender" or "separately for each gender". Because x_i takes on only two values, we can be more explicit and write the statistical model as

$$y_i | x_i = 0 \ \sim \ N(\mu_0, \sigma_0^2)$$
$$y_i | x_i = 1 \ \sim \ N(\mu_1, \sigma_1^2)$$

Each conditional distribution has its own conditional expectation, or conditional population mean,

$$\mu_{x_i} \equiv E(y_i | x_i)$$

denoted μ_1 for men and μ_0 for women ($\equiv$ stands for "defined as"). Each conditional distribution also has its own conditional variance

$$\sigma_{x_i}^2 \equiv \mathrm{Var}(y_i | x_i)$$

denoted σ_1^2 for men and σ_0^2 for women. A final assumption is that y_i is conditionally independent of $y_{i'}$, given the values of x_i and $x_{i'}$, for different professors i and i'. This means that knowing one professor's salary does not help us predict another professor's salary if we already know that other professor's gender and the corresponding gender-specific mean salary.

By modeling y_i conditional on x_i, we are treating y_i as a *response variable* (sometimes also called dependent variable or criterion variable) and x_i as an *explanatory variable* or *covariate* (sometimes referred to as independent variable, predictor, or regressor).

The normality assumptions stated above are usually assessed by inspecting the conditional sample distributions for the men and women, using box plots,

```
. graph box salary, over(male) ytitle(Academic salary) asyvars
```

(see left panel of figure 1.1), or histograms,

```
. histogram salary, by(male, rows(2)) xtitle(Academic salary)
```

(see left panel of figure 1.2). We see that both distributions are somewhat positively skewed.

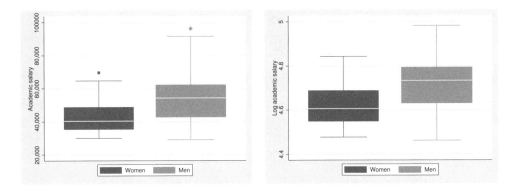

Figure 1.1: Box plots of salary and log salary by gender

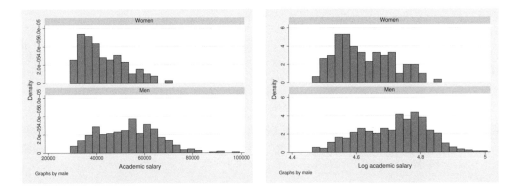

Figure 1.2: Histograms of salary and log salary by gender

A logarithmic transformation of salary makes the distributions more symmetric (like normal distributions)

```
. generate lsalary = log10(salary)
. graph box lsalary, over(male) ytitle(Log academic salary) asyvars
. histogram lsalary, by(male, rows(2)) xtitle(Log academic salary)
```

(see right panels of figures 1.1 and 1.2).

We have specified a full statistical model for the data and will now turn to making inferences about the means μ_0 and μ_1. Note that the methods of inference discussed in this chapter do not rely on the normality assumptions unless the sample is small.

We use the `ttest` command to estimate the population means, produce confidence intervals, and test the null hypothesis that the two population means are equal,

$$H_0: \mu_0 = \mu_1 \qquad \text{or} \qquad H_0: \mu_0 - \mu_1 = 0$$

against the two-sided alternative that they are different,

$$H_a: \mu_0 \neq \mu_1 \qquad \text{or} \qquad H_a: \mu_0 - \mu_1 \neq 0$$

The most popular version of the t test makes the additional assumption that the conditional variances are equal, $\sigma_0^2 = \sigma_1^2$, which we denote by dropping the subscript of $\sigma_{x_i}^2$, that is,

$$\text{Var}(y_i | x_i) = \sigma^2$$

so that the model becomes

$$y_i | x_i \sim N(\mu_{x_i}, \sigma^2)$$

The equal-variance assumption seems more reasonable for the log-transformed salaries than for the salaries on their original scale. However, salary in dollars is more interpretable than its log transformation, so for simplicity, we will work with the untransformed variable in this chapter (but see exercise 1.5).

We can perform t tests with the assumption of equal variances using

```
. ttest salary, by(male)
Two-sample t test with equal variances
```

Group	Obs	Mean	Std. Err.	Std. Dev.	[95% Conf. Interval]	
Women	128	42916.6	809.7795	9161.61	41314.2	44519.01
Men	386	53499.24	640.4822	12583.48	52239.96	54758.52
combined	514	50863.87	558.972	12672.77	49765.72	51962.03
diff		-10582.63	1206.345		-12952.63	-8212.636

```
    diff = mean(Women) - mean(Men)                              t =  -8.7725
Ho: diff = 0                                  degrees of freedom =      512

    Ha: diff < 0                 Ha: diff != 0                  Ha: diff > 0
 Pr(T < t) = 0.0000       Pr(|T| > |t|) = 0.0000           Pr(T > t) = 1.0000
```

and without the assumption of equal variances by specifying the `unequal` option:

```
. ttest salary, by(male) unequal
Two-sample t test with unequal variances
```

Group	Obs	Mean	Std. Err.	Std. Dev.	[95% Conf. Interval]	
Women	128	42916.6	809.7795	9161.61	41314.2	44519.01
Men	386	53499.24	640.4822	12583.48	52239.96	54758.52
combined	514	50863.87	558.972	12672.77	49765.72	51962.03
diff		-10582.63	1032.454		-12614.48	-8550.787

```
         diff = mean(Women) - mean(Men)                           t = -10.2500
Ho: diff = 0                      Satterthwaite's degrees of freedom =  297.227

    Ha: diff < 0                  Ha: diff != 0                   Ha: diff > 0
 Pr(T < t) = 0.0000          Pr(|T| > |t|) = 0.0000          Pr(T > t) = 1.0000
```

For both versions of the t test, the population means are estimated by the sample means,

$$\widehat{\mu}_0 = \overline{y}_0, \quad \widehat{\mu}_1 = \overline{y}_1$$

where the "hat" ($\,\widehat{}\,$) denotes an estimator. The t statistic is given by

$$t = \frac{\widehat{\mu}_0 - \widehat{\mu}_1}{\widehat{\text{SE}}(\widehat{\mu}_0 - \widehat{\mu}_1)}$$

the estimated difference in population means (the difference in sample means) divided by the estimated standard error of the estimated difference in population means. Under the null hypothesis of equal population means, the statistic has a t distribution with df degrees of freedom, where df $= n - 2$ (sample size n minus 2 for two estimated means $\widehat{\mu}_0$ and $\widehat{\mu}_1$) for the test assuming equal variances.

The 95% confidence interval for the difference in population means $\mu_0 - \mu_1$ is

$$\widehat{\mu}_0 - \widehat{\mu}_1 \ \pm \ t_{0.975,\text{df}} \ \widehat{\text{SE}}(\widehat{\mu}_0 - \widehat{\mu}_1)$$

where $t_{0.975,\text{df}}$ is the 97.5th percentile of the t distribution with df degrees of freedom.

The standard error is estimated as \$1,206.345 under the equal-variance assumption $\sigma_0^2 = \sigma_1^2$ and as \$1,032.454 without the equal-variance assumption; the degrees of freedom also differ. In both cases, the two-tailed p-value (given under `Ha: diff != 0`) is less than 0.0005 (typically reported as $p < 0.001$), leading to rejection of the null hypothesis at, say, the 5% level. For example, with the equal-variance assumption, $t = -8.77$, df $= 512$, $p < 0.001$. The 95% confidence intervals for the difference in population mean salary for men and women are from $-\$12,953$ to $-\$8,213$ under the equal-variance assumption and from $-\$12,615$ to $-\$8,551$ without the equal-variance assumption. Repeating the analysis for log-salary (not shown) also gives $p < 0.001$ and a smaller relative difference between the estimated standard errors for the two versions of the t test.

It is important to note that the normality assumption is not necessary for valid estimation of the model parameters (the population means and population standard deviations) and the standard error(s) used for inference. Normality ensures that the t statistic has a t distribution under the null hypothesis, but this null distribution is also approximately correct in large samples if normality is violated. These remarks apply to the rest of this volume.

1.4 One-way analysis of variance

The model underlying the t test with equal variances is also called a one-way analysis-of-variance (ANOVA) model.

Analysis of variance involves partitioning the *total sum of squares* (TSS), the sum of squared deviations of the y_i from their overall mean,

$$\text{TSS} = \sum_{i=1}^{n}(y_i - \overline{y})^2$$

into the *model sum of squares* (MSS) and the *sum of squared errors* (SSE).

The MSS, also known as regression sum of squares, is

$$\text{MSS} = \sum_{i=1}^{n}(\widehat{y}_i - \overline{y})^2 = \sum_{i,\,x_i=0}(\overline{y}_0 - \overline{y})^2 + \sum_{i,\,x_i=1}(\overline{y}_1 - \overline{y})^2 = n_0(\overline{y}_0 - \overline{y})^2 + n_1(\overline{y}_1 - \overline{y})^2$$

the sum of squared deviations of the sample means from the overall mean, interpretable as the between-group sum of squares (here "$i, x_i{=}0$" and "$i, x_i{=}1$" mean that the sums are taken over females and males, respectively).

The SSE, also known as residual sum of squares, is

$$\text{SSE} = \sum_{i=1}^{n}(y_i - \widehat{y}_i)^2 = \sum_{i,\,x_i=0}(y_i - \overline{y}_0)^2 + \sum_{i,\,x_i=1}(y_i - \overline{y}_1)^2 = (n_0 - 1)s_0^2 + (n_1 - 1)s_1^2$$

the sum of squared deviations of responses from their respective sample means, interpretable as the within-group sum of squares (s_0 and s_1 are the within-group sample standard deviations).

The group-specific sample means can be viewed as predictions, $\widehat{y}_i = \widehat{\mu}_{x_i} = \overline{y}_{x_i}$, representing the best guess of the salary when all that is known about the professor is the gender. These predictions, $\overline{y}_0$ and $\overline{y}_1$, minimize the SSE and are therefore referred to as *ordinary least-squares* (OLS) estimates. When evaluating the quality of the predictions $\widehat{y}_i$, the SSE is interpreted as the sum of the squared prediction errors $y_i - \widehat{y}_i$.

The total sum of squares equals the model sum of squares plus the sum of squared errors

$$\text{TSS} = \text{MSS} + \text{SSE}$$

The deviations contributing to each of these sums of squares are shown in figure 1.3 for an observation y_i (shown as •) in a hypothetical dataset. These deviations add up in the same way as the corresponding sums of squares. For example, for men, $y_i - \overline{y} = (\overline{y}_1 - \overline{y}) + (y_i - \overline{y}_1)$.

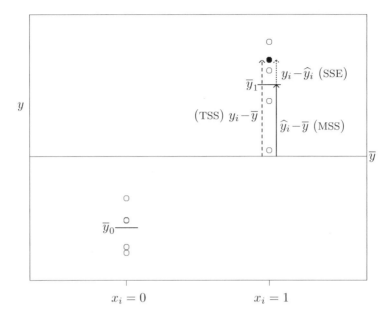

Figure 1.3: Illustration of deviations contributing to total sum of squares (TSS), model sum of squares (MSS), and sum of squared errors (SSE)

The model mean square (MMS) and mean squared error (MSE) can be obtained from the corresponding sums of squares by dividing by the appropriate degrees of freedom as shown in table 1.1 for the general case of g groups (for comparing males and females, $g = 2$).

Table 1.1: Sums of squares (SS) and mean squares (MS) for one-way ANOVA

Source	SS	df	MS $= \frac{\text{SS}}{\text{df}}$	
Model	MSS	$g-1$	MMS	Between group
Error	SSE	$n-g$	MSE	Within group
Total	TSS	$n-1$		

The MSE is the pooled within-group sample variance, which is an estimator for the population variance parameter σ^2:

$$\widehat{\sigma^2} = \text{MSE}$$

The F statistic for the null hypothesis that the population means are the same (against the two-sided alternative) then is

$$F = \frac{\text{MMS}}{\text{MSE}}$$

Under the null hypothesis, this statistic has an F distribution with $g-1$ and $n-g$ degrees of freedom. When $g=2$ as in our example, the F statistic is the square of the t statistic from the independent-samples t test under the equal-variance assumption and the p-values from both tests are identical.

We can perform a one-way ANOVA in Stata using

```
. anova salary male
```

			Number of obs =	514	R-squared	= 0.1307
			Root MSE	= 11827.4	Adj R-squared =	0.1290
Source	Partial SS	df	MS	F	Prob > F	
Model	1.0765e+10	1	1.0765e+10	76.96	0.0000	
male	1.0765e+10	1	1.0765e+10	76.96	0.0000	
Residual	7.1622e+10	512	139887048			
Total	8.2387e+10	513	160599133			

This gives a statistically significant difference in mean salaries for men and women as before $[F(1,512) = 76.96, p < 0.001]$. The estimate $\sqrt{\widehat{\sigma^2}} = \$11{,}827$ is given under Root MSE in the output. Estimates of μ_0 and μ_1 are not shown but can be obtained using the postestimation command `margins` (available as of Stata 11):

```
. margins male
```

Adjusted predictions   Number of obs   =   514
Expression   : Linear prediction, predict()

		Delta-method				
	Margin	Std. Err.	z	P>|z|	[95% Conf. Interval]	
male						
0	42916.6	1045.403	41.05	0.000	40867.65	44965.56
1	53499.24	601.9981	88.87	0.000	52319.34	54679.13

1.5 Simple linear regression

Salaries can vary considerably between academic departments. Some disciplines are more marketable than others, perhaps because there are highly paid jobs in those disciplines outside academia or because there is a low supply of qualified graduates. The dataset contains a variable, `market`, for the marketability of the discipline, defined as the mean U.S. faculty salary in that discipline divided by the mean salary across all disciplines.

Let us now investigate the relationship between salaries and marketability of the discipline. Marketability ranges from 0.71 to 1.33 in this sample, taking on 46 different values. A one-way ANOVA model, allowing for a different mean salary μ_{x_i} for each value of marketability x_i would have a large number of parameters and many groups containing only one individual and would hence be *overparameterized*. There are two popular ways of dealing with this problem: 1) categorize the continuous explanatory variable into intervals, thus producing fewer and larger groups or 2) assume a parametric, typically linear, relationship between μ_{x_i} and x_i.

Taking the latter approach, a *simple linear regression model* can be written as

$$y_i | x_i \ \sim \ N(\mu_{x_i}, \sigma^2)$$

where

$$\mu_{x_i} \equiv E(y_i | x_i) \ = \ \beta_1 + \beta_2 x_i$$

Alternatively, it can be written as

$$y_i \ = \ \beta_1 + \beta_2 x_i + \epsilon_i, \qquad \epsilon_i | x_i \ \sim \ N(0, \sigma^2)$$

where ϵ_i is the residual or error term for the ith professor, assumed to be independent of the residuals for other professors. Here β_1 (the Greek letter β is pronounced "beta") is called the *intercept* (often denoted β_0 or α) and represents the conditional expectation of y_i when $x_i = 0$:

$$E(y_i | x_i = 0) \ = \ \beta_1$$

β_2 is called the *slope*, or *regression coefficient*, of x_i and represents the increase in conditional expectations when x_i increases one unit, from some value a to $a+1$:

$$E(y_i | x_i = a+1) - E(y_i | x_i = a) \ = \ [\beta_1 + \beta_2(a+1)] - [\beta_1 + \beta_2 a] = \beta_2$$

We refer to $\beta_1 + \beta_2 x_i$ as the *fixed part* and ϵ_i as the *random part* of the model.

In addition to assuming that the conditional expectations fall on a straight line, the model assumes that the conditional variances, or residual variances, of the y_i are equal for all x_i,

$$\mathrm{Var}(y_i | x_i) = \mathrm{Var}(\epsilon_i | x_i) \ = \ \sigma^2$$

known as the *homoskedasticity* assumption (in contrast to *heteroskedasticity*).

A graphical illustration of the simple linear regression model is given in figure 1.4. Here the line represents the conditional expectation $E(y_i | x_i)$ as a function of x_i, and the density curves represent the conditional distributions of $y_i | x_i$ shown only for some values of x_i.

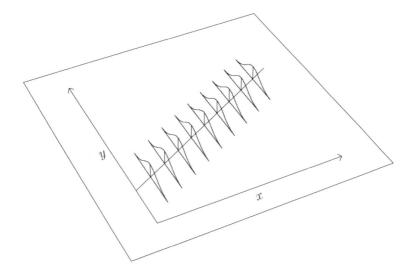

Figure 1.4: Illustration of simple linear regression model

We have described the regression coefficients as changes in conditional expectations without making any causal claims. Sometimes, especially in econometrics, regression coefficients are interpreted as causal effects; this necessitates further assumptions known as exogeneity assumptions, which are discussed in section 1.13. We will occasionally use the term *effect* in this book, but this does not necessarily mean that we are making causal claims.

For OLS estimates, we can partition the TSS into MSS and SSE exactly as in section 1.4, the only difference being the form of the predicted value of y_i:

$$\widehat{y}_i \;=\; \widehat{\beta}_1 + \widehat{\beta}_2 x_i \tag{1.1}$$

The contribution to each sum of squares is shown in figure 1.5 for an observation (shown as •) in a hypothetical dataset. The OLS estimates of β_1 and β_2 are obtained by minimizing the SSE, and the estimate of σ^2 is again the MSE, where the error degrees of freedom are $n-2$ (number of observations n minus 2 estimated regression coefficients $\widehat{\beta}_1$ and $\widehat{\beta}_2$). Maximum likelihood estimation, used for more complex models in later chapters, gives the same estimates of the regression coefficients as OLS but a smaller estimate of the residual variance because the latter is given by the SSE divided by n instead of divided by $n-2$ as for OLS.

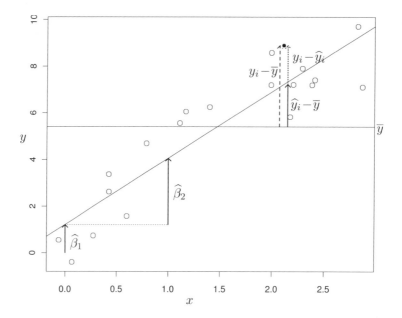

Figure 1.5: Illustration of sums of squares for simple linear regression

The *coefficient of determination* R^2 is defined as

$$R^2 \equiv \frac{\text{MSS}}{\text{TSS}} = \frac{\text{TSS} - \text{SSE}}{\text{TSS}}$$

and can be motivated as the proportion of the total variability (TSS) that is "explained" by the model that includes the covariate x_i (MSS). Alternatively, if TSS is viewed as the sum of squared prediction errors when the predictions are simply $\widehat{y}_i = \overline{y}$ (not using x_i to make predictions) whereas SSE is the sum of squared prediction errors taking into account x_i as in (1.1), then R^2 can be interpreted as the proportional reduction in prediction error variability due to knowing x_i.

R^2 is not a measure of model fit, but rather, as the name *coefficient of determination* implies, a measure of how well x_i predicts y_i. If the true model is a regression model with a large residual variance, then R^2 will tend to be small, and it would therefore be erroneous to interpret a small R^2 as indicating model misspecification. R^2 is also not a measure of the magnitude of the effect of x_i (or effect size), which is estimated by $\widehat{\beta}_2$. For a given estimated effect size $\widehat{\beta}_2$, R^2 will decrease as the estimated residual variance $\widehat{\sigma}^2$ increases (leading to a larger SSE) or the sample variance of x_i decreases (leading to a larger MSS).

The Stata command for the simple linear regression model is

```
. regress salary market

      Source |       SS       df       MS              Number of obs =     514
-------------+------------------------------           F(  1,   512) =  101.77
       Model |  1.3661e+10      1  1.3661e+10           Prob > F      =  0.0000
    Residual |  6.8726e+10    512   134231433           R-squared     =  0.1658
-------------+------------------------------           Adj R-squared =  0.1642
       Total |  8.2387e+10    513   160599133           Root MSE      =   11586

------------------------------------------------------------------------------
      salary |      Coef.   Std. Err.      t    P>|t|     [95% Conf. Interval]
-------------+----------------------------------------------------------------
      market |   34545.22   3424.333    10.09   0.000     27817.75    41272.69
       _cons |   18096.99   3288.009     5.50   0.000     11637.35    24556.64
------------------------------------------------------------------------------
```

The estimated coefficient of marketability is given next to `market` in the regression table, and the estimated intercept is given next to `_cons`. The reason for the label `_cons` is that the intercept can be thought of as the coefficient of a variable that is equal to 1 for each observation, referred to as a constant (because it does not vary).

The estimated regression coefficients, their estimated standard errors, and the estimated residual standard deviations are given for all models fit in this chapter in table 1.2, starting with the estimates for the model above in the first two columns under "Section 1.5".

Table 1.2: Ordinary least-squares (OLS) estimates for salary data (in U.S. dollars)

	Section 1.5				Section 1.6				Section 1.7		Section 1.8	
	Est	(SE)	Est	(SE)	Est	(SE)	Est	(SE)	Est	(SE)	Est	(SE)
[_cons]	18,097	(3,288)	50,864	(511)	42,917	(1,045)	44,324	(983)	34,834	(734)	36,774	(1,072)
[male]					10,583	(1,206)	8,708	(1,139)	2,040	(783)	−593	(1,320)
[market]	34,545	(3,424)										
[marketc]			34,545	(3,424)			29,972	(3,302)	38,402	(2,172)	38,437	(2,161)
[yearsdg]									949	(36)	763	(83)
[male×yearsdg]											227	(92)
[associate]												
[full]												
[male×associate]												
[male×full]												
[marketc×yearsdg]												
[yearsdg2]												
σ	11,586		11,586		11,827		10,986		7,148		7,112	

	Section 1.9				Section 1.10				Section 1.11	
	Est	(SE)	Est	(SE)	Est	(SE)	Est	(SE)	Est	(SE)
[_cons]	39,866	(746)	37,493	(989)	36,397	(886)	36,044	(712)	38,838	(1,027)
[male]			−1,043	(1,215)	465	(1,119)	926	(712)	−1,182	(1,252)
[market]										
[marketc]			36,987	(1,975)	36,951	(1,985)	46,906	(3,456)	46,578	(3,544)
[yearsdg]			405	(87)	552	(52)	541	(51)	39	(145)
[male×yearsdg]			184	(85)					178	(89)
[associate]	7,285	(1,026)	3,349	(872)	3,008	(1,307)	3,303	(860)	4,812	(968)
[full]	21,267	(966)	11,168	(1,168)	9,099	(1,894)	11,573	(1,166)	12,791	(1,230)
[male×associate]					284	(1,575)				
[male×full]					2539	(1,920)				
[marketc×yearsdg]							−750	(214)	−726	(222)
[yearsdg2]									10	(4)
σ	8,917		6,482		6,506		6,434		6,353	

We obtain the estimates $\widehat{\beta}_1 = \$18{,}097$, $\widehat{\beta}_2 = \$34{,}545$, and $\sqrt{\widehat{\sigma^2}} = \$11{,}586$. The estimated intercept $\widehat{\beta}_1$ is the estimated population mean salary when marketability is zero, a value that does not occur in this sample and is meaningless. Before interpreting the estimates, we therefore refit the model after mean-centering `market`:

```
. egen mn_market = mean(market)

. generate marketc = market - mn_market

. regress salary marketc
```

Source	SS	df	MS			
Model	1.3661e+10	1	1.3661e+10			
Residual	6.8726e+10	512	134231433			
Total	8.2387e+10	513	160599133			

Number of obs =   514
F(1,   512) = 101.77
Prob > F   = 0.0000
R-squared   = 0.1658
Adj R-squared = 0.1642
Root MSE   =   11586

salary	Coef.	Std. Err.	t	P>|t|	[95% Conf. Interval]	
marketc	34545.22	3424.333	10.09	0.000	27817.75	41272.69
_cons	50863.87	511.029	99.53	0.000	49859.9	51867.85

Here `mn_market`, the sample mean of `market`, was subtracted from `market` so that `marketc` equals zero when `market` is equal to its sample mean. The only estimate that is affected by the centering is the intercept that now represents the estimated population mean salary when marketability is equal to its sample mean (when `marketc` is zero). The estimated standard error of the intercept is considerably smaller after mean-centering because we are no longer extrapolating outside the range of the data.

For each unit increase in marketability, the population mean salary is estimated to increase by \$34,545 (with 95% confidence interval from \$27,818 to \$41,273). Because marketability ranges from 0.71 to 1.33, with no two people differing by as much as 1 unit, we could consider the effect of a 0.1 point increase in marketability, which is associated with an estimated increase in population mean salary of \$3,454 ($= \widehat{\beta}_2/10$). Alternatively, we could standardize marketability (giving it a standard deviation of one in addition to a mean of zero), in which case the estimated regression parameter would be interpreted as the estimated increase in population mean salary when marketability increases by one sample standard deviation.

If we standardize both the response variable `salary` and the covariate `market`, the estimated regression coefficient becomes a *standardized regression coefficient*, interpreted as the estimated number of standard deviations that salary increases, on average, when marketability increases by one standard deviation. Standardized regression coefficients can also be obtained by using the `beta` option in the `regress` command. However, they should be used with caution because they depend on sample-specific standard deviations, which invalidates comparisons across samples. See Greenland, Schlesselman, and Criqui (1986) for further discussion. A similar issue arises for mean-centering based on sample means. It might have been preferable to subtract 1 (where the mean salary for the discipline equals the mean salary across disciplines) from `marketability` instead of subtracting the sample mean 0.9485214.

To test the null hypothesis that the regression coefficient of marketability is zero, H_0: $\beta_2 = 0$, against the two-sided alternative H_a: $\beta_2 \neq 0$, we again use a t statistic, now given by

$$t = \frac{\widehat{\beta}_2}{\widehat{SE}(\widehat{\beta}_2)}$$

If the null hypothesis is true, this statistic has a t distribution with degrees of freedom given by the error degrees of freedom, denoted `Residual df` in the regression output. Here $t = 10.09$, df $= 512$, and $p < 0.001$, so we can reject the null hypothesis at the 5% level of significance. The null hypothesis for the intercept H_0: $\beta_1 = 0$ is usually irrelevant, and p-values for intercepts are thus ignored.

To visualize the fitted model, we can calculate predicted values $\widehat{y}_i$ using the postestimation command `predict` with the `xb` option (were `xb` is short for "x's times betas" and stands for $\widehat{\beta}_1 + \widehat{\beta}_2 x_i$ here):

```
. predict yhat, xb
```

We can then produce a scatterplot of the data points (y_i, x_i) together with a line connecting the predicted points $(\widehat{y}_i, x_i)$ with the following command:

```
. twoway (scatter salary market) (line yhat market, sort),
> ytitle(Academic salary) xtitle(Marketability)
```

The graph in figure 1.6 shows that a straight line appears to fit reasonably well and that the constant-variance assumption does not appear to be violated.

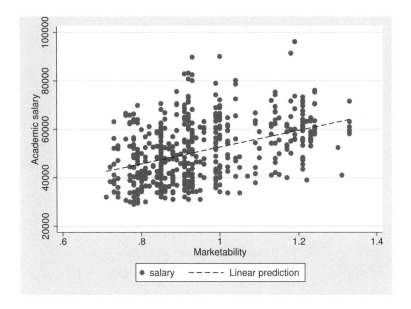

Figure 1.6: Scatterplot with predicted line from simple regression

There is considerable scatter around the regression line with only 16.6% (`R-squared` in regression output) of the variability in salaries being explained by marketability.

1.6   Dummy variables

Instead of using a t test to compare the population mean salaries between men and women, we can use simple linear regression. This becomes obvious by considering the diagram in figure 1.7, where we have simply used the variable x_i, equal to 1 for men and 0 for women, and connected the corresponding conditional expectations of y_i by a straight line.

We are not making any assumption regarding the relationship between the conditional means and x_i here because any two means can be connected by a straight line. (In contrast, assuming in the previous section that the conditional means for the 46 values of marketability lay on a straight line was a strong assumption.) We are, however, assuming equal conditional variances for the two populations because of the homoskedasticity assumption discussed earlier.

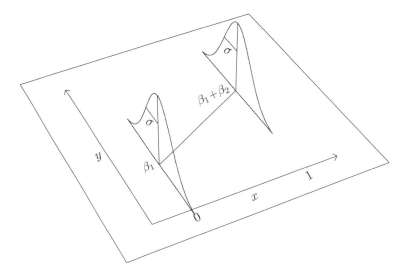

Figure 1.7: Illustration of simple linear regression with a dummy variable

The model can be written as

$$\mu_{x_i} \equiv E(y_i|x_i) \; = \; \beta_1 + \beta_2 x_i, \qquad y_i|x_i \; \sim \; N(\mu_{x_i}, \sigma^2)$$

so that

$$\mu_0 \; = \; \beta_1 + \beta_2 \times 0 = \beta_1$$
$$\mu_1 \; = \; \beta_1 + \beta_2 \times 1 = \beta_1 + \beta_2$$

The intercept β_1 can now be interpreted as the expectation for the group coded 0, the *reference group* (here women), whereas the slope β_2 represents the difference in expectations $\beta_2 = \mu_1 - \mu_0$ between the group coded 1 (here men) and the reference group.

When a dichotomous variable, coded 0 and 1, is used in a regression model like this, it is referred to as a *dummy variable* or *indicator variable*. A useful convention is to give the dummy variable the name of the group for which it is 1 and to describe it as a dummy variable for being in that group, here a dummy variable for being male.

The null hypothesis that $\beta_2 = 0$ is equivalent to the null hypothesis that the population means are the same, $\mu_1 - \mu_0 = 0$. We can therefore use simple regression to obtain the same result as previously produced for the two-sided independent-samples t test with the equal-variance assumption:

```
. regress salary male
```

Source	SS	df	MS			Number of obs =	514
						F(1, 512) =	76.96
Model	1.0765e+10	1	1.0765e+10			Prob > F =	0.0000
Residual	7.1622e+10	512	139887048			R-squared =	0.1307
						Adj R-squared =	0.1290
Total	8.2387e+10	513	160599133			Root MSE =	11827

salary	Coef.	Std. Err.	t	P>|t|	[95% Conf. Interval]	
male	10582.63	1206.345	8.77	0.000	8212.636	12952.63
_cons	42916.6	1045.403	41.05	0.000	40862.8	44970.41

The t statistic has the same value as that previously shown for the independent-samples t test (with the equal-variance assumption) apart from the sign, which depends on whether $\mu_0 - \mu_1$ or $\mu_1 - \mu_0$ is estimated and is thus arbitrary.

We can also relax the homoskedasticity (and normality) assumption in linear regression by replacing the conventional *model-based estimator* for the standard errors with the so-called *sandwich estimator*. Simply add the vce(robust) option (where vce stands for "variance–covariance matrix of estimates") to the regress command:

```
. regress salary male, vce(robust)
Linear regression
```

					Number of obs =	514
					F(1, 512) =	105.26
					Prob > F =	0.0000
					R-squared =	0.1307
					Root MSE =	11827

salary	Coef.	Robust Std. Err.	t	P>|t|	[95% Conf. Interval]	
male	10582.63	1031.462	10.26	0.000	8556.213	12609.05
_cons	42916.6	808.184	53.10	0.000	41328.84	44504.37

The resulting t statistic is almost identical to that from the ttest command with the unequal option in section 1.3 (10.26 compared with 10.25). In the rest of this chapter, we will use model-based standard errors to facilitate comparisons between estimation methods. (Robust standard errors can perform worse than model-based standard errors in small samples.)

In this section, we entered a dummy variable into the model in the same way as we would enter a continuous covariate. However, it usually does not make sense to evaluate dummy variables at their mean or to report standardized regression coefficients for dummy variables, because a standard-deviation change in a dummy variable, such as male, is meaningless.

The utility of using a regression model with a dummy variable instead of the t test to compare two groups becomes evident in the next section, where we want to control or adjust for other variables, which is straightforward in a multiple regression model.

1.7   Multiple linear regression

An important question when investigating gender discrimination is whether the men and women being compared are similar in the variables that justifiably affect salaries. As we have seen, there is some variability in marketability, and marketability has an effect on salary. Could the lower mean salary for women be due to women tending to work in disciplines with lower marketability? This is a possible explanation only if women tend to work in disciplines with lower marketability than do men, which is indeed the case:

```
. tabstat marketc, by(male) statistics(mean sd)
Summary for variables: marketc
      by categories of: male

   male |       mean        sd
--------+---------------------
  Women | -.0469589   .1314393
    Men |  .0155718   .1518486
--------+---------------------
  Total | -2.96e-08    .14938
```

A variable like marketability that is associated with both the covariate of interest (here gender) and the response variable is often called a *confounder*. The impact of ignoring one or more confounders on the estimated gender effect is called confounding.

We could render the comparison of salaries more fair by matching each woman to a man with the same value of marketability; however, this would be cumbersome, and we may not find matches for everyone. Instead, we can assume that marketability has the same, linear effect on salary for both genders and check whether gender has any additional effect after allowing for the effect of marketability.

This can be accomplished by specifying a multiple linear regression model,

$$y_i \;=\; \beta_1 + \beta_2 x_{2i} + \beta_3 x_{3i} + \epsilon_i, \qquad \epsilon_i | x_{2i}, x_{3i} \;\sim\; N(0, \sigma^2)$$

where we have multiple covariates or explanatory variables: a dummy variable x_{2i} for being a man and mean-centered marketability x_{3i}. (We number covariates beginning with 2 to correspond to the regression coefficients.)

For disciplines with mean marketability ($x_{3i} = 0$), the model specifies that the expected salary is β_1 for women ($x_{2i} = 0$) and $\beta_1 + \beta_2$ for men ($x_{2i} = 1$). Therefore, β_2 can be interpreted as the difference in population mean salary between men and women in disciplines with mean marketability. Fortunately, β_2 has an even more general interpretation as the difference in population mean salary between men and women in disciplines with any level of marketability, as long as both genders have the same value for marketability, $x_{3i} = a$:

$$E(y_i | x_{2i} = 1, x_{3i} = a) - E(y_i | x_{2i} = 0, x_{3i} = a) \;=\; (\beta_1 + \beta_2 + \beta_3 a) - (\beta_1 + \beta_3 a) \;=\; \beta_2$$

When comparing genders, we are now controlling for, adjusting for, partialling out, or keeping constant marketability.

Figure 1.8 shows model-implied regression lines for this model and how the lines depend on the coefficients. We see that β_2 determines the difference in means between men and women (vertical distance between the gender-specific regression lines) for any value of `marketc`.

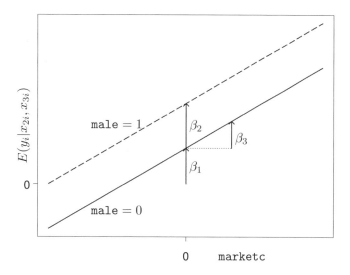

Figure 1.8: Illustration of multiple regression with a dummy variable for `male` (x_{2i}) and a continuous covariate, `marketc` (x_{3i})

The Stata command for the model is

```
. regress salary male marketc
```

Source	SS	df	MS
Model	2.0711e+10	2	1.0356e+10
Residual	6.1676e+10	511	120696838
Total	8.2387e+10	513	160599133

Number of obs =	514
F(2, 511) =	85.80
Prob > F =	0.0000
R-squared =	0.2514
Adj R-squared =	0.2485
Root MSE =	10986

salary	Coef.	Std. Err.	t	P>|t|	[95% Conf. Interval]	
male	8708.423	1139.411	7.64	0.000	6469.917	10946.93
marketc	29972.6	3301.766	9.08	0.000	23485.89	36459.3
_cons	44324.09	983.3533	45.07	0.000	42392.17	46256

The difference in population mean salaries between men and women, controlling for marketability, is estimated as \$8,708, considerably smaller than the unadjusted difference of \$10,583 that we estimated in the previous section. The estimated coefficient of `marketc` is also reduced and is now interpretable as the effect of marketability for a given gender or the within-gender effect of marketability. The coefficient of determination is now $R^2 = 0.251$ compared with $R^2 = 0.166$ for the model containing only `marketc` (R^2 cannot decrease when more covariates are added).

We can obtain predicted salaries $\widehat{y}_i$ and plot them with the observed salaries y_i against mean-centered marketability x_{3i} separately for each gender x_{2i} using

```
. predict yhat2, xb

. twoway (scatter salary marketc if male==1, msymbol(o))
>        (line yhat2 marketc if male==1, sort lpatt(dash))
>        (scatter salary marketc if male==0, msymbol(oh))
>        (line yhat2 marketc if male==0, sort lpatt(solid)),
>        ytitle(Academic salary) xtitle(Mean-centered marketability)
>        legend(order(1 " " 2 "Men" 3 " " 4 "Women"))
```

which produces the graph in figure 1.9.

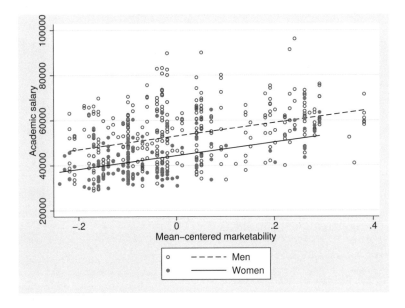

Figure 1.9: Scatterplot with predicted lines from multiple regression

In figure 1.9, The vertical distance between the regression lines is $\widehat{\beta}_2$, and the slope of both regression lines is $\widehat{\beta}_3$ (see also figure 1.8). The figure suggests that there is considerable overlap in the distributions of marketability for men and women. If there were little overlap (with men having higher values of marketability than women), the estimate of the coefficient of `male` would rely on extrapolating the regression line for males to low values of marketability and the regression line for females to high values of marketability. Such extrapolation beyond the range of data for each gender would hinge completely on the linearity assumption and would therefore be problematic. The degree of overlap, or *common support*, can be better assessed by plotting estimated probability density functions of `marketc` for both genders on the same graph,

```
. twoway (kdensity marketc if male==0) (kdensity marketc if male==1),
>        legend(order(2 "Men" 1 "Women")) xtitle(Mean-centered marketability)
>        ytitle(Estimated density)
```

where `kdensity` uses a kernel density smoother to obtain a smooth version of a histogram. We see in figure 1.10 that there is very good overlap between the densities except beyond `marketc` equal to 0.3, which occurs only for a few men.

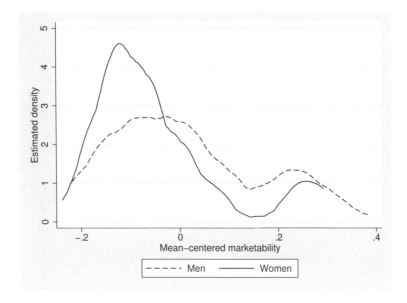

Figure 1.10: Estimated densities of `marketc` for men and women

Figure 1.11 illustrates the ideas of confounding using a hypothetical population model and simulated data. (To exaggerate the confounding, the data were simulated with little common support, but here we know that the relationships are linear.) There are two treatment groups, whose responses are represented by hollow and filled circles. The lines in the top panel represent the fixed part of the true regression model, which includes a dummy variable for treatment and a continuous covariate x_i that is correlated with treatment. The coefficient of treatment is given by the vertical distance between the regression lines. The OLS estimator of the regression coefficient for treatment in the model including both treatment and the covariate (that is, the true model) is an unbiased estimator of this coefficient.

In the bottom panel, the vertical distance between the lines instead represents the difference in marginal population means for the treatments. This difference might be the parameter of interest. However, if we are interested in the regression coefficient of treatment in the true model, it is clear from the figure that x_i is a confounder, because x_i is associated with both the treatment (the mean of x_i is larger for the treatment represented by hollow circles) and the response (the mean of y_i is larger for larger values of x_i). Thus the conditional difference in population means, given x_i, represented by the vertical distance between the population regression lines in the top panel, is different from the unconditional counterpart in the lower panel. If the regression lines in the top panel had coincided (with no vertical distance between them), then the association between the response variable and the treatment would be said to be *spurious*. For a further discussion of confounding—relating the idea to the concepts of causality and exogeneity, and giving an expression for omitted variable bias—see section 1.13.

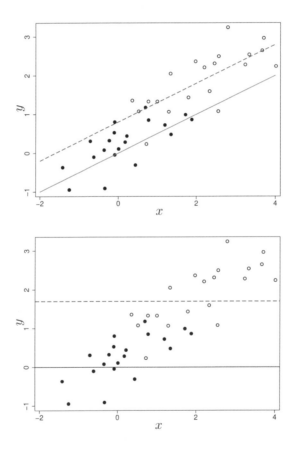

Figure 1.11: Illustration of confounding: Top panel shows conditional population means for the treatment groups, given x (true, data-generating model); bottom panel shows population means for two treatment groups, not conditioning on x

The model used in this section is sometimes called an analysis of covariance (ANCOVA) model because in addition to the categorical explanatory variable or *factor* (gender) used in one-way ANOVA, there is a continuous *covariate* marketability. Departing from the ANOVA/ANCOVA terminology, we use the word *covariate* for any observed explanatory variable, also including dummy variables, throughout this book.

Another covariate we should perhaps control for is time since the degree (in years), yearsdg, and we can do so using

```
. regress salary male marketc yearsdg
```

Source	SS	df	MS		Number of obs =	514
					F(3, 510) =	367.56
Model	5.6333e+10	3	1.8778e+10		Prob > F =	0.0000
Residual	2.6054e+10	510	51087083.4		R-squared =	0.6838
					Adj R-squared =	0.6819
Total	8.2387e+10	513	160599133		Root MSE =	7147.5

salary	Coef.	Std. Err.	t	P>|t|	[95% Conf. Interval]	
male	2040.211	783.122	2.61	0.009	501.6684	3578.753
marketc	38402.39	2171.689	17.68	0.000	34135.83	42668.95
yearsdg	949.2583	35.94867	26.41	0.000	878.6326	1019.884
_cons	34834.3	733.7898	47.47	0.000	33392.68	36275.93

For a given marketability and time since degree, the estimated population mean salary for men is $2,040 greater than for women, a substantial reduction in the estimated gender gap due to controlling for yearsdg. For a given gender and time since degree, the estimated effect of marketability is $38,402 extra mean salary per unit increase in marketability. Comparing professors of the same gender from disciplines with the same marketability, the estimated effect of time since degree is $949 extra mean salary per year since degree. It is tempting to interpret this coefficient as an effect of experience, but those with, for instance, yearsdg=40 differ from those with yearsdg=0 not only in terms of experience but also because they were recruited in a different epoch. As we will discuss in *Part III: Introduction to models for longitudinal and panel data*, we cannot separate such cohort effects from age or experience effects by using cross-sectional data.

We will now use Stata's margins command to obtain different kinds of predicted salaries for males and females based on the multiple regression model. To do this, we must first rerun the regression, this time declaring male as a categorical variable by using i.male (see section 1.8 for more information on factor variables). To suppress output, we use the quietly prefix.

```
. quietly regress salary i.male marketc yearsdg
```

We obtain predicted mean estimates for the two genders for certain values of the other covariates in the model, for instance, marketc=0 and yearsdg=10, by using

```
. margins male, at(marketc=0 yearsdg=10)
Adjusted predictions                                Number of obs   =        514
Model VCE     : OLS

Expression    : Linear prediction, predict()
at            : marketc         =              0
                yearsdg         =             10
```

		Delta-method				
	Margin	Std. Err.	z	P>|z|	[95% Conf. Interval]	
male						
0	44326.89	639.7602	69.29	0.000	43072.98	45580.79
1	46367.1	444.0554	104.42	0.000	45496.77	47237.43

The difference in adjusted means is equal to the estimated coefficient of the male dummy variable.

Alternatively, we can predict the mean salaries that males and females would have if both genders had the same distribution of the covariates `marketc` and `yearsdg`, namely, the distribution of the combined sample of males and females:

```
. margins male
Predictive margins                                  Number of obs   =        514
Model VCE     : OLS

Expression    : Linear prediction, predict()
```

		Delta-method				
	Margin	Std. Err.	z	P>|z|	[95% Conf. Interval]	
male						
0	49331.73	667.2756	73.93	0.000	48023.89	50639.57
1	51371.94	370.7068	138.58	0.000	50645.37	52098.51

One way of obtaining this *predictive margin* for females ourselves would be to set `male` to 0 for the entire sample, obtain the predicted value $\hat{y}_i$, and average it over the sample; we would do similarly for males. Because the predictions are linear in the covariates, the same results would be obtained by substituting the mean of `marketc` and `yearsdg` into the prediction formula, which can be achieved by using the command `margins male, atmeans`. Such predictions are also often referred to as *adjusted means*. We see that the difference between males and females is again equal to the estimated coefficient of the dummy variable for males.

1.8   Interactions

The models considered in the previous section assumed that the effects of different covariates were additive. For instance, if the dummy variable x_{2i} changes from 0 (women) to 1 (men), the mean salary increases by an amount β_2 regardless of the values of the other covariates (`marketc` and `yearsdg`).

However, this is a strong assumption that can be violated. The gender difference may depend on `yearsdg` if, for instance, starting salaries are similar for men and women but men receive larger or more frequent increases. We can investigate this possibility by including an *interaction* between gender and time since degree. An interaction between two variables implies that the effect of each variable depends on the value of the other variable: in our case, the effect of gender depends on time since degree and the effect of time since degree depends on gender.

We can incorporate the interaction in the regression model (with the usual assumptions) by simply including the product of `male` (x_{2i}) and `yearsdg` (x_{4i}) as a further covariate with regression coefficient β_5:

$$
\begin{aligned}
y_i &= \beta_1 + \beta_2\texttt{male}_i + \beta_3 x_{3i} + \beta_4\texttt{yearsdg}_i + \beta_5\texttt{male}_i \times \texttt{yearsdg}_i + \epsilon_i \\
&= \beta_1 + (\beta_2 + \beta_5\texttt{yearsdg}_i)\texttt{male}_i + \beta_3 x_{3i} + \beta_4\texttt{yearsdg}_i + \epsilon_i & (1.2) \\
&= \beta_1 + \beta_2\texttt{male}_i + \beta_3 x_{3i} + (\beta_4 + \beta_5\texttt{male}_i)\texttt{yearsdg}_i + \epsilon_i & (1.3)
\end{aligned}
$$

From (1.2), we see that the effect of `male` (also called the gender gap) is given by $\beta_2 + \beta_5\texttt{yearsdg}$ and hence depends on time since degree if $\beta_5 \neq 0$. From (1.3), we see that the effect of `yearsdg` is given by $\beta_4 + \beta_5\texttt{male}$ and hence depends on gender if $\beta_5 \neq 0$. We can describe time since degree as a *moderator* or an *effect modifier* of the effect of gender or vice versa.

When including an interaction between two variables, it is usually essential to keep both variables in the model. For instance, dropping `male` or setting $\beta_2 = 0$ would force the gender gap to be exactly 0 when time since degree is 0, which is a completely arbitrary constraint unless it corresponds to a specific research question.

An illustration of the model is given in figure 1.12 (with `marketc` set to zero; $x_{3i} = 0$). If β_5 were 0, we would obtain two parallel regression lines with vertical distance β_2. We see that β_5 represents the additional slope for men compared with women, or the additional gender gap when `yearsdg` increases by one unit. We also see that β_2 is the gender gap when `yearsdg` $= 0$ and β_4 is the slope of `yearsdg` when `male` $= 0$.

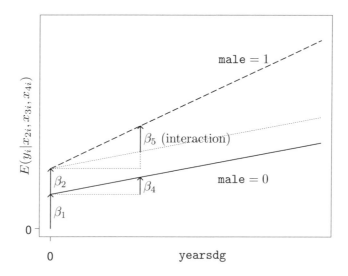

Figure 1.12: Illustration of interaction between `male` (x_{2i}) and `yearsdg` (x_{4i}) for `marketc` (x_{3i}) equal to 0 (not to scale)

To fit this model in Stata, we can generate the interaction as the product of the dummy variable for being male and years since degree:

```
. generate male_years = male*yearsdg
```

We can then include the interaction in the regression:

```
. regress salary male marketc yearsdg male_years
```

Source	SS	df	MS
Model	5.6641e+10	4	1.4160e+10
Residual	2.5746e+10	509	50581607.4
Total	8.2387e+10	513	160599133

Number of obs =	514
F(4, 509) =	279.95
Prob > F =	0.0000
R-squared =	0.6875
Adj R-squared =	0.6850
Root MSE =	7112.1

salary	Coef.	Std. Err.	t	P>|t|	[95% Conf. Interval]
male	-593.3088	1320.911	-0.45	0.654	-3188.418 2001.8
marketc	38436.65	2160.963	17.79	0.000	34191.14 42682.15
yearsdg	763.1896	83.4169	9.15	0.000	599.3057 927.0734
male_years	227.1532	91.99749	2.47	0.014	46.41164 407.8947
_cons	36773.64	1072.395	34.29	0.000	34666.78 38880.51

Here it is natural to interpret the interaction in terms of the effect of gender. When time since degree is 0 years, the population mean salary for men minus the population mean salary for women (after adjusting for marketability) is estimated as −$593. For every additional year since completing the degree, we add $227 to the difference, giving a difference of $0 after a little over 2 years; a difference of about $−593.31 + $227.15 \times 10 = $1,678 after 10 years; a difference of $−593.31 + $227.15 \times 20 = $3,949 after 20 years; and a difference of $−593.31 + $227.15 \times 30 = $6,221 after 30 years. Although the estimated gender gap at 0 years is not statistically significant at the 5% level ($t = −0.45$, df $= 509$, $p = 0.65$), the change in gender gap with years since degree (or interaction) is significant ($t = 2.47$, df $= 509$, $p = 0.01$).

We might wonder if the gender gap is statistically significant for faculty with 10 years of experience (adjusting for marketability), hence testing the null hypothesis H_0: $\beta_2 + \beta_5 \times 10 = 0$ against the two-sided alternative. This null hypothesis involves a linear combination of coefficients, and we can use the `lincom` command (which stands for "linear combination") to perform the test,

```
. lincom male + male_years*10
 ( 1)   male + 10*male_years = 0
```

salary	Coef.	Std. Err.	t	P>|t|	[95% Conf. Interval]
(1)	1678.223	792.9094	2.12	0.035	120.4449	3236.001

giving $t = 2.12$, df $= 509$, and $p = 0.04$. We also obtain a 95% confidence interval for the adjusted difference in population mean salaries 10 years since the degree that ranges from $120 to $3,236.

The regression model now includes three covariates, and it is difficult to represent it using a two-dimensional graph. However, as in figure 1.12, we can hold `marketc` constant and display the estimated population mean salary as a function of the other variables when `marketc` is zero. We could do this by setting `marketc` to zero and then using `predict`, or we can use Stata's `twoway function` command to produce figure 1.13:

```
. twoway (function Women = 36773 + 763.19*x, range(0 41) lpatt(dash))
>        (function Men = 36773 + -593.31 + (763.19 + 227.15)*x,
>        range(0 41) lpatt(solid)),
>        xtitle(Time since degree (years)) ytitle(Mean salary)
```

Here we have typed in the predicted regression lines for women and men as a function of `yearsdg` (here referred to as `x`), using (1.3) with `male` $= 0$ for women, `male` $= 1$ for men, and `marketc` $= 0$ (that is, $x_{3i} = 0$).

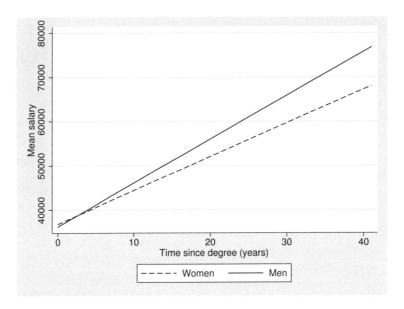

Figure 1.13: Estimated effect of gender and time since degree on mean salary for disciplines with mean marketability

Because we have just fit the model in Stata, we can also refer to the regression coefficients as _b[_cons] for the intercept, _b[yearsdg] for the coefficient of yearsdg, etc., saving us from having to type in all the coefficients, which is error-prone. The Stata command then looks like this:

```
twoway (function Women =_b[_cons] + _b[yearsdg]*x, range(0 41) lpatt(dash))
    (function Men =_b[_cons] + _b[male] + (_b[yearsdg] + _b[male_years])*x,
    range(0 41) lpatt(solid)), xtitle(Time since degree (years)) ytitle(Mean salary)
```

Instead of creating the interaction variable male_years first and then including it as a covariate in the regression model, we can also use *factor variables* to specify the interactions within the regress command (see also help fvvarlist for more information). To introduce an interaction between two variables, simply bind them together with a hash, #, making sure to declare their type: use the prefix i. for categorical variables (where i. stands for indicators, another term for dummy variables), and use c. for continuous variables. The command becomes

```
. regress salary male marketc yearsdg i.male#c.yearsdg
```

Source	SS	df	MS		
Model	5.6641e+10	4	1.4160e+10		
Residual	2.5746e+10	509	50581607.4		
Total	8.2387e+10	513	160599133		

```
Number of obs =      514
F( 4,  509) = 279.95
Prob > F      =  0.0000
R-squared     =  0.6875
Adj R-squared =  0.6850
Root MSE      =  7112.1
```

salary	Coef.	Std. Err.	t	P>|t|	[95% Conf. Interval]
male	-593.3088	1320.911	-0.45	0.654	-3188.418 2001.8
marketc	38436.65	2160.963	17.79	0.000	34191.14 42682.15
yearsdg	763.1896	83.4169	9.15	0.000	599.3057 927.0734
male# c.yearsdg 1	227.1532	91.99749	2.47	0.014	46.41164 407.8947
_cons	36773.64	1072.395	34.29	0.000	34666.78 38880.51

By binding the two variables together with two hashes, **##**, we specify that in addition to the interaction, we would also like to include each variable on its own:

```
. regress salary marketc i.male##c.yearsdg
```

Source	SS	df	MS		
Model	5.6641e+10	4	1.4160e+10		
Residual	2.5746e+10	509	50581607.4		
Total	8.2387e+10	513	160599133		

```
Number of obs =      514
F( 4,  509) = 279.95
Prob > F      =  0.0000
R-squared     =  0.6875
Adj R-squared =  0.6850
Root MSE      =  7112.1
```

salary	Coef.	Std. Err.	t	P>|t|	[95% Conf. Interval]
marketc	38436.65	2160.963	17.79	0.000	34191.14 42682.15
1.male	-593.3088	1320.911	-0.45	0.654	-3188.418 2001.8
yearsdg	763.1896	83.4169	9.15	0.000	599.3057 927.0734
male# c.yearsdg 1	227.1532	91.99749	2.47	0.014	46.41164 407.8947
_cons	36773.64	1072.395	34.29	0.000	34666.78 38880.51

In the output, the prefix "1." in 1.male stands for a dummy variable for the group of professors having the value 1 for the variable male, that is, men. The number 1 also appears in the interaction term to signify that it is an interaction between the dummy variable for male=1 and the continuous variable yearsdg.

In the lincom command, the coefficients are referred to by the names shown in the regression output, except that the number shown next to the interaction term (here 1) comes before the categorical variable, followed by a "."—for example, 1.male#c.yearsdg. So, 10 years after the degree, the gender difference is estimated as

```
. lincom 1.male + 1.male#c.yearsdg*10
( 1)  1.male + 10*1.male#c.yearsdg = 0
```

salary	Coef.	Std. Err.	t	P>|t|	[95% Conf. Interval]
(1)	1678.223	792.9094	2.12	0.035	120.4449 3236.001

Similarly, in the `twoway` command for producing figure 1.13, we refer to the coefficients as `_b[1.male]` and `_b[1.male#c.yearsdg]`.

We could also include an interaction between `marketc` and `male` and, as will be discussed in section 1.10.1, an interaction between the two continuous covariates `marketc` and `yearsdg`. In addition to these *two-way interactions*, we could consider the *three-way interaction* represented by the product of all three covariates. However, such higher-order interactions are rarely used because they are difficult to interpret.

1.9   Dummy variables for more than two groups

Another important explanatory variable for salary is academic rank, coded in the variable `rank` as 1 for assistant professor, 2 for associate professor, and 3 for full professor.

Using `rank` as a continuous covariate in a simple regression model would force a constraint on the population mean salaries for the three groups, namely, that the mean salary of associate professors is halfway between the mean salaries of assistant and full professors:

$$E(y_i|x_i) = \beta_1 + \beta_2 x_i = \begin{cases} \beta_1 + \beta_2 \times 1 & \text{for assistant professors} \\ \beta_1 + \beta_2 \times 2 & \text{for associate professors} \\ \beta_1 + \beta_2 \times 3 & \text{for full professors} \end{cases}$$

Such linearity is a strong assumption for an ordinal variable, such as `rank`, and a meaningless assumption for unordered categorical covariates, such as ethnicity, where the ordering of the values assigned to categories is arbitrary. It thus makes sense to estimate the mean of each rank freely by treating one of the ranks, for instance, assistant professor, as the reference category and using dummy variables for the other two ranks.

We can create the dummy variables by typing

```
. generate associate = rank==2 if rank < .
. generate full = rank==3 if rank < .
```

Here the logical expression `rank==2` evaluates to 1 if it is true and zero otherwise. This expression yields a 0 when `rank` is 1, 3, or missing, but we do not want to interpret a missing value; therefore, the `if` condition is necessary to ensure that missing values in `rank` translate to missing values in the dummy variables. (We specify `rank < .` because Stata interprets all missing values, ., as very large numbers.)

Our preferred method for producing dummy variables is using the `tabulate` command with the `generate()` option, as follows:

```
. drop associate full
. tabulate rank, generate(r)
. rename r2 associate
. rename r3 full
```

The `generate(r)` option produces dummy variables for each unique value of `rank`, here named `r1`, `r2`, and `r3` for the values 1, 2, and 3. (The naming of the dummy variables would have been the same if the unique values had been 0, 1, and 4.) An advantage of using `tabulate` is that it places missing values into dummy variables whenever the original variable is missing—as shown above, this requires extra caution when using the `generate` command.

Denoting these dummy variables x_{2i} and x_{3i}, respectively, we specify the model

$$E(y_i|x_{2i}, x_{3i}) = \beta_1 + \beta_2 x_{2i} + \beta_3 x_{3i} = \begin{cases} \beta_1 & \text{for assistant professors} \\ \beta_1 + \beta_2 & \text{for associate professors} \\ \beta_1 + \beta_3 & \text{for full professors} \end{cases}$$

showing that the intercept β_1 represents the population mean salary for the reference category (assistant professors), β_2 represents the difference in mean salaries between associate and assistant professors, and β_3 represents the difference in mean salaries between full and assistant professors. Hence, the coefficient of each dummy variable represents the population mean of the corresponding group minus the population mean of the reference group. Figure 1.14 illustrates how the model-implied means for the three ranks are determined by the regression coefficients.

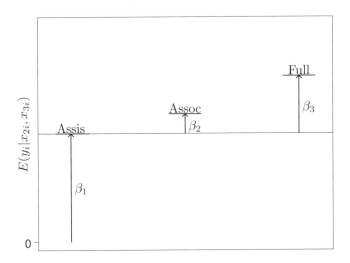

Figure 1.14: Illustration: Interpretations of coefficients of dummy variables x_{2i} and x_{3i} for associate and full professors, with assistant professors as the reference category

Estimates for the regression model with dummy variables for academic rank are obtained by using

```
. regress salary associate full
      Source |       SS       df       MS              Number of obs =     514
-------------+------------------------------           F(  2,   511) =  262.54
       Model |  4.1753e+10      2  2.0877e+10           Prob > F      =  0.0000
    Residual |  4.0634e+10    511  79518710.1           R-squared     =  0.5068
-------------+------------------------------           Adj R-squared =  0.5049
       Total |  8.2387e+10    513   160599133           Root MSE      =  8917.3

------------------------------------------------------------------------------
      salary |      Coef.   Std. Err.      t    P>|t|     [95% Conf. Interval]
-------------+----------------------------------------------------------------
   associate |   7285.121    1026.19     7.10   0.000     5269.049    9301.192
        full |   21267.11   965.8886    22.02   0.000     19369.51    23164.71
       _cons |   39865.86   745.7043    53.46   0.000     38400.84    41330.88
------------------------------------------------------------------------------
```

We see that the estimated difference in population means between associate professors and assistant professors is $7,285 and that the estimated difference in population means between full professors and assistant professors is $21,267. The mean salary for assistant professors is $39,866. We can obtain the estimated population mean salary for associate professors, $\widehat{\beta}_1 + \widehat{\beta}_2$, by using

```
. lincom _cons + associate
( 1)  associate + _cons = 0
```

salary	Coef.	Std. Err.	t	P>|t|	[95% Conf. Interval]	
(1)	47150.98	704.9766	66.88	0.000	45765.97	48535.99

We can obtain the estimated difference in population mean salary between full and associate professors, $(\widehat{\beta}_3 + \widehat{\beta}_1) - (\widehat{\beta}_2 + \widehat{\beta}_1) = \widehat{\beta}_3 - \widehat{\beta}_2$, by using

```
. lincom full - associate
( 1)  - associate + full = 0
```

salary	Coef.	Std. Err.	t	P>|t|	[95% Conf. Interval]	
(1)	13981.99	934.8036	14.96	0.000	12145.45	15818.52

The difference in estimated population mean salaries between full and associate professors ($13,982) is considerably larger than the difference between associate and assistant professors ($7,285), suggesting that the constraint imposed if academic rank were treated as a continuous covariate is unreasonable.

We can fit the same regression model without forming dummy variables ourselves by using factor variables, that is, by simply preceding the categorical covariate(s) with the i. prefix:

```
. regress salary i.rank
```

Source	SS	df	MS		Number of obs =	514
					F(2, 511) =	262.54
Model	4.1753e+10	2	2.0877e+10		Prob > F =	0.0000
Residual	4.0634e+10	511	79518710.1		R-squared =	0.5068
					Adj R-squared =	0.5049
Total	8.2387e+10	513	160599133		Root MSE =	8917.3

salary	Coef.	Std. Err.	t	P>|t|	[95% Conf. Interval]	
rank						
2	7285.121	1026.19	7.10	0.000	5269.049	9301.192
3	21267.11	965.8886	22.02	0.000	19369.51	23164.71
_cons	39865.86	745.7043	53.46	0.000	38400.84	41330.88

Here the 2 and 3 listed under **rank** denote the dummy variables for rank 2 (associate professor) and rank 3 (full professor), respectively. We can refer to the corresponding coefficients as _b[2.rank] and _b[3.rank]. The syntax for the lincom command to compare these coefficients is

```
. lincom 3.rank - 2.rank
( 1)   - 2.rank + 3.rank = 0
```

salary	Coef.	Std. Err.	t	P>|t|	[95% Conf. Interval]
(1)	13981.99	934.8036	14.96	0.000	12145.45 15818.52

The `i.` prefix treats the lowest value of a categorical variable as the reference category. For `rank`, the lowest value is 1 (assistant professor), and we wanted to treat assistant professors as the reference category. If we had instead wanted to treat the value 3 (full professors) as the reference category, we could have replaced `i.rank` by `ib3.rank` or `b3.rank`. Here the "b" stands for base level, another term for reference category. Although factor variables are very convenient, an advantage of constructing your own dummy variables is that you can give them meaningful names.

Instead of using regression with dummy variables, we could use one-way ANOVA with $g = 3$ groups (see table 1.1 on page 18). The one-way ANOVA model with g groups is sometimes written as

$$y_{ij} = \beta + \alpha_j + \epsilon_{ij}, \qquad \sum_{j=1}^{g} \alpha_j = 0, \quad \epsilon_{ij} \sim N(0, \sigma^2)$$

which corresponds to the model considered in section 1.3 with $\mu_j = \beta + \alpha_j$. The Stata command to fit this model is

```
. anova salary rank
```

	Number of obs =	514	R-squared	= 0.5068
	Root MSE = 8917.33		Adj R-squared =	0.5049

Source	Partial SS	df	MS	F	Prob > F
Model	4.1753e+10	2	2.0877e+10	262.54	0.0000
rank	4.1753e+10	2	2.0877e+10	262.54	0.0000
Residual	4.0634e+10	511	79518710.1		
Total	8.2387e+10	513	160599133		

This command produces the same sums of squares, mean squares, and F statistic as given at the top of the regression output, but no estimates of population means or their differences. The F test is a test of the null hypothesis that all three population means are the same, or in other words, that the coefficients β_2 and β_3 of the dummy variables are both zero. The alternative hypothesis is that at least one of the coefficients differs from zero. Such joint or simultaneous hypotheses can also be tested by using `testparm` after fitting the regression model (see below).

Adding the dummy variables for academic rank to the regression model from the previous section, we obtain

```
. regress salary i.male##c.yearsdg marketc i.rank
```

Source	SS	df	MS			Number of obs =	514
						F(6, 507) =	242.32
Model	6.1086e+10	6	1.0181e+10			Prob > F =	0.0000
Residual	2.1301e+10	507	42014709.8			R-squared =	0.7414
						Adj R-squared =	0.7384
Total	8.2387e+10	513	160599133			Root MSE =	6481.9

salary	Coef.	Std. Err.	t	P>|t|	[95% Conf.	Interval]
1.male	-1043.394	1215.034	-0.86	0.391	-3430.516	1343.727
yearsdg	405.2749	86.72844	4.67	0.000	234.8835	575.6663
male#						
c.yearsdg						
1	184.3764	85.06732	2.17	0.031	17.24853	351.5042
marketc	36987.08	1974.888	18.73	0.000	33107.11	40867.06
rank						
2	3349.005	871.6155	3.84	0.000	1636.582	5061.428
3	11168.26	1167.809	9.56	0.000	8873.923	13462.6
_cons	37493.09	988.658	37.92	0.000	35550.72	39435.46

The estimated coefficients of **associate** and **full** are considerably lower than before because they now represent the estimated adjusted or partial differences in population means, holding the other covariates in the model constant. However, interpreting the effect of rank, adjusted for **yearsdg**, may be problematic because rank and years since degree are inherently strongly associated. Therefore, estimating the adjusted difference between full and assistant professors effectively requires extrapolation for full professors to unrealistically low values of **yearsdg** and for assistant professors to unrealistically high values (there is little common support).

We can test the null hypothesis that the coefficients of these dummy variables are both zero by using the **testparm** command:

```
. testparm i.rank

 ( 1)   2.rank = 0
 ( 2)   3.rank = 0

       F( 2,   507) =   52.89
            Prob > F =    0.0000
```

The F statistic is equal to the difference in MSS between the models that do and do not contain the two dummy variables for academic rank (but contain all the other terms of the model), divided by the product of the difference in model degrees of freedom and the MSE of the larger model. (Had we used the dummy variables **associate** and **full** instead of the factor variable **i.rank**, the syntax for the above test would have been **testparm associate full**.)

After controlling for academic rank (and the other covariates), the difference in population mean salary between men and women is estimated as $-\$1043.39 + \$184.38 \times$

yearsdg, which is lower for every year since degree than the estimate of $-\$593.31 + \$227.15 \times$ yearsdg before adjusting for academic rank. For example, at 10 years since degree, the difference in mean salary is now estimated as about \$800:

```
. lincom 1.male + 1.male#c.yearsdg*10
( 1)  1.male + 10*1.male#c.yearsdg = 0
```

salary	Coef.	Std. Err.	t	P>|t|	[95% Conf. Interval]	
(1)	800.3697	727.9737	1.10	0.272	−629.8468	2230.586

The corresponding estimate not adjusting for academic rank was about \$1,678.

We see that the estimated gender effect is smaller when it is adjusted for academic rank. However, adjusting for academic rank could be problematic if rank is a mediating or intervening variable on the causal pathway from gender to salary, that is, if gender affects promotions, which in turn affect salary. After controlling for rank, we obtain an estimate of the direct effect of gender on salary, not mediated by rank (although other intervening variables may be involved). Here we must decide whether we are interested in the direct effect or the total effect of gender on salary (the sum of the direct effect and the indirect effect mediated by rank). If we are interested in the direct effect, we should control for rank, whereas we should not control for rank if we are interested in the total effect. Boudreau et al. (1997) discuss a study arguing that gender does not affect academic rank at Bowling Green State University, in which case the direct effect should be the same as the total effect.

1.10 Other types of interactions

1.10.1 Interaction between dummy variables

Could the salary difference between ranks be gender specific? Equivalently, could the gender gap in salaries depend on academic rank? These questions can be answered by including the two interaction terms male×associate $(x_{2i}x_{5i})$ and male×full $(x_{2i}x_{6i})$ in the model. (We omit the male by yearsdg interaction here for simplicity.)

$$
\begin{aligned}
y_i &= \beta_1 + \beta_2\text{male}_i + \cdots + \beta_5\text{associate}_i + \beta_6\text{full}_i + \beta_7\text{male}_i \times \text{associate}_i \\
&\quad + \beta_8\text{male}_i \times \text{full}_i + \epsilon_i \\
&= \beta_1 + \beta_2\text{male}_i + \cdots + (\beta_5 + \beta_7\text{male}_i)\text{associate}_i + (\beta_6 + \beta_8\text{male}_i)\text{full}_i + \epsilon_i \quad (1.4) \\
&= \beta_1 + (\beta_2 + \beta_7\text{associate}_i + \beta_8\text{full}_i)\text{male}_i + \cdots + \beta_5\text{associate}_i + \beta_6\text{full}_i + \epsilon_i \quad (1.5)
\end{aligned}
$$

If the other terms denoted "$\cdots$" above are omitted, this model becomes a two-way ANOVA model with main effects and an interaction between academic rank and gender.

An interaction between dummy variables can be interpreted as a difference of a difference. For instance, we see from (1.4) that β_7 represents the difference between men and women of the difference between the mean salaries of associate and assistant professors.

We now construct the interactions

```
. generate male_assoc = male*associate
. generate male_full = male*full
```

and fit the regression model including these interactions:

```
. regress salary male marketc yearsdg associate full male_assoc male_full

      Source |       SS         df       MS              Number of obs =      514
-------------+------------------------------             F(  7,   506) =   205.77
       Model |  6.0969e+10       7   8.7099e+09          Prob > F       =   0.0000
    Residual |  2.1418e+10     506   42328437.6          R-squared      =   0.7400
-------------+------------------------------             Adj R-squared =   0.7364
       Total |  8.2387e+10     513   160599133           Root MSE       =     6506

------------------------------------------------------------------------------
      salary |     Coef.   Std. Err.      t    P>|t|     [95% Conf. Interval]
-------------+----------------------------------------------------------------
        male |  465.2322   1118.953     0.42   0.678    -1733.133    2663.598
     marketc |  36950.82   1985.138    18.61   0.000     33050.69    40850.95
     yearsdg |  552.3409   51.93243    10.64   0.000     450.3112    654.3706
   associate |  3008.126   1306.744     2.30   0.022     440.8131    5575.438
        full |  9098.926   1894.294     4.80   0.000     5377.275    12820.58
  male_assoc |  284.0408    1574.62     0.18   0.857    -2809.557    3377.639
   male_full |  2539.387   1919.637     1.32   0.186    -1232.053    6310.826
       _cons |  36397.49   885.9509    41.08   0.000      34656.9    38138.09
------------------------------------------------------------------------------
```

From (1.5), we see that the estimated difference in population mean salaries between men and women is $\widehat{\beta}_2 + \widehat{\beta}_7 \texttt{associate} + \widehat{\beta}_8 \texttt{full}$. In other words, the estimated coefficient $\widehat{\beta}_7$ of $\texttt{male\_assoc}$ can be interpreted as the difference in estimated gender gap between associate and assistant professors, and similarly for $\texttt{male\_full}$. Neither interaction coefficient is significant at the 5% level. However, it is not considered good practice to include only some of the interaction terms for a group of dummy variables representing one categorical variable; hence, we should test both coefficients simultaneously:

```
. testparm male_assoc male_full

 ( 1)   male_assoc = 0
 ( 2)   male_full = 0

       F(  2,   506) =     0.95
            Prob > F =   0.3864
```

There is little evidence for an interaction between gender and academic rank $[F(2, 506) = 0.95, p = 0.39]$.

Using factor variables instead of constructing interaction terms ourselves, the `regress` command becomes

```
. regress salary marketc yearsdg i.male##i.rank
```

Source	SS	df	MS		Number of obs =	514
					F(7, 506) =	205.77
Model	6.0969e+10	7	8.7099e+09		Prob > F =	0.0000
Residual	2.1418e+10	506	42328437.6		R-squared =	0.7400
					Adj R-squared =	0.7364
Total	8.2387e+10	513	160599133		Root MSE =	6506

salary	Coef.	Std. Err.	t	P>|t|	[95% Conf. Interval]	
marketc	36950.82	1985.138	18.61	0.000	33050.69	40850.95
yearsdg	552.3409	51.93243	10.64	0.000	450.3112	654.3706
1.male	465.2322	1118.953	0.42	0.678	-1733.133	2663.598
rank						
2	3008.126	1306.744	2.30	0.022	440.8131	5575.438
3	9098.926	1894.294	4.80	0.000	5377.275	12820.58
male#rank						
1 2	284.0408	1574.62	0.18	0.857	-2809.557	3377.639
1 3	2539.387	1919.637	1.32	0.186	-1232.053	6310.826
_cons	36397.49	885.9509	41.08	0.000	34656.9	38138.09

and the `testparm` command for testing the gender by rank interaction becomes

```
. testparm i.male#i.rank

 ( 1)  1.male#2.rank = 0
 ( 2)  1.male#3.rank = 0

       F( 2,   506) =    0.95
            Prob > F =    0.3864
```

The output above also shows how to refer to the two coefficients representing the interaction between gender and rank, namely, `_b[1.male#2.rank]` and `_b[1.male#3.rank]`.

1.10.2   Interaction between continuous covariates

The effect of marketability, `marketc` (x_{3i}), could increase or decrease with time since degree, `yearsdg` (x_{4i}). We can include an interaction between these two continuous covariates in a regression model:

$$y_i = \beta_1 + \beta_2 x_{2i} + \beta_3 \text{marketc}_i + \beta_4 \text{yearsdg}_i + \cdots + \beta_7 \text{marketc}_i \times \text{yearsdg}_i + \epsilon_i$$

$$= \beta_1 + \beta_2 x_{2i} + (\beta_3 + \beta_7 \text{yearsdg}_i) \text{marketc}_i + \beta_4 \text{yearsdg}_i + \cdots + \epsilon_i \qquad (1.6)$$

$$= \beta_1 + \beta_2 x_{2i} + \beta_3 \text{marketc}_i + (\beta_4 + \beta_7 \text{marketc}_i) \text{yearsdg}_i + \cdots + \epsilon_i \qquad (1.7)$$

We can then fit this model in Stata using the commands

```
. generate market_yrs = marketc*yearsdg
. regress salary male marketc yearsdg associate full market_yrs
```

Source	SS	df	MS		Number of obs =	514
					F(6, 507) =	247.16
Model	6.1397e+10	6	1.0233e+10		Prob > F =	0.0000
Residual	2.0990e+10	507	41401072		R-squared =	0.7452
					Adj R-squared =	0.7422
Total	8.2387e+10	513	160599133		Root MSE =	6434.4

salary	Coef.	Std. Err.	t	P>|t|	[95% Conf. Interval]	
male	926.1298	712.2859	1.30	0.194	-473.2657	2325.525
marketc	46905.65	3455.747	13.57	0.000	40116.31	53695
yearsdg	540.73	51.40369	10.52	0.000	439.7395	641.7205
associate	3303.134	859.6452	3.84	0.000	1614.229	4992.04
full	11573.46	1165.936	9.93	0.000	9282.799	13864.12
market_yrs	-750.4151	214.1251	-3.50	0.000	-1171.097	-329.7334
_cons	36044.19	711.6195	50.65	0.000	34646.1	37442.28

Using (1.6), we see that the estimated effect of marketability, $\widehat{\beta}_3 + \widehat{\beta}_7 \texttt{yearsdg}$, decreases from \$46,906 for faculty who have just completed their degree to $46,906 - \$750.41 \times 30 = \$24,394$ for faculty who completed their degree 30 years ago.

Using factor variables, the model can be fit like this:

```
. regress salary i.male c.marketc##c.yearsdg i.rank
```

Source	SS	df	MS		Number of obs =	514
					F(6, 507) =	247.16
Model	6.1397e+10	6	1.0233e+10		Prob > F =	0.0000
Residual	2.0990e+10	507	41401072		R-squared =	0.7452
					Adj R-squared =	0.7422
Total	8.2387e+10	513	160599133		Root MSE =	6434.4

salary	Coef.	Std. Err.	t	P>|t|	[95% Conf. Interval]	
1.male	926.1298	712.2859	1.30	0.194	-473.2657	2325.525
marketc	46905.65	3455.747	13.57	0.000	40116.31	53695
yearsdg	540.73	51.40369	10.52	0.000	439.7395	641.7205
c.marketc#						
c.yearsdg	-750.415	214.1251	-3.50	0.000	-1171.097	-329.7334
rank						
2	3303.134	859.6452	3.84	0.000	1614.229	4992.04
3	11573.46	1165.936	9.93	0.000	9282.799	13864.12
_cons	36044.19	711.6195	50.65	0.000	34646.1	37442.28

1.11   Nonlinear effects

We have assumed that the relationship between population mean salary and each of the continuous covariates `marketc` and `yearsdg` is linear after controlling for the other variables. However, the difference in population mean salary for each extra year is likely to increase with time since degree (for instance, if percentage increases are constant over time).

Such a nonlinear relationship can be modeled by including the square of `yearsdg` in the model in addition to `yearsdg` itself (adding the interaction `male_years` back in):

```
. generate yearsdg2 = yearsdg^2
. regress salary male marketc yearsdg male_years associate full
> market_yrs yearsdg2
```

Source	SS	df	MS		Number of obs =	514
					F(8, 505) =	192.04
Model	6.2005e+10	8	7.7507e+09		Prob > F =	0.0000
Residual	2.0382e+10	505	40360475		R-squared =	0.7526
					Adj R-squared =	0.7487
Total	8.2387e+10	513	160599133		Root MSE =	6353

salary	Coef.	Std. Err.	t	P>|t|	[95% Conf. Interval]	
male	-1181.825	1251.647	-0.94	0.346	-3640.901	1277.251
marketc	46578.2	3544.482	13.14	0.000	39614.45	53541.95
yearsdg	39.04014	144.8758	0.27	0.788	-245.5933	323.6736
male_years	177.7129	89.36428	1.99	0.047	2.141361	353.2845
associate	4811.533	967.9374	4.97	0.000	2909.853	6713.213
full	12791	1230.256	10.40	0.000	10373.95	15208.05
market_yrs	-726.3863	222.4265	-3.27	0.001	-1163.382	-289.3911
yearsdg2	10.1092	3.970824	2.55	0.011	2.307829	17.91057
_cons	38837.69	1027.381	37.80	0.000	36819.23	40856.16

The estimated coefficient of `yearsdg2` is significantly different from 0 at, say, the 5% level ($t = 2.55$, df = 505, $p = 0.01$), whereas the coefficient of `yearsdg` is no longer statistically significant. It should nevertheless be retained to form a flexible quadratic curve because the minimum of the curve is otherwise arbitrarily forced to occur when `yearsdg` = 0.

After setting the covariates `marketc` (and hence `market_yrs`), `associate`, and `full` to zero, we can visualize the relationship between salary and time since degree for male and female assistant professors in disciplines with mean marketability by using the `twoway function` command, which produces figure 1.15:

```
. twoway (function Women = _b[_cons] + _b[yearsdg]*x + _b[yearsdg2]*x^2,
>         range(0 41) lpatt(dash))
>         (function Men = _b[_cons] + _b[male] + (_b[yearsdg] + _b[male_years])*x
>         + _b[yearsdg2]*x^2, range(0 41) lpatt(solid)),
>         xtitle(Time since degree (years)) ytitle(Mean salary)
```

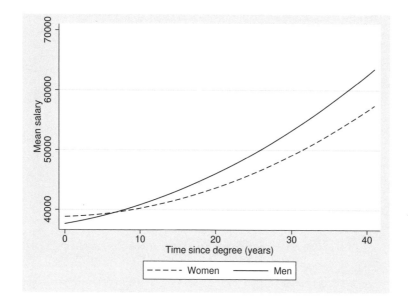

Figure 1.15: Estimated effects of gender and time since degree on mean salary for assistant professors in disciplines with mean marketability

The quadratic term for `yearsdg` can also be included in the regression by using factor variables, simply by using the expression `c.yearsdg#c.yearsdg`. We can therefore use the following command to fit the same model as before:

```
regress salary i.male##c.yearsdg c.marketc c.marketc#c.yearsdg i.rank
    c.yearsdg#c.yearsdg
```

Here we use a double-hash, `##`, in `i.male##c.yearsdg` to include `i.male`, `c.yearsdg`, and `i.male#c.yearsdg`. To avoid duplicating the term `c.yearsdg`, we use a single hash for `c.marketc#c.yearsdg` and then include the missing term `c.marketc`. In fact, it is not necessary to avoid duplication because Stata will drop any redundant terms. However, the terms are then listed in the output together with the label `omitted`. A preferred approach therefore is to factorize out all terms that interact with `c.yearsdg` using

```
. regress salary (i.male c.marketc c.yearsdg)##c.yearsdg i.rank
```

Source	SS	df	MS
Model	6.2005e+10	8	7.7507e+09
Residual	2.0382e+10	505	40360475.1
Total	8.2387e+10	513	160599133

Number of obs =	514
F(8, 505) =	192.04
Prob > F =	0.0000
R-squared =	0.7526
Adj R-squared =	0.7487
Root MSE =	6353

salary	Coef.	Std. Err.	t	P>|t|	[95% Conf. Interval]	
1.male	-1181.825	1251.647	-0.94	0.346	-3640.901	1277.251
marketc	46578.2	3544.482	13.14	0.000	39614.45	53541.95
yearsdg	39.04014	144.8758	0.27	0.788	-245.5933	323.6736
male# c.yearsdg 1	177.7129	89.36428	1.99	0.047	2.141358	353.2845
c.marketc# c.yearsdg	-726.3863	222.4265	-3.27	0.001	-1163.382	-289.3911
c.yearsdg# c.yearsdg	10.1092	3.970824	2.55	0.011	2.307829	17.91057
rank 2	4811.533	967.9374	4.97	0.000	2909.853	6713.213
3	12791	1230.256	10.40	0.000	10373.95	15208.05
_cons	38837.69	1027.381	37.80	0.000	36819.23	40856.16

Stata now makes sure not to duplicate any terms. The estimated regression coefficients can be referred to as _b[1.male], _b[1.male#c.yearsdg], _b[c.marketc#c.yearsdg], _b[c.yearsdg#c.yearsdg], _b[2.rank], and _b[3.rank].

A more flexible curve can be produced by using *higher-order polynomials*, also including yearsdg cubed, which can be expressed as c.yearsdg#c.yearsdg#c.yearsdg, and possibly higher powers. Unless there are specific hypotheses about the shape of the curve, the coefficients of lower powers should be kept in the model. See section 7.3 for a further discussion of polynomials and for an alternative approach to modeling nonlinearity based on linear splines.

Finally, we note that the effect of gender becomes nonsignificant at the 5% level after adding yearsrank, the number of years spent at the current academic rank, to the final regression model presented here.

1.12 Residual diagnostics

Predicted residuals are defined as the differences between the observed responses y_i and the predicted responses $\widehat{y}_i$:

$$\widehat{\epsilon}_i \;=\; y_i - \underbrace{(\widehat{\beta}_1 + \widehat{\beta}_2 x_{2i} + \cdots + \widehat{\beta}_p x_{pi})}_{\widehat{y}_i}$$

Predicted standardized residuals are obtained as

$$r_i \;=\; \frac{\widehat{\epsilon}_i}{s_{ri}}$$

where s_{ri} is the estimated standard error of the residual.

Predicted residuals or standardized residuals can be used to investigate whether model assumptions, such as homoskedasticity and normally distributed errors, are violated. Predicted standardized residuals have the advantage that they have an approximate standard normal distribution if the model assumptions are true. For instance, a value greater than 3 should occur only about 0.1% of the time and may therefore be an outlier.

The postestimation command `predict` with the `residual` option provides predicted residuals for the last regression model that was fit.

```
. predict res, residual
```

Standardized residuals can be obtained by using the `rstandard` option of `predict`.

A histogram of the predicted residuals with an overlayed normal distribution is produced by the `histogram` command with the `normal` option,

```
. histogram res, normal
```

and is presented in figure 1.16.

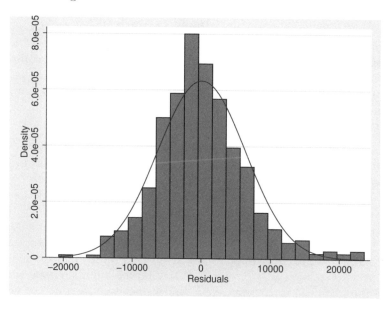

Figure 1.16: Predicted residuals with overlayed normal distribution

The distribution is somewhat skewed, suggesting that salary should perhaps be log-transformed as discussed in section 1.3. Again it may be advisable to use the `vce(robust)` option, which provides standard errors that are not just robust to heteroskedasticity but also to other violations of the distributional assumptions.

1.13 ❖ Causal and noncausal interpretations of regression coefficients

1.13.1 Regression as conditional expectation

Consider the regression model

$$y_i = \beta_1 + \beta_2 T_i + \epsilon_i$$

where y_i is income and T_i is a dummy variable for having a bachelor's degree (taking the value 1 for those with a degree and the value 0 for those without a degree). More generally, $T_i = 1$ could represent "treatment" and $T_i = 0$ could represent "control".

So far, we have followed the conventions of traditional statistics by not making causal interpretations of the regression coefficient β_2. β_2 is merely interpreted as the difference in the conditional expectation of income between those with and those without a degree:

$$\beta_2 = E(y_i|T_i = 1) - E(y_i|T_i = 0)$$

This difference can be due to the causal effect of having a degree on income or due to inherent differences that affect income, such as intelligence, between those who do and do not pursue and earn a degree. The reason for such differences is that subjects are not randomly assigned to getting a degree or not. Instead, there are various mechanisms at play that affect selection into the treatment, having a degree. Variables that affect both selection and income, such as intelligence, are called confounders or lurking variables. If there are confounders, then it is evident that β_2 is not a causal effect.

Some of the combined effect of the confounders on income contributes to β_2, and the remainder contributes to the error term ϵ_i. The part that contributes to β_2 is called the *linear projection* onto T_i, whereas the part that contributes to the error term is called orthogonal to (or uncorrelated with) T_i. In other words, β_2 absorbs all aspects of unobserved variables that are correlated with T_i, rendering ϵ_i uncorrelated with T_i by definition. This is the case for the true model, and consistent estimation refers to estimating the conflated parameter. Why would we want to estimate such a conflated parameter? The parameters β_1 and β_2 would give the best (smallest mean-squared prediction error) prediction $\widehat{y}_i = \beta_1 + \beta_2 T_i$ of the income y_i for someone who happens to have a degree ($T_i = 1$) or someone who happens not to have a degree ($T_i = 0$). In that sense, β_2 is a measure of association.

In many disciplines, confounding is implicitly viewed as inevitable, and the term "causation" is avoided at all cost. Many introductory statistics textbooks state that regression or correlation does not imply causation, and that is the first and last time causality is mentioned.

1.13.2   Regression as structural model

In contrasts to traditional statistics, the regression coefficient of T_i is usually interpreted as causal in econometrics, and the assumptions under which the causal effect can be estimated consistently (or unbiasedly) are stated explicitly.

We will write the causal or *structural* model as

$$y_i = \beta_1^c + \beta_2^c T_i + \epsilon_i^c$$

where β_2^c is now the *causal effect* of having a degree on income, that is, the mean change in income produced by changing degree status, keeping all else constant. The error term ϵ_i^c represents the combined effects of all omitted variables, not just the component that is uncorrelated with T_i.

For unbiased estimation of β_2^c, the *strict exogeneity* assumption $E(\epsilon_i^c | T_i) = 0$ is required. This assumption implies lack of correlation between T_i and ϵ_i^c and this latter, weaker assumption is often loosely referred to as exogeneity. Endogeneity is due to confounders, such as intelligence that have been omitted from the model and are hence part of the error term ϵ_i^c and that are correlated with T_i. If assignment to the treatment and control groups is random, as in a randomized experiment, then T_i is strictly exogenous by design.

The top panel of figure 1.17 illustrates violation of exogeneity. Here x_i is intelligence, the hollow and filled circles represent observations from the group with a degree ($T_i = 1$) and group without a degree ($T_i = 0$), respectively, and the dashed and solid lines represent the corresponding fixed part of the structural model. The vertical distance between these lines is the causal treatment effect β_2^c. We see that the residuals tend to be positive for the degree group and negative for the nondegree group. This positive correlation between the degree dummy variable T_i and the error term ϵ_i is due to the omitted variable, intelligence x_i, which has a higher mean in the degree group than in the nondegree group and is positively correlated with income y_i.

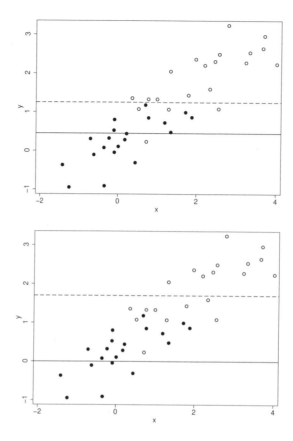

Figure 1.17: Illustration of violation of exogeneity: Top panel shows the structural model, where the errors are correlated with the treatments; bottom panel shows the estimated regression model when x_i is omitted

The bottom panel shows the least-squares regression lines when x_i is omitted. Here the vertical distance between the lines is the estimated effect of treatment, which is too large because of the positive correlation between the error term and the degree dummy variable. It can be shown that for given values of x_i and T_i, under repeated sampling of y_i from the model $y_i = \beta_1^c + \beta_2^c T_i + \beta_x^c x_i + \epsilon_i^c$, the bias of the OLS estimator of β_2^c for the model not containing x_i is given by $r_{Tx}\beta_x^c s_x/s_T$, where r_{Tx} is the correlation between x_i and T_i, β_x^c is the structural parameter of x_i, and s_x and s_T are the standard deviations of x_i and T_i, respectively.

1.14 Summary and further reading

In this chapter, we have shown how linear regression can be used to model the relationship between a continuous response variable and explanatory variables of different types, including continuous, dichotomous, and ordered or unordered polytomous (multicategory) variables. Special cases of such regression models are one-way ANOVA and the model underlying an independent-samples t test.

We have also demonstrated how dummy variables can be used to represent categories of categorical explanatory variables and how products of variables can be used to model interactions, where the effect of each variable is moderated by the other variable. We have shown how nonlinear relationships between the response and a continuous covariate can be modeled using polynomials. Specification of such models is greatly facilitated by Stata's factor variables, introduced in Stata release 11.

We have introduced the idea of a structural model representing the causal effect of covariates in contrast to a regression model for conditional means that makes no causal claims. Exogeneity assumptions that are required for using regression models to estimate causal effects have been briefly discussed. These issues become more complex in multilevel models and are revisited in chapter 3. We refer the interested reader to Morgan and Winship (2007) for a readable introduction to the modern literature on causal inference.

A good elementary introduction to linear regression models is provided by Agresti and Finlay (2007). More advanced, but accessible, introductions that also include regression for other response types, such as logistic regression, include DeMaris (2004) for social science, Vittinghoff et al. (2005) for biomedicine, and Stock and Watson (2011) and Wooldridge (2009) for economics.

The next section, 1.15, contains exercises designed to help reinforce the material discussed in this chapter. Exercises 1.1 to 1.5 involve analysis of four different datasets in Stata, whereas exercises 1.6 and 1.7 concern interpretation of estimated regression coefficients in models that include interactions. Exercise 1.5 concerns the interpretation of regression coefficients when the response variable is log-transformed, a topic not discussed in this chapter. Finally, exercise 1.8 is a self-assessment exercise (with solutions provided) that reviews many of the ideas discussed in this chapter. As indicated by the label $\boxed{\text{Solutions}}$, solutions to exercise 1.1 can be found on the website for this book (http://www.stata-press.com/books/mlmus3).

Brief introductions to logistic regression for dichotomous, ordinal, and nominal responses; discrete-time hazard models for survival or duration data; and Poisson regression for counts are given in the beginning of some chapters in volume 2, before we embark on multilevel versions of these models.

1.15 Exercises

1.1 High-school-and-beyond data $\boxed{\text{Solutions}}$

The data considered here are from the High School and Beyond Survey. They are discussed and analyzed in Raudenbush et al. (2004) and accompany the HLM program (Raudenbush et al. 2004).

The variables in the dataset `hsb.dta` that we will use here are

- `schoolid`: school identifier
- `mathach`: a measure of mathematics achievement
- `ses`: socioeconomic status based on parental education, occupation, and income
- `minority`: dummy variable for student being nonwhite

1. Keep only data on the five schools with the lowest values of `schoolid` (1224, 1288, 1296, 1308, and 1317). Also drop the variables not listed above.

2. Obtain the means and standard deviations for the continuous variables and frequency tables for the categorical variables. Also obtain the mean and standard deviation of the continuous variables for each of the five schools (using the `table` or `tabstat` command).

3. Produce a histogram and a box plot of `mathach`.

4. Produce a scatterplot of `mathach` versus `ses`. Also produce a scatterplot for each school (using the `by()` option).

5. Treating `mathach` as the response variable y_i and `ses` as an explanatory variable x_i, consider the linear regression of y_i on x_i:

$$y_i = \beta_1 + \beta_2 x_i + \epsilon_i, \quad \epsilon_i | x_i \sim N(0, \sigma^2)$$

   a. Fit the model.
   b. Report and interpret the estimates of the three parameters of this model.
   c. Interpret the confidence interval and p-value associated with β_2.

6. Using the `predict` command, create a new variable `yhat` that is equal to the predicted values $\widehat{y}_i$ of `mathach`:

$$\widehat{y}_i = \widehat{\beta}_1 + \widehat{\beta}_2 x_i$$

7. Produce a scatterplot of `mathach` versus `ses` with the regression line (`yhat` versus `ses`) superimposed. Produce the same scatterplot by school. Does it appear as if schools differ in their mean math achievement after controlling for `ses`?

8. Extend the regression model from step 5 by including dummy variables for four of the five schools.

   a. Fit the model with and without factor variables.

> b. Describe what the coefficients of the school dummies represent.
>
> c. Test the null hypothesis that the population coefficients of all four dummy variables are zero (use `testparm`).

9. Add interactions between the school dummies and `ses` using factor variables, and interpret the estimated coefficients.

See also exercise 3.7 for random-intercept models applied to this dataset.

1.2 Anorexia data

Hand et al. (1994) provide data on young girls with anorexia who were randomly assigned to three different treatments: cognitive behavioral therapy, family therapy, and control (treatment as usual). The response variable is the girls' weight in kilograms after treatment, and we also have their weights before treatment.

The variables in `anorexia.dta` are

- `treat`: treatment group (1: cognitive behavioral therapy; 2: control; 3: family therapy)
- `weight1`: weight before treatment (in kilograms)
- `weight2`: weight after treatment (in kilograms)

1. Produce a table of the means and standard deviations of `weight2`, as well as sample sizes by treatment group.
2. Plot box plots and histograms of `weight2` by group as shown in section 1.3.
3. Create dummy variables `cbt` and `ft` for cognitive behavioral therapy and family therapy, respectively.
4. Fit a regression model with `weight2` as the response variable and `weight1`, `cbt`, and `ft` as covariates.
5. Interpret the estimated regression coefficients.
6. For which of the three pairs of treatment groups is there any evidence, at the 5% level, that one treatment is better than the other?
7. Fit the model again, this time relaxing the homoskedasticity assumption. Does this alter your answer for step 6?
8. Plot a histogram of the estimated residuals for this model with a normal density curve superimposed.

1.3 Smoking-and-birthweight data

Here we consider a subset of data on birth outcomes, provided by Abrevaya (2006), which is analyzed in chapter 3. The data were derived from birth certificates by the U.S. National Center for Health Statistics.

The variables in `smoking.dta` that we will consider here are

- `momid`: mother identifier
- `idx`: chronological numbering of multiple children to the same mother in the database (1: first child; 2: second child; 3: third child)

- `birwt`: birthweight (in grams)
- `mage`: mother's age at the birth of the child (in years)
- `smoke`: dummy variable for mother smoking during pregnancy (1: smoking; 0: not smoking)
- `male`: dummy variable for baby being male (1: male; 0: female)
- `hsgrad`: dummy variable for mother having graduated from high school
- `somecoll`: dummy variable for mother having some college education (but no degree)
- `collgrad`: dummy variable for mother having graduated from college
- `black`: dummy variable for mother being black (1: black; 0: white)

1. Keep only the data on each mother's first birth, that is, where `idx` is 1.

2. Create the variable `education`, taking the value 1 if `hsgrad` is 1, the value 2 if `somecoll` is 1, the value 3 if `collgrad` is 1, and the value 0 otherwise.

3. Produce a table of the means and standard deviations of `birwt` for all the subgroups defined by `smoke`, `education`, `male`, and `black`. Hint: Use the `table` command with `smoke` as *rowvar*, `education` as *colvar*, and `male` and `black` as *superrowvars*; see `help table`.

4. Produce box plots for the same groups. Hint: Use the `asyvars` option and the `over()` option for each grouping variable except the last (starting with `over(education)`), and use the `by()` option for the last grouping variable. Use the `nooutsides` option to suppress the display of outliers, making the graph easier to interpret. What do you observe?

5. Regress `birwt` on `smoke` and interpret the estimated regression coefficients.

6. Add `mage`, `male`, `black`, `hsgrad`, `somecoll`, and `collgrad` to the model in step 5.

7. Interpret each of the estimated regression coefficients from step 6.

8. Discuss the difference in the estimated coefficient of `smoke` from steps 5 and 6.

9. Use the `margins` command to produce a table of estimated population means for girls born to white mothers of average age by smoking status and education. (This requires you to run the `regress` command again with the factor variables `i.smoke` and `i.education`.)

10. Extend the model from step 6 to investigate whether the adjusted difference in mean birthweight between boys and girls differs between black and white mothers. Is there any evidence at the 5% level that it does?

1.4 Class-attendance data

This dataset on college students is taken from Wooldridge (2010). The variables in `attend.dta` are

- `stndfnl`: standardized final exam score
- `atndrte`: percent of lectures attended
- `frosh`: dummy variable for being a freshman
- `soph`: dummy variable for being a sophomore
- `priGPA`: prior cumulative GPA (grade-point average)
- `ACT`: score on ACT test (a test used for admission to college)

The questions below are adapted from Wooldridge (2010, 86).

1. To assess the effect of lecture attendance on final exam performance, regress `stndfnl` on `atndrte` and the dummy variables `frosh` and `soph`. Interpret the estimated regression coefficient of `atndrte`.
2. Is the model suitable for estimating the causal effect of attendance? Explain.
3. Add `priGPA` and `ACT` to the regression model in step 1. Now what is the estimated coefficient of `atndrte`? Discuss how the estimate differs from that in step 1.
4. Add the squares of `priGPA` and `ACT` to the model in step 3 by using factor variables. What happens to the estimated coefficient of `atndrte`? Are the two quadratic terms jointly significant at the 5% level?
5. Retain the squares of `priGPA` and `ACT` only if they are jointly significant at the 5% level, and then add the square of `atndrte` to the model. What do you conclude?

1.5 ❖ Faculty salary data

In this exercise, we will revisit the data analyzed in this chapter. Instead of treating salary as the response variable, we will use the log-transform of salary as response. As shown in display 1.1 below, the exponentiated regression coefficients can then be interpreted as multiplicative effects on the mean salary.

1. Generate a variable equal to log-salary, and regress the new variable on a dummy variable for being male.
2. Calculate the exponential of the estimated coefficient of the male dummy variable and interpret it.
3. Add mean-centered marketability, `marketc`, as a further covariate and fit the extended model.
4. Interpret the exponentiated regression coefficients for the male dummy and for `marketc`.

Taking the exponential (antilogarithm) of the log-linear model

$$\ln(y_{ij}) = \beta_1 + \beta_2 x_{2ij} + \cdots + \beta_p x_{pij} + \zeta_j + \epsilon_{ij}$$

we obtain the multiplicative model

$$y_{ij} = \exp(\beta_1) \times \exp(\beta_2)^{x_{2ij}} \times \cdots \times \exp(\beta_p)^{x_{pij}} \times \exp(\zeta_j + \epsilon_{ij})$$

Taking the conditional expectation of y_{ij}, conditional on the covariates $\mathbf{x}_{ij} \equiv (x_{2ij}, \ldots, x_{pij})'$, we get

$$E(y_{ij}|\mathbf{x}_{ij}) = \exp(\beta_1) \times \exp(\beta_2)^{x_{2ij}} \times \cdots \times \exp(\beta_p)^{x_{pij}} \times E\{\exp(\zeta_j + \epsilon_{ij})|\mathbf{x}_{ij}\}$$

Using the fact that

$$E\{\exp(\zeta_j + \epsilon_{ij})|\mathbf{x}_{ij}\} = \exp\{(\psi + \theta)/2\}$$

for the log-normal distribution, we obtain

$$E(y_{ij}|\mathbf{x}_{ij}) = \exp\{\beta_1 + (\psi+\theta)/2\} \times \exp(\beta_2)^{x_{2ij}} \times \cdots \times \exp(\beta_p)^{x_{pij}}$$

Display 1.1: Log-linear models and multiplicative effects

1.6 Interaction I

The table below gives estimates for a multiple regression model fit to data from the 2000 Program for International Student Assessment (PISA). Specifically, the reading score for students from three of the countries was regressed on dummy variables for two of the countries (country=2 for the United Kingdom, country=3 the United States, and country=1 for Germany), a dummy variable test_lan for the test language (English or German depending on the country) being spoken at home, and the interactions between the country and test language dummy variables. The following output was obtained:

Source	SS	df	MS				
Model	1368555.21	5	273711.043				
Residual	39308808	4522	8692.79257				
Total	40677363.2	4527	8985.50104				

Number of obs = 4528
F(5, 4522) = 31.49
Prob > F = 0.0000
R-squared = 0.0336
Adj R-squared = 0.0326
Root MSE = 93.235

wleread	Coef.	Std. Err.	t	P>|t|	[95% Conf. Interval]	
country						
2	114.7891	17.88846	6.42	0.000	79.71894	149.8592
3	42.36503	14.83595	2.86	0.004	13.27931	71.45075
1.test_lan	98.53028	11.30737	8.71	0.000	76.36232	120.6982
country#test_lan						
2 1	-98.53611	18.17351	-5.42	0.000	-134.1651	-62.90716
3 1	-35.66298	15.32532	-2.33	0.020	-65.7081	-5.617867
_cons	423.4007	11.06498	38.26	0.000	401.7079	445.0935

1. Interpret each estimated coefficient.
2. Plot the estimated relationship between mean reading score and test_lan for each of the countries. You may do this using the twoway function command in Stata.
3. Write down the Stata command to fit this model (using factor variables).

1.7 Interaction II

The following estimates were obtained by regressing a continuous measure y_i of fear of crime on the variables age_i, fem_i, and their interaction (using data from the British Crime Survey, 2001–2002):

$$\widehat{y}_i = -0.19 + 0.66fem_i - 0.02age_i - 0.07fem_i \times age_i$$

Here fem_i is a dummy variable for being female, and age_i is the number of 10-year intervals since age 16, that is, $(age - 16)/10$.

1. What is the predicted fear of crime for males and females at age 16?
2. What is the predicted fear of crime for males and females at age 80?
3. Interpret the estimated coefficient of $fem_i \times age_i$.
4. Plot the predicted relationship between fear of crime and age (for the age range 16–90 years) using a separate line for each gender. You may use Stata's twoway function command.
5. Adding age_i^2 to the model gave these estimates:

$$\widehat{y}_i = -0.24 + 0.66fem_i + 0.02age_i - 0.006age_i^2 - 0.06fem_i \times age_i$$

Use Stata to plot the predicted relationship between fear of crime and age (for the age range 16–90 years) using a separate curve for each gender.

1.8 Self-assessment exercise

1. In the simple linear regression model shown below, salary (y_i) is regressed on years of experience (x_i):

$$y_i = \beta_1 + \beta_2 x_i + \epsilon_i, \quad \epsilon_i | x_i \sim N(0, \sigma^2)$$

   a. What are the usual terms used to describe β_1, β_2, ϵ_i, and σ^2?
   b. State the assumptions of this model in words.
   c. If the variables are called `salary` (y_i) and `yearsexp` (x_i) in Stata, write down the command for fitting the model.

2. In a simple linear regression model for salary, the coefficient of a dummy variable for being male is estimated as $3,623.

   a. Interpret the estimated coefficient.
   b. If a dummy variable for being female had been used instead of a dummy variable for being male, what would have been the value of the estimated regression coefficient?
   c. The estimated standard error for the estimated coefficient of the male dummy variable was $1,912. Can you reject the null hypothesis that the population mean salary is the same for men and women by using a two-sided test at the 5% level?

3. What is the relationship between one-way ANOVA and multiple linear regression? Is one model a special case of the other? Explain.

4. In the output for a multiple linear regression model, an F test is given with $F(4, 268) = 12.63$. State the null and alternative hypotheses being tested.

5. In a regression of salary (in U.S. dollars) on age (in years), the intercept is estimated as $-$2,000. Explain how this is possible given that salary must be positive.

6. Using the United States sample of the 2000 Programme for International Student Assessment (PISA) study, the difference in population mean English reading score between those who do and do not speak English at home is estimated as 63 with a standard error of 10. When controlling for socioeconomic status, the adjusted difference in population mean reading score is estimated as 49 with a standard error of 10.

   a. What does it mean to "control" or "adjust for" socioeconomic status?
   b. Under what circumstances does controlling for a variable, x_1, alter the estimated regression coefficient of another variable, x_2?
   c. In the context of this example, explain why the adjusted estimate differs from the unadjusted estimate, paying attention to the direction of the difference.

7. The salaries of a company's employees were regressed on a dummy variable for being male, `male`, and years of experience, `yearsexp`, and their interaction, giving the following results:

$$\widehat{y_i} = \$30{,}000 + \$2{,}000 \, \mathtt{male}_i + \$600 \, \mathtt{yearsexp}_i - \$100 \, \mathtt{male}_i \times \mathtt{yearsexp}_i$$

  a. Interpret each estimated coefficient.
  b. What is the estimated difference in population mean salary between men and women who have 10 years of experience?

8. Regression output for data from the 2000 PISA study is given below. The sample analyzed here included children from the United States, the United Kingdom, and Germany. `wleread` is the reading score, `usa` is a dummy variable for the child being from the United States, `uk` is a dummy variable for the child being from the United Kingdom, and `female` is a dummy variable for the child being female.

```
. regress wleread usa uk female
```

Source	SS	df	MS			Number of obs =	4528
						F(3, 4524) =	44.21
Model	1158538.51	3	386179.502			Prob > F =	0.0000
Residual	39518824.7	4524	8735.37239			R-squared =	0.0285
						Adj R-squared =	0.0278
Total	40677363.2	4527	8985.50104			Root MSE =	93.463

wleread	Coef.	Std. Err.	t	P>|t|	[95% Conf. Interval]	
usa	4.00653	3.715402	1.08	0.281	-3.277473	11.29053
uk	20.30115	3.159347	6.43	0.000	14.10728	26.49501
female	26.13154	2.781327	9.40	0.000	20.67878	31.5843
_cons	504.7174	2.672635	188.85	0.000	499.4778	509.9571

  a. Write down the linear regression model being fit.
  b. Interpret the estimated coefficient of `uk`.
  c. Estimate the difference in population mean reading scores between the United States and the United Kingdom.
  d. What percentage of the variability in reading scores is explained by nationality and gender?
  e. What is the magnitude of the estimated residual variance?
  f. Write down the necessary Stata commands for investigating whether the difference in population mean reading scores between girls and boys differs between countries. Assume that the data contain a variable, `country`, taking the value 1 for Germany, 2 for United States, and 3 for United Kingdom. Give the commands with and without factor variables.

Solutions for self-assessment exercise

1. a. β_1 is the intercept, β_2 is the slope or coefficient of x_i, e_i is the residual or error term for subject i, and σ^2 is the residual variance.

   b. For given x_i, the expectation of y_i is linearly related to x_i, the residuals e_i are normally distributed with zero mean and variance σ^2, and the residuals e_i are independent for different units i. The residual variance is constant for all values of x_i, an assumption called homoskedasticity.

   c. regress salary yearsexp

2. a. The estimated difference in population mean salaries between men and women is $3,623.

   b. $3,623

   c. No, the t statistic is $t = 3,623/1,912 = 1.89$. The smallest t statistic that yields a two-sided p-value of $p < 0.05$ is 1.96, when the degrees of freedom are very large (larger t statistics are required for smaller degrees of freedom), so the test is not significant at the 5% level.

3. One-way ANOVA is a special case of multiple linear regression where the only explanatory variables are the dummy variables for each category of a categorical variable, except the reference category.

4. The null hypothesis is that all true regression coefficients, except the intercept, are zero. The alternative hypothesis is that at least one of the true regression coefficients (except the intercept) is nonzero.

5. The intercept is the estimated population mean salary, or predicted salary, at age 0. This is an extrapolation well outside the range of the data. Even if it were possible to have a salary at age 0, there would be no reason to believe that the relationship between salary and age is linear beginning with age 0.

6. a. Controlling for socioeconomic status means attempting to estimate the difference in mean reading score for two (sub)populations (native and nonnative speakers) having the same socioeconomic status.

   b. When x_1 is associated with both x_2 and y.

   c. Native speakers may have higher mean socioeconomic status, and socioeconomic status may be positively correlated with reading scores, so some of the apparent advantage of native speakers is actually due to their higher mean socioeconomic status.

7. a. The estimated population mean salary is $30,000 for females with no experience and $2,000 greater for males with no experience. For each extra year of experience, the estimated population mean salary increases $600 for females and $100 less (that is, $500) for males.

   b. The difference in population means for males and females after 10 years is estimated as $2,000 - $100 \times 10 = $1,000$.

8. a. $y_i = \beta_1 + \beta_2 \; \mathtt{usa}_i + \beta_3 \; \mathtt{uk}_i + \beta_4 \; \mathtt{female}_i + e_i$ where, for given covariates, e_i has a normal distribution with mean 0 and constant variance σ^2, and is independent of $e_{i'}$ for another student i'.

   b. The estimated coefficient of $\mathtt{uk}$ represents the estimated difference in population mean reading scores between children in the United Kingdom and children in Germany.

   c. The estimated difference in population mean reading scores between children in the United States and children in the United Kingdom is the difference between the estimated coefficients of $\mathtt{usa}$ and $\mathtt{uk}$, giving $4.01 - 20.30 = -16.29$.

d. 2.85%

e. $\widehat{\sigma^2} = 8735.37$ (under `MS` for `Residual`).

f. Without factor variables:

```
generate fem_usa = female*usa
generate fem_uk = female*uk
regress wleread usa uk female fem_usa fem_uk
testparm fem_usa fem_uk
```

With factor variables:

```
regress wleread i.country##i.female
testparm i.country#i.female
```

Part II

Two-level models

2 Variance-components models

2.1 Introduction

Units of observation often fall into groups or clusters. For example, individuals could be nested in families, hospitals, schools, neighborhoods, or firms. Longitudinal data also consist of clusters of observations made at different occasions for the same individual or cluster. For two examples of clustered data, the nesting structure is depicted in figure 2.1.

In clustered data, it is usually important to allow for dependence or correlations among the responses observed for units belonging to the same cluster. For example, the adult weights of siblings are likely to be correlated because siblings are genetically related to each other and have usually been raised within the same family. Variance-components models are designed to model and estimate such within-cluster correlations.

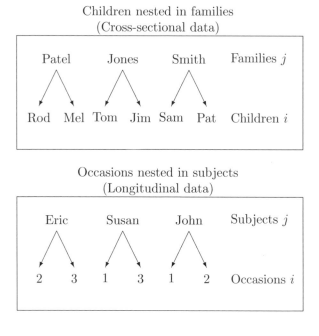

Figure 2.1: Examples of clustered data

In this chapter, we consider the simple situation of clustered data without covariates or explanatory variables. This situation is important in its own right and is also useful for introducing and motivating the notions of random effects and variance components. We also describe basic principles of estimation and prediction in this simple setting. However, this means that some parts of the chapter may be a bit demanding, and you might want to skip sections 2.10 and 2.11 on first reading.

In addition to describing variance-components models, we introduce the Stata commands `xtreg` and `xtmixed`, which will be used throughout volume 1.

2.2   How reliable are peak-expiratory-flow measurements?

The data come from a reliability study conducted by Professor Martin Bland using 17 of his family and colleagues as subjects. The purpose was to illustrate a way of assessing the quality of two instruments for measuring people's peak-expiratory-flow rate (PEFR). The PEFR, which is roughly speaking how strongly subjects can breathe out, is a central clinical measure in respiratory medicine.

The subjects had their PEFR measured twice (in liters per minute) using the standard Wright peak-flow meter and twice using the new Mini Wright peak-flow meter. The methods were used in random order to avoid confounding practice (prior experience) effects with method effects. If the new method agrees sufficiently well with the old, the old method may be replaced with the more convenient Mini meter. Interestingly, the paper reporting this study (Bland and Altman 1986) is the most cited paper in *The Lancet*, one of the most prestigious medical journals.

The data are presented in table 2.1 and are stored in `pefr.dta` in the same form as in the table, with the following variable names:

- `id`: subject identifier
- `wp1`: Wright peak-flow meter, occasion 1
- `wp2`: Wright peak-flow meter, occasion 2
- `wm1`: Mini Wright flow meter, occasion 1
- `wm2`: Mini Wright flow meter, occasion 2

Table 2.1: Peak-expiratory-flow rate measured on two occasions using both the Wright and the Mini Wright peak-flow meters

Subject id	Wright peak-flow meter		Mini Wright peak-flow meter	
	First wp1	Second wp2	First wm1	Second wm2
1	494	490	512	525
2	395	397	430	415
3	516	512	520	508
4	434	401	428	444
5	476	470	500	500
6	557	611	600	625
7	413	415	364	460
8	442	431	380	390
9	650	638	658	642
10	433	429	445	432
11	417	420	432	420
12	656	633	626	605
13	267	275	260	227
14	478	492	477	467
15	178	165	259	268
16	423	372	350	370
17	427	421	451	443

We load the data into Stata using the command

```
. use http://www.stata-press.com/data/mlmus3/pefr
```

In this chapter, we analyze the two sets of measurements from the Mini Wright peak-flow meter only. Analyses comparing the standard Wright and Mini Wright peak-flow meters are discussed in chapter 8.

2.3 Inspecting within-subject dependence

The first and second recordings on the Mini Wright peak-flow meter can be plotted against the subject identifier with a horizontal line representing the overall mean by using

```
. generate mean_wm = (wm1+wm2)/2
. summarize mean_wm
```

Variable	Obs	Mean	Std. Dev.	Min	Max
mean_wm	17	453.9118	111.2912	243.5	650

```
. twoway (scatter wm1 id, msymbol(circle))
>        (scatter wm2 id, msymbol(circle_hollow)),
>        xtitle(Subject id)  xlabel(1/17) ytitle(Mini Wright measurements)
>        legend( order(1 "Occasion 1" 2 "Occasion 2")) yline(453.9118)
```

The resulting graph is shown in figure 2.2.

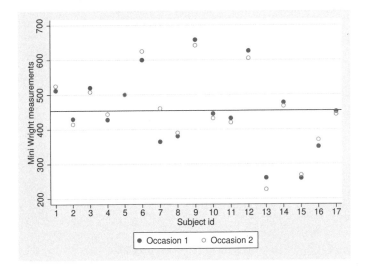

Figure 2.2: First and second measurements of peak-expiratory-flow using Mini Wright meter versus subject number (the horizontal line represents the overall mean)

It may be tempting to model the response y_{ij} of unit (here measurement occasion) i in cluster (here subject) j using a standard regression model without covariates

$$y_{ij} = \beta + \xi_{ij} \tag{2.1}$$

where ξ_{ij} are residuals or error terms that are uncorrelated over both subjects and occasions (the Greek letter ξ is pronounced "xi").

However, it is clear from the figure that repeated measurements on the same subject tend to be closer to each other than to the measurements on a different subject. Indeed, if this were not the case, the Mini Wright peak-flow meter would be useless as a tool for discriminating between the subjects in this sample. Because there are large differences between subjects (for example, compare subjects 9 and 15) and only small differences within subjects, the responses for occasions 1 and 2 on the same subject tend to lie on the same side of the overall mean, shown as a horizontal line in the figure, and are therefore positively correlated (that is, they have a positive covariance, defined as the expectation of products of deviations from the mean). See also section 2.4.4.

We can also see that there is within-subject dependence by considering prediction of a subject's response at occasion 2 if we only know all the subjects' responses at occasion 1. If the response for a given subject at occasion 2 were independent of his or her response at occasion 1, a good prediction would be the mean response at occasion 1 across all subjects. However, it is clear that a much better prediction here is the subject's own response at occasion 1 because the responses are highly dependent within subject.

The within-subject dependence is due to between-subject heterogeneity. If all subjects were more or less alike (for example, pick subjects 2, 4, 10, 11, 14, and 17), there would be much less within-subject dependence.

2.4 The variance-components model

2.4.1 Model specification

As we saw in the previous section and in figure 2.2, it is unreasonable to assume that the deviations ξ_{ij} of y_{ij} from the population mean β are uncorrelated within subjects in the regression model

$$y_{ij} = \beta + \xi_{ij} \tag{2.2}$$

We can model the within-subject dependence by splitting the residual ξ_{ij} into two uncorrelated components: a permanent component ζ_j (ζ is pronounced "zeta"), which is specific to each subject j and constant across occasions i; and an idiosyncratic component ϵ_{ij}, which is specific to each occasion i for each subject j. We then obtain a variance-components model,

$$y_{ij} = \beta + \zeta_j + \epsilon_{ij} \tag{2.3}$$

as shown for subject j in figure 2.3.

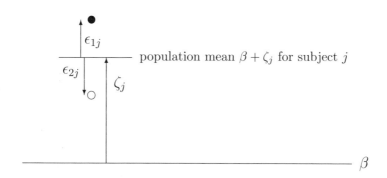

Figure 2.3: Illustration of variance-components model for a subject j

Here ζ_j is the random deviation of subject j's mean measurement (over a hypothetical population of measurement occasions) from the overall mean β. The component

ζ_j, often called a random effect of subject or a *random intercept*, has zero population mean and is uncorrelated across subjects. ζ_j can be viewed as representing individual differences due to personal characteristics not included as variables in the model. The component ϵ_{ij}, often called the level-1 residual or within-subject residual, is the random deviation of y_{ij} from subject j's mean. This residual has zero population mean and is uncorrelated across occasions and subjects.

In classical psychometric test theory, (2.3) represents a measurement model where $\beta + \zeta_j$ is the *true score* for subject j, defined as the long-term mean measurement, and ϵ_{ij} is the measurement error at occasion i for subject j.

The random intercept ζ_j has variance ψ (pronounced "psi"), interpretable as the between-subject variance, and the residual ϵ_{ij} has constant variance θ (pronounced "theta"), interpretable as the within-subject variance.

The model is a simple example of a two-level model, where occasions are level-1 units and subjects are level-2 units or clusters. The random intercept ζ_j is then referred to as the level-2 residual with level-2 (between-subject) variance ψ; ϵ_{ij} is referred to as the level-1 residual with level-1 (between-occasion, within-subject) variance θ.

2.4.2   Path diagram

We can display the random part of the model (every term except β) by using a path diagram or a directed acyclic graph (DAG), as shown in figure 2.4. Here the rectangles represent the observed responses y_{1j} and y_{2j} for each subject j, where the j subscript is implied by the label "subject j" inside the frame surrounding the diagram. The long arrows from ζ_j to the responses represent regressions with slopes equal to 1. The short arrows pointing at the responses from below represent the additive level-1 residuals ϵ_{1j} and ϵ_{2j}.

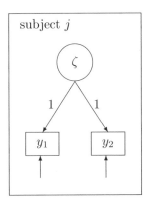

Figure 2.4: Path diagram of random part of random-intercept model

The path diagram makes it clear that the dependence between the two responses is solely due to the shared random intercept. The responses are *conditionally independent* given ζ_j because they are regressed on ζ_j and there is no arrow directly connecting them.[1] (There is also no two-way arrow connecting the level-1 errors ϵ_{1j} and ϵ_{2j} to indicate that they are correlated.) It follows that the responses are conditionally uncorrelated given ζ_j:

$$\text{Cor}(y_{ij}, y_{i'j} | \zeta_j) = 0$$

This can also be seen by imagining that the data in figure 2.2 were generated by the model depicted in figure 2.3, where the dependence is solely due to the measurements being shifted up or down by the shared random intercept ζ_j for each cluster j. One way of conditioning on ζ_j is to imagine a dataset consisting of just one cluster (or consisting of a subset of clusters with identical values of ζ_j). For that dataset, the responses would be uncorrelated. Another way of understanding conditional independence is to consider predicting the response y_{2j} at occasion 2 for a subject. Given that we know ζ_j (and β), knowing y_{1j} would not improve our prediction.

The (marginal) within-subject correlation is induced by ζ_j because this is shared by all responses for the same subject (see section 2.4.4). As we will see in later chapters, path diagrams are useful for conveying the structure of complex models involving several random effects.

2.4.3   Between-subject heterogeneity

Each response differs from the overall mean β by a total residual or error ξ_{ij}, the sum of two error terms or *error components*: ζ_j and ϵ_{ij}

$$\xi_{ij} \equiv \zeta_j + \epsilon_{ij}$$

The random intercept ζ_j is shared between measurement occasions for the same subject j, whereas ϵ_{ij} is unique for each occasion i (and subject).

The variance of the responses becomes $\psi + \theta$

$$\text{Var}(y_{ij}) = E\{(y_{ij} - \underbrace{\beta}_{E(y_{ij})})^2\} = E\{(\zeta_j + \epsilon_{ij})^2\} = E(\zeta_j^2) + 2 \underbrace{E(\zeta_j \epsilon_{ij})}_{\text{Cov}(\zeta_j, \epsilon_{ij})=0} + E(\epsilon_{ij}^2)$$

$$= \psi + \theta$$

which is the sum of *variance components* representing between-subject and within-subject variances. The proportion of the total variance that is between subjects is

$$\rho = \frac{\text{Var}(\zeta_j)}{\text{Var}(y_{ij})} = \frac{\psi}{\psi + \theta} \tag{2.4}$$

---

1. If the arrows between ζ_j and the y_{ij} were reversed, ζ_j would become a so-called "collider" and conditional independence would not hold (for example, Morgan and Winship [2007, chap. 3]).

The coefficient ρ is similar to the coefficient of determination R^2 in linear regression discussed in section 1.5, because it expresses how much of the total variability is "explained" by subjects.

In the measurement context, ψ is the variance of subjects' true scores $\beta + \zeta_j$, θ is the *measurement error variance* (the squared *standard error of measurement*), and ρ is a *reliability*, here a test–retest reliability. Note that the reliability is not just a characteristic of the method; it also depends on the between-subject variance, ψ, which can differ between populations.

2.4.4   Within-subject dependence

Intraclass correlation

The marginal (not conditional on ζ_j) covariance between the measurements on two occasions i and i' for the same subject j is defined as

$$\text{Cov}(y_{ij}, y_{i'j}) = E[\{y_{ij} - E(y_{ij})\}\{y_{i'j} - E(y_{i'j})\}]$$

The corresponding marginal correlation is the above covariance divided by the product of the standard deviations:

$$\text{Cor}(y_{ij}, y_{i'j}) = \frac{\text{Cov}(y_{ij}, y_{i'j})}{\sqrt{\text{Var}(y_{ij})}\sqrt{\text{Var}(y_{i'j})}} \tag{2.5}$$

It follows from the variance-components model that the population means at both occasions are constrained to be equal to β and the standard deviations are constrained to be equal to $\sqrt{\psi + \theta}$. For the variance-components model, the marginal (not conditional on ζ_j) covariance between the measurements therefore equals ψ:

$$\text{Cov}(y_{ij}, y_{i'j}) = E\{(y_{ij} - \underbrace{\beta}_{E(y_{ij})})(y_{i'j} - \underbrace{\beta}_{E(y_{i'j})})\} = E\{(\zeta_j + \epsilon_{ij})(\zeta_j + \epsilon_{i'j})\}$$

$$= E(\zeta_j^2) + \underbrace{E(\zeta_j \epsilon_{i'j})}_{0} + \underbrace{E(\epsilon_{ij}\zeta_j)}_{0} + \underbrace{E(\epsilon_{ij}\epsilon_{i'j})}_{0} = E(\zeta_j^2) = \psi$$

The corresponding correlation, called the *intraclass correlation*, becomes

$$\text{Cor}(y_{ij}, y_{i'j}) = \frac{\text{Cov}(y_{ij}, y_{i'j})}{\sqrt{\text{Var}(y_{ij})}\sqrt{\text{Var}(y_{i'j})}} = \frac{\psi}{\sqrt{\psi + \theta}\sqrt{\psi + \theta}} = \frac{\psi}{\psi + \theta} = \rho$$

Thus ρ, previously given in (2.4), also represents the within-cluster correlation, which cannot be negative in the variance-components model because $\psi \geq 0$. We see that between-cluster heterogeneity and within-cluster correlations are different ways of describing the same phenomenon; both are zero when there is no between-cluster variance ($\psi = 0$), and both increase when the between-cluster variance increases relative to the within-cluster variance.

The intraclass correlation is estimated by simply plugging in estimates for the unknown parameters:

$$\widehat{\rho} = \frac{\widehat{\psi}}{\widehat{\psi} + \widehat{\theta}}$$

Figure 2.5 shows data with an estimated intraclass correlation of $\widehat{\rho} = 0.58$ (left panel) and data with an estimated intraclass correlation of $\widehat{\rho} = 0.87$ (right panel).

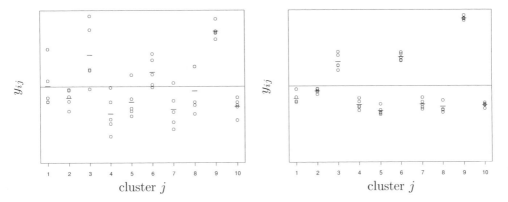

Figure 2.5: Illustration of lower intraclass correlation (left) and higher intraclass correlation (right)

Intraclass correlation versus Pearson correlation

In contrast to the estimated intraclass correlation, the Pearson correlation r is obtained by plugging in separate sample means $\overline{y}_{i\cdot}$ and $\overline{y}_{i'\cdot}$ and sample standard deviations s_{y_i} and $s_{y_{i'}}$ for the two occasions in the estimate of the marginal correlation (2.5),

$$r = \frac{\frac{1}{J-1}\sum_{j=1}^{J}(y_{ij} - \overline{y}_{i\cdot})(y_{i'j} - \overline{y}_{i'\cdot})}{s_{y_i}s_{y_{i'}}}$$

where J is the number of clusters. Here it is not assumed that the population means and standard deviations are constant across occasions.

To give more insight into the interpretation of the estimated intraclass correlation and Pearson correlation, consider what happens if we alter the second Mini Wright peakflow measurements by adding 100 to them, as shown in figure 2.6. (Such a systematic increase could, for instance, be due to a practice effect.) For the variance-components model, it is obvious that the within-cluster variance has increased, giving a much smaller intraclass correlation than for the original data (estimated as 0.63 instead of 0.97). In contrast, the Pearson correlation r is 0.97 in both cases (figures 2.2 and 2.6) because it is based on deviations of the first and second measurements from their *respective* means. In contrast, the intraclass correlation is based on deviations from the *overall* or pooled mean.

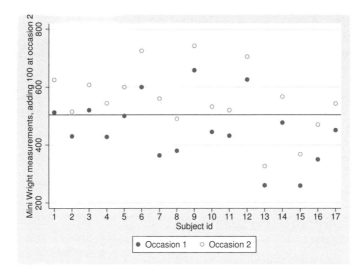

Figure 2.6: First recording of Mini Wright meter and second recording plus 100 versus subject number (the horizontal line represents the overall mean)

The Pearson correlation can be thought of as a measure of *relative agreement*, which refers to how well rankings of subjects based on each measure agree, and is therefore not affected by linear transformations of the measurements. In contrast, the intraclass correlation is a measure of *absolute agreement*.

The intraclass correlation is useful when the units i are *exchangeable* with identical means and standard deviations. For instance, for twin data, there may not even be such a thing as the first and second twin (presuming that birth order is either irrelevant or unknown). Whereas the Pearson correlation can only be obtained by making an arbitrary assignment to y_{1j} and y_{2j} for each twin-pair, the intraclass correlation does not require this. Twins are an example of exchangeable dyads, where the intraclass correlation is more appropriate; married couples are an example of nonexchangeable or distinguishable dyads, where the Pearson correlation between husbands y_{1j} and wives y_{2j} is more appropriate because it is usually difficult to justify that husbands and wives have the same population mean and the same variance. Another difference between the estimated intraclass correlation and the Pearson correlation is that the latter is only defined for pairs of variables, whereas the former summarizes dependence for clusters of size 2 and larger, and clusters of variable sizes; see, for example, exercise 2.4.

2.5   Estimation using Stata

In Stata, maximum likelihood estimates for variance-components models can be obtained using `xtreg` with the `mle` option or `xtmixed` with the `mle` option (the default for `xtmixed` since Stata 12). Restricted maximum likelihood (REML) estimates can be obtained using `xtmixed, reml`, and feasible generalized least-squares (FGLS) estimates can be obtained using `xtreg, re` (the default method). See sections 2.10.2 and 3.10.1 for information on these estimation methods.

The `xtreg` command is the most computationally efficient for variance-components models. However, the postestimation command `predict` for `xtmixed` is more useful than `predict` for `xtreg`. The user-written command `gllamm` can also be used for maximum likelihood estimation of linear variance-components models and the other models discussed in this volume. However, we do not generally recommend using `gllamm` for these models because `xtreg` and `xtmixed` are more computationally efficient, and `gllamm` is sometimes less accurate for linear models for continuous responses. For readers interested in learning to use `gllamm`, either to apply it to noncontinuous responses or to use its extended modeling framework, a `gllamm` companion for this book is available from the `gllamm` website.

2.5.1 Data preparation: Reshaping to long form

We now set up the data for estimation in Stata. Currently, the responses for occasions 1 and 2 are in *wide form* as two separate variables, `wp1` and `wp2` for the Wright peak-flow meter, and `wm1` and `wm2` for the Mini Wright peak-flow meter

```
. list if id < 6, clean noobs
    id    wp1    wp2    wm1    wm2    mean_wm
     1    494    490    512    525      518.5
     2    395    397    430    415      422.5
     3    516    512    520    508        514
     4    434    401    428    444        436
     5    476    470    500    500        500
```

For model fitting, we need to stack the occasion 1 and 2 measurements using a given meter into one variable. We can use the `reshape` command to obtain such a *long form* with one variable, `wp`, for both Wright peak-flow meter measurements; one variable, `wm`, for both Mini Wright peak-flow meter measurements; and a variable, `occasion` (equal to 1 and 2), for the measurement occasion:

```
. reshape long wp wm, i(id) j(occasion)
(note: j = 1 2)
Data                                wide   ->   long

Number of obs.                        17   ->     34
Number of variables                    6   ->      5
j variable (2 values)                      ->   occasion
xij variables:
                                 wp1 wp2   ->   wp
                                 wm1 wm2   ->   wm
```

Note that `i()` is used to specify clusters, denoted j in this book, and `j()` is used to specify units within clusters, denoted i in this book.

The data for the first five subjects now look like this:

```
. list if id < 6, clean noobs
    id   occasion     wp     wm   mean_wm
     1          1     494    512     518.5
     1          2     490    525     518.5
     2          1     395    430     422.5
     2          2     397    415     422.5
     3          1     516    520       514
     3          2     512    508       514
     4          1     434    428       436
     4          2     401    444       436
     5          1     476    500       500
     5          2     470    500       500
```

2.5.2   Using xtreg

We can estimate the parameters of the variance-components model (2.3) using the xtreg command with the mle option, which stands for maximum likelihood estimation (see section 2.10.2).

Before using xtreg and many of the commands starting with xt, the data should be declared as clustered data (referred to as panel data in Stata documentation because longitudinal or panel data are a common example of clustered data) using the xtset command. Here it is sufficient to declare that id is the cluster identifier $j = 1, \ldots, 17$:

```
. xtset id
       panel variable:  id (balanced)
```

The output states that our data are balanced, meaning that the cluster size is constant (here two measurements per subject).

We are now ready to use the xtreg command. As in the regress command, the response variable wm and covariates are listed after the command name. In variance-components models, the fixed part is just the intercept β, which is included by default, so we do not specify any covariates. The random part includes a random intercept ζ_j for the clusters defined in the xtset command. The level-1 residual ϵ_{ij} need not be specified because it is always included. Therefore the command is

```
. xtreg wm, mle

Random-effects ML regression          Number of obs      =        34
Group variable: id                    Number of groups   =        17

Random effects u_i ~ Gaussian         Obs per group: min =         2
                                                     avg =       2.0
                                                     max =         2

                                      Wald chi2(0)       =      0.00
Log likelihood  = -184.57839          Prob > chi2        =         .

─────────────┬──────────────────────────────────────────────────────────
          wm │     Coef.   Std. Err.      z    P>|z|     [95% Conf. Interval]
─────────────┼──────────────────────────────────────────────────────────
       _cons │  453.9118   26.18616    17.33   0.000     402.5878   505.2357
─────────────┼──────────────────────────────────────────────────────────
     /sigma_u │  107.0464   18.67858                      76.0406   150.6949
     /sigma_e │  19.91083   3.414659                      14.2269    27.8656
         rho │  .9665602   .0159494                     .9210943   .9878545
─────────────┴──────────────────────────────────────────────────────────
Likelihood-ratio test of sigma_u=0: chibar2(01)=   46.27 Prob>=chibar2 = 0.000
```

We see in the output that there are 34 observations belonging to 17 groups (the clusters, here subjects) and that there are 2 observations in each group (the minimum, maximum, and hence average number are all 2). In the Stata output and in the Stata documentation for xtreg, the i subscript is used for clusters (instead of j used in this book), u_i is used for the random intercept (instead of ζ_j), and the t subscript is used for occasions (instead of i used in this book).

The estimate of the overall population mean β, given next to _cons in the output, is 453.91. The estimate of the between-subject standard deviation $\sqrt{\psi}$ of the random intercepts of subjects, referred to as /sigma_u, is 107.05, and the estimate of the within-subject standard deviation $\sqrt{\theta}$, referred to as /sigma_e, is 19.91. It follows that the intraclass correlation is estimated as

$$\widehat{\rho} = \frac{\widehat{\psi}}{\widehat{\psi} + \widehat{\theta}} = \frac{107.0464^2}{19.91083^2 + 107.0464^2} = 0.97$$

which is referred to as rho in the output. This estimate is close to 1, indicating that the Mini Wright peak-flow meter is very reliable. The parameter estimates are also given in table 2.2 below.

To be explicit about the structure of the model, we will run the xtset command every time we run xtreg although this is not required if the data have already been xtset in the current Stata session.

2.5.3   Using xtmixed

The variance-components model considered here is a simple special case of a linear mixed-effects model that can be fit using the xtmixed command (available as of Stata 9). The xtmixed command can be used for models with random slopes as well as models with more than one clustering variable (for example, three-level models). The structure of the model is completely specified in the xtmixed command instead of using the xtset command to define any aspect of the model as for `xtreg`. A nice consequence is that it is completely transparent what model is being specified in `xtmixed`.

The fixed part of the model, here β, is specified as in any estimation command in Stata (the response variable followed by a list of covariates). The random part, except the residual ϵ_{ij}, is specified after two vertical bars (or pipes), ||. The cluster identifier, here `id`, is first given to define the clusters j over which ζ_j varies. This is followed by a colon and nothing, because a random intercept ζ_j is included by default (it can be excluded using the `noconstant` option). Finally, we request maximum likelihood estimation by using the `mle` option (the default as of Stata 12).

```
. xtmixed wm || id:, mle

Mixed-effects ML regression                 Number of obs      =        34
Group variable: id                          Number of groups   =        17

                                            Obs per group: min =         2
                                                           avg =       2.0
                                                           max =         2

                                            Wald chi2(0)       =         .
Log likelihood = -184.57839                 Prob > chi2        =         .
```

wm	Coef.	Std. Err.	z	P>|z|	[95% Conf. Interval]
_cons	453.9118	26.18617	17.33	0.000	402.5878 505.2357

Random-effects Parameters	Estimate	Std. Err.	[95% Conf. Interval]
id: Identity			
sd(_cons)	107.0464	18.67858	76.04062 150.695
sd(Residual)	19.91083	3.414678	14.22687 27.86564

```
LR test vs. linear regression: chibar2(01) =    46.27 Prob >= chibar2 = 0.0000
```

The table of estimates for the fixed part of the model has the same form as that for `xtreg` and all Stata estimation commands. The random part is given under the `Random-effects Parameters` header. Here `sd(_cons)` is the estimate of the random-intercept standard deviation $\sqrt{\psi}$, and `sd(Residual)` is the estimate of the standard deviation $\sqrt{\theta}$ of the level-1 residuals. All of these estimates are identical to the estimates using `xtreg` given in table 2.2. We could also obtain estimated variances (instead of standard deviations) with their standard errors by using the `variance` option.

Table 2.2: Maximum likelihood estimates for Mini Wright peak-flow meter

	Est	(SE)
Fixed part		
β	453.91	(26.18)
Random part		
$\sqrt{\psi}$	107.05	
$\sqrt{\theta}$	19.91	
Log likelihood	-184.58	

2.6 Hypothesis tests and confidence intervals

2.6.1 Hypothesis test and confidence interval for the population mean

In the regression tables produced by `xtreg` and `xtmixed`, z statistics are reported for β instead of the t statistics given by the `regress` command.

Like the t statistic in ordinary linear regression, the z statistic for the null hypothesis

$$H_0: \beta = 0 \quad \text{against} \quad H_a: \beta \neq 0$$

is given by

$$z = \frac{\widehat{\beta}}{\widehat{SE}(\widehat{\beta})}$$

(where the standard error takes a different form than in standard linear regression, as discussed in section 2.10.3).

The reason this statistic is called z instead of t is that a standard normal sampling distribution is assumed under the null hypothesis instead of a t distribution. The t distribution is a finite-sample distribution whose shape depends on the degrees of freedom. For the variance-components model, the finite-sample distribution does not have a simple form, so Stata's commands use the asymptotic (large-sample) sampling distribution. (Some other software packages approximate the finite-sample distribution by a t distribution where the degrees of freedom are some function of the data.) The null hypothesis that the population mean β is zero is not of interest in the peak-expiratory-flow example.

Squaring the z statistics gives the Wald statistic, an approximation to the likelihood-ratio statistic, described in section 2.6.2 for testing the between-cluster variance.

An asymptotic 95% confidence interval for β is given by

$$\widehat{\beta} \pm z_{0.975}\widehat{SE}(\widehat{\beta})$$

where $z_{0.975}$ is the 97.5th percentile of the standard normal distribution, that is, $z_{0.975} = 1.96$. This kind of confidence interval based on assuming a normal sampling distribution is often called a Wald confidence interval. In the Mini Wright application, the 95% Wald confidence interval for the population mean β is from 402.59 to 505.24, as shown, for instance, in the output from xtreg on page 85.

As for linear regression, there are two versions of estimated standard errors: a model-based version and a robust version based on the so-called sandwich estimator. The latter can be obtained for the maximum likelihood estimator in xtmixed and for the feasible generalized least-squares (FGLS) estimator in xtreg, re with the vce(robust) option. However, it should be noted that robust standard errors are known to perform poorly in small samples (samples with a small number of clusters).

2.6.2   Hypothesis test and confidence interval for the between-cluster variance

We now consider testing hypotheses regarding the between-cluster variance, ψ. In particular, we are often interested in the hypothesis

$$H_0: \psi = 0 \quad \text{against} \quad H_a: \psi > 0$$

This null hypothesis is equivalent to the hypothesis that $\zeta_j = 0$ or that there is no random intercept in the model. If the null hypothesis is true, we can use ordinary regression instead of a variance-components model.

The test we will be using most in this book for testing variance components is the likelihood-ratio test.

Likelihood-ratio test

A likelihood-ratio test can be used by fitting the model with and without the random intercept. The likelihood-ratio test statistic then is

$$L = 2(l_1 - l_0)$$

where l_1 is the maximized log likelihood for the variance-components model (which includes ζ_j) and l_0 is the maximized log likelihood for the model without ζ_j. Importantly, the distribution of L under H_0 is not χ^2 with 1 degree of freedom (df) as usual. This is because the null hypothesis is on the boundary of the parameter space since $\psi \geq 0$, violating the regularity conditions of standard statistical test theory.

For datasets simulated under the null hypothesis, without the random intercept, we would expect positive within-cluster correlations in about half the datasets and negative within-cluster correlations in the other half. Thus ψ would be estimated as positive half the time and as zero (because ψ would have to be negative to produce negative correlations but is constrained to be nonnegative) the other half the time. The correct asymptotic sampling distribution under the null hypothesis hence takes a simple form, being a 50:50 mixture of a spike at 0 and a χ^2 with 1 df, often written as $0.5\chi^2(0) + 0.5\chi^2(1)$, where $\chi^2(0)$ is spike of height 1 at 0. The correct p-value can be obtained by simply dividing the "naïve" p-value, based on the χ^2 with 1 df, by 2.

This p-value is given at the bottom of the `xtreg` and `xtmixed` output, where the correct sampling distribution is referred to as `chibar2(01)` (click on `chibar2(01)`, which is shown in blue in the Stata Results window to find an explanation). We can also perform the likelihood-ratio test ourselves by fitting the variance-components model, storing the estimates, then fitting the model without the random intercept, and finally comparing the models using the `lrtest` command:

```
. quietly xtmixed wm || id:, mle
. estimates store ri
. quietly xtmixed wm, mle
. lrtest ri .

Likelihood-ratio test                               LR chi2(1)  =      46.27
(Assumption: . nested in ri)                        Prob > chi2 =     0.0000

Note: The reported degrees of freedom assumes the null hypothesis is not on
      the boundary of the parameter space.  If this is not true, then the
      reported test is conservative.
```

Here the `quietly` prefix command is used to suppress output from `xtmixed`. In the `lrtest` command, `ri` refers to the estimates stored under that name, and "." refers to the current (or last) estimates. As the note in the output and the notation `LR chi2(1)` imply, we now have to divide the p-value by 2. We see that the test of the null hypothesis $\psi = 0$ has a very small p-value, and the null hypothesis is rejected at standard significance levels.

Recall that the number of clusters is only 17, perhaps too few to rely on the asymptotic distribution of the likelihood-ratio statistic. The same comment applies to the score test described below.

It does of course not make sense to test the null hypothesis that $\theta = 0$, or in other words that all $\epsilon_{ij} = 0$, because this would force all responses y_{ij} for the same cluster j to be identical.

❖ Score test

There are two approximations to the likelihood-ratio statistic: the Wald statistic and the score or Lagrange multiplier statistic (see display 2.1 below if you are interested in the details). For variance parameters, Wald tests do not perform well but score tests can be used. Breusch and Pagan's Lagrange multiplier test is a *score test* based on a quadratic approximation of the likelihood at $\psi = 0$. The implementation in Stata is based on the FGLS estimator (see section 2.10.2), obtained using the `re` option of `xtreg`:

```
. quietly xtset id
. quietly xtreg wm, re
```

The test is then performed using the postestimation command `xttest0` for `xtreg`

```
. xttest0
Breusch and Pagan Lagrangian multiplier test for random effects
        wm[id,t] = Xb + u[id] + e[id,t]
        Estimated results:
                            |     Var     sd = sqrt(Var)
                   ---------+------------------------------
                        wm  |   12214.63        110.5198
                         e  |   396.4412        19.91083
                         u  |   12187.51        110.397
        Test:   Var(u) = 0
                            chibar2(01) =     15.88
                            Prob > chi2 =     0.0000
```

leading to the same conclusion as before. Here the *p*-value has already been divided by 2, taking into account that the null hypothesis is on the boundary of the parameter space.

The graph below shows twice the log likelihood, $2l(\psi)$, as a function of the parameter ψ. (If there are other parameters, this is twice the profile likelihood, maximized with respect to the other parameters.) The maximum of this curve is at the maximum likelihood estimate $\widehat{\psi}$. We describe three statistics for testing H_0: $\psi = 0$:

- The *likelihood-ratio statistic* is given by $2\{l(\widehat{\psi}) - l(0)\}$ and is represented by the arrow labeled "L" pointing from $2l(\widehat{\psi})$ to $2l(0)$.

- The *Wald statistic* is based on approximating the function $2l(\psi)$ by the dotted quadratic curve at the maximum likelihood estimate $\widehat{\psi}$. The value of this curve at $\psi = 0$ is an approximation of $2l(0)$, and the Wald statistic is the decrease in the quadratic curve from $\widehat{\psi}$ to $\psi = 0$, as shown by the arrow labeled "Wald". Because the quadratic approximation to twice the log likelihood at the mode is $2l(\widehat{\psi}) - \frac{(\psi - \widehat{\psi})^2}{\mathrm{SE}(\widehat{\psi})^2}$, it follows that the Wald statistic is $\{\widehat{\psi}/\mathrm{SE}(\widehat{\psi})\}^2$.

- The *score statistic* is based on approximating $2l(\psi)$ by the dashed quadratic curve at the null hypothesis value, $\psi = 0$. The maximum value of this curve is an approximation of $2l(\widehat{\psi})$; the score statistic is the difference between that maximum and $2l(0)$, as shown by the arrow labeled "Score".

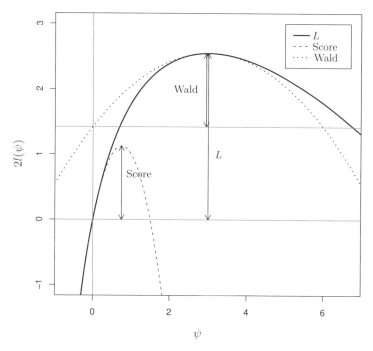

The Wald statistic is obtained by fitting only the unconstrained model, whereas the score statistic is obtained by fitting only the constrained model. A nice feature of the score statistic is that several model extensions can be tested by fitting only one model.

Adapted from Brian Ripley's notes for a course on Applied Statistics at the University of Oxford in 2005.

Display 2.1: Wald and score statistics as approximations to likelihood-ratio statistic

F test

We can also base the test for unexplained between-cluster heterogeneity on a regression model that includes dummy variables for clusters instead of including a random intercept for clusters. Such a model can be thought of as a one-way ANOVA model with clusters as a factor. As will be discussed in section 3.7.2, the model is often called a fixed-effects model because clusters are represented by fixed rather than random effects. The natural test in this setting is an F test for the joint null hypothesis that all clusters have the same mean (or that the coefficients of the 16 dummy variables are all zero). This F test can be obtained using `xtreg` with the `fe` (where `fe` stands for "fixed effects") option:

```
. quietly xtset id

. xtreg wm, fe
Fixed-effects (within) regression          Number of obs     =        34
Group variable: id                         Number of groups  =        17

R-sq:  within  = 0.0000                     Obs per group: min =         2
       between =      .                                    avg =       2.0
       overall =      .                                    max =         2

                                            F(0,17)           =      0.00
corr(u_i, Xb)  =      .                     Prob > F          =         .

------------------------------------------------------------------------------
          wm |      Coef.   Std. Err.      t    P>|t|     [95% Conf. Interval]
-------------+----------------------------------------------------------------
       _cons |   453.9118   3.414679   132.93   0.000     446.7074    461.1161
-------------+----------------------------------------------------------------
     sigma_u |  111.29118
     sigma_e |  19.910831
         rho |  .96898482   (fraction of variance due to u_i)
------------------------------------------------------------------------------
F test that all u_i=0:     F(16, 17) =     62.48             Prob > F = 0.0000
```

We see from the bottom of the output that the null hypothesis is clearly rejected. (Use of the `xtreg` command with the `fe` option is discussed in more detail in section 3.7.2.)

If the uncertainty in the estimated level-1 residual variances is ignored and $\widehat{\theta}$ is treated as the true θ, then the F test can be replaced by a χ^2 test. A similar χ^2 test is described in Raudenbush and Bryk (2002) and implemented in the HLM software of Raudenbush et al. (2004). Instead of basing the χ^2 statistic on the estimated mean $\widehat{\beta}$ from the fixed-effects model, they base it on the estimated mean from the variance-components model.

Confidence intervals

Both `xtmixed` and `xtreg` report confidence intervals for the random-intercept standard deviation. These intervals are obtained by exponentiating the limits of the Wald confidence interval for the log standard deviation. The reason for this is that the sampling distribution of $\log(\widehat{\psi})$ approaches normality faster than that of $\widehat{\psi}$ as the number of clusters increases. The intervals may be adequate, as long as the lower limit is not too close to 0.

The estimated standard errors reported for the estimated between-cluster and within-cluster standard deviations $\sqrt{\psi}$ and $\sqrt{\theta}$ by `xtreg` and `xtmixed` and for the corresponding variances ψ and θ by `xtmixed` with the `variance` option should not be used to construct test statistics or confidence intervals. In particular, when the estimates are small or there are few clusters, the sampling distributions of the estimators may be very different from normal.

2.7 Model as data-generating mechanism

Figure 2.7 shows how the responses y_{ij} can be viewed as resulting from sequential (or hierarchical) sampling, first of ζ_j and then of y_{ij} given ζ_j. For concreteness, we consider normal distributions for ζ_j and ϵ_{ij}, but these distributional assumptions are usually not important for inferences. As seen in the top of the figure, the random intercept ζ_j has a normal distribution with mean zero (and variance ψ). Drawing a realization from this distribution for subject j determines the mean $\beta + \zeta_j$ of the distribution (with variance θ) from which responses y_{ij} for this subject are subsequently drawn. At a given measurement occasion i, a response y_{ij} is therefore sampled from a normal distribution with mean $\beta + \zeta_j$ and variance θ, $y_{ij} \sim N(\beta + \zeta_j, \theta)$ (see the bottom distribution in the figure). Equivalently, a residual (or measurement error) ϵ_{ij} is drawn from a normal distribution with mean zero and variance θ. The hierarchical sampling perspective is the reason why multilevel models are sometimes called hierarchical models.

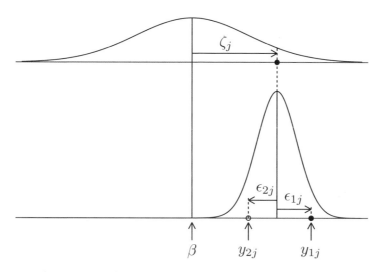

Figure 2.7: Illustration of hierarchical sampling in variance-components model

In this description, we have viewed the variance-components model as the data-generating mechanism for y_{ij} for given occasions and subjects.

The variance-components model is often motivated in terms of two-stage survey sampling, consisting of random sampling of clusters, such as schools, followed by random sampling of units (for example, students) from clusters. In this view, the top distribution of figure 2.7 represents the distribution of cluster means $\beta + \zeta_j$ in the population of clusters, where each cluster comes with a realized value of ζ_j. In stage 1, cluster j is sampled, and the bottom distribution represents the population of units in that cluster, where each unit in the cluster comes with a realized value of ϵ_{ij} and hence y_{ij}. We then randomly draw units from the cluster population, which determines the y_{ij} in our sample.

However, when motivating the model through survey sampling, it is important to remember that the sampling itself does not produce the within-cluster dependence. Such dependence must already exist in the population from which the sample was drawn, which we shall refer to as the finite population (because it is not infinite). Two-stage sampling merely guarantees that the sample contains multiple units per cluster, making it possible to separately estimate the between-cluster and within-cluster variance components ψ and θ. In contrast, a simple random sample of, say, 1,000 students from all U.S. high school students would be unlikely to contain any two students belonging to the same school, so we could only estimate the total variance, $\psi + \theta$, not the separate variance components.

Because the dependence preexists in the finite population, it is more useful to think of the variance-components model as the data-generating mechanism that generated the responses y_{ij} for the finite population. Furthermore, it is the model parameters—β, ψ, and θ—of this underlying variance-components model that we wish to estimate, not any finite population characteristic. Even if the sample contained the entire finite population, that is, all high school students in the U.S., we would still estimate the parameters with error, because all we have observed is one realization from the model, albeit for a large number of clusters and units. This imprecision of parameter estimates is even more pronounced if the finite population is small, for instance, all high school students in Monaco.

When the model is viewed as a data-generating mechanism, randomness comes from drawing the response from a distribution [the ζ_j from $N(0, \psi)$ and the ϵ_{ij} from $N(0, \theta)$, resulting in $y_{ij} = \beta + \zeta_j + \epsilon_{ij}$], not only from sampling units from a finite population. In the survey sampling literature, inference about the data-generating mechanism is referred to as superpopulation inference because data on the finite population (for example, all U.S. high school students) still represent only a sample from the model. Remembering that we wish to make inferences regarding the data-generating mechanism is particularly important in multilevel modeling, where it is not unusual to have data on the entire finite population of clusters (for example, all U.S. states) or the entire finite population of units within clusters (for example, both eyes on each head).

We can think of the randomness in the data as arising from two sources: the data-generating mechanism that produced the y_{ij} for the finite population and survey sampling of a subset of the finite population into the sample. When the parameters of interest are finite population characteristics, such as the proportion of eligible U.S. voters intending to vote for a given presidential candidate, the only randomness we need to care about when making inferences is randomness due to survey sampling, that is, sampling individuals from the finite population (as in an opinion poll). Such inferences, taking into account design features such as stratification, primary sampling units, and sample selection probabilities, are called design-based inferences. (In Stata, design-based inference is performed using the `svyset` command and the `svy` prefix command.)

When estimating model parameters of the data-generating mechanism (or superpopulation or infinite population parameters) rather than finite population characteristics, randomness due to drawing units from the finite population is often ignored, and the inferences are called "model based". For single-level data, ignoring randomness due to survey sampling is legitimate for simple (equal probability) random sampling (conditional on the covariates) because under such a design, the finite population distribution also holds in the sample. This means that the responses in the sample can be viewed as directly drawn from the model. For multilevel data, another design that preserves the distribution is two-stage sampling if simple random sampling (conditional on the covariates) is employed in each stage. Such sampling designs are called "ignorable".

To summarize, in this book we make inferences regarding the parameters of statistical models or data-generating mechanisms under the assumption that the sampling design is ignorable. We see no problem with applying this approach to data that contain the entire population of clusters or the entire population of units within clusters.

2.8   Fixed versus random effects

In the peak-expiratory-flow data, each subject j has a different effect ζ_j on the measured peak-expiratory-flow rates. In analysis of variance (ANOVA) terminology (see sections 1.4 and 1.9), the subjects can therefore be thought of as the levels of a *factor* or categorical explanatory variable. Because the effects of subjects are random, the variance-components model is therefore sometimes referred to as a one-way random-effects ANOVA model.

The one-way random-effects ANOVA model can be written as

$$y_{ij} \;=\; \beta + \zeta_j + \epsilon_{ij}, \qquad E(\epsilon_{ij}|\zeta_j)=0,\; E(\zeta_j)=0,\; \text{Var}(\epsilon_{ij}|\zeta_j)=\theta,\; \text{Var}(\zeta_j)=\psi \qquad (2.6)$$

where ζ_j is a random intercept. In contrast, the one-way fixed-effects ANOVA model is

$$y_{ij} \;=\; \beta + \alpha_j + \epsilon_{ij}, \qquad E(\epsilon_{ij})=0,\; \text{Var}(\epsilon_{ij})=\theta,\; \sum_{j=1}^{J}\alpha_j = 0 \qquad (2.7)$$

where α_j are unknown, fixed, cluster-specific parameters. In the random-effects model, the random intercepts are uncorrelated across clusters and uncorrelated with the level-1 residuals. In both models, the level-1 residuals are uncorrelated across units. Both random-effects models and fixed-effects models include cluster-specific intercepts—ζ_j and α_j, respectively—to account for unobserved heterogeneity. Thus a natural question is whether to use a random- or fixed-effects approach.

One way of answering this question is by being explicit about the target of inference, namely, whether interest concerns the *population* of clusters or the particular clusters in the *dataset*. Here the "population of clusters" refers to the infinite population, or the data-generating mechanism for the clusters. If we are interested in the population of clusters, the random-effects model is appropriate. In that model, β represents the population mean for the population of clusters (and for each cluster, the population of units in the cluster) and ψ represents the variance for the population of clusters. The model specifies how the cluster-specific means $\beta + \zeta_j$ are generated. In the variance-components model, ψ represents between-cluster variability due to cluster-level random (unexplained) processes that affect the response variable. As we will see in later chapters, the data-generating model for the cluster means can also contain cluster-level covariates to explain between-cluster variability.

If we do not wish to generalize beyond the particular clusters in the sample, the fixed-effects model is appropriate. In that model, β represents the mean for the sample of clusters (and for each cluster, the population of units in the cluster). The model allows each cluster to have a different mean $\beta + \alpha_j$ but does not specify how the means are generated. If the cluster means were generated by a random process, we merely condition on their realized values and do not learn about the process. It is not possible to include cluster-level covariates in fixed-effects models, so in this approach, no attempt is made to explain between-cluster variability.

Standard errors, confidence intervals, and p-values are based on the notion of repeated samples of the data from a model. For instance, the standard error for $\widehat{\beta}$ is the standard deviation of the estimates over repeated samples. In the random-effects approach, the random intercepts ζ_j change in repeated samples (in addition to ϵ_{ij}). In the fixed-effects approach, the fixed intercepts α_j remain constant in repeated samples (only ϵ_{ij} changes). As we will see in section 2.10.3, this leads to a larger standard error for $\widehat{\beta}$ in the random-effects approach (compared with the fixed-effects approach) because we are generalizing to the population of clusters and not just making inferences for the particular clusters in the data.

As we will see in section 2.11, the random-effects approach allows the ζ_j to be predicted after estimating the model parameters. In that sense, we can make inferences regarding the effects of clusters in the sample. These predictions can have better properties than the estimates of α_j in the fixed-effects approach, and this is sometimes the reason for adopting a random-effects approach.

A necessary assumption when treating the cluster effects as random is that they are exchangeable in the sense that their joint distribution (across clusters) does not change if the cluster labels j are permuted. In other words, there is no a priori ordering or grouping of the clusters. This assumption may appear unreasonable if the clusters are, for instance, countries. Sweden seems very different from Nigeria. While the same could be said about two individual people, an important difference is that the distinct nature of different countries is known to us a priori. In this case, we can include country-specific covariates in the model, as we will see in the next chapter, and exchangeability is then assumed conditional on the covariates. A fixed-effects approach should be used if it does not make sense at all to think of the clusters in the sample merely as examples of possible clusters—for example, males and females as examples of possible genders—or if the clusters themselves are of such intrinsic interest that we do not want to model them as exchangeable even after conditioning on covariates.

A random-effects approach should be used only if there is a sufficient number of clusters in the sample, typically more than 10 or 20. The reason for this is that the between-cluster variance ψ is poorly estimated if there are few clusters. Poor estimation of ψ translates to poor estimation of the standard error of $\widehat{\beta}$. In two-level models, large-sample properties or asymptotics, such as consistency of point estimates and standard errors, rely on the number of clusters going to infinity, possibly for fixed cluster sizes. In contrast, the fixed-effects approach does not rely on a large number of clusters as long as the total sample size is large.

Regarding cluster sizes, these should be large in the fixed-effects approach if the α_j are of interest. This is also the case in the random-effects approach if prediction of the random effects is of interest, but prediction of ζ_j performs better with small cluster sizes than estimation of α_j because of "shrinkage" (see section 2.11). For parameter estimation in random-effects models, it is only required that there are a good number of clusters of size 2 or more; it does not matter if there are also "clusters" of size 1. Such singleton clusters do not provide information on the within-cluster correlation or on how the total variance is partitioned into ψ and θ, but they do contribute to the estimation of β and $\psi + \theta$.

In the peak-expiratory-flow application, the one-way fixed-effects ANOVA model has 19 parameters ($\beta, \alpha_1, \ldots, \alpha_{17}, \theta$) and one constraint ($\sum_j \alpha_j = 0$). The one-way random-effects ANOVA model is thus much more parsimonious, having only 3 parameters (β, ψ, θ).

2.9   Crossed versus nested effects

So far, we have considered the random or fixed effects of a single cluster variable or factor, subjects. Another potential factor in the peak-expiratory-flow dataset is the measurement occasion with 2 levels, occasions 1 and 2. In the variance-components model, occasion was allowed to have an effect on the response variable only via the residual term ϵ_{ij}, which takes on a different value for each combination of subject and occasion and has mean zero for each occasion across subjects. We have therefore implicitly treated occasions as *nested* within subjects, meaning that occasion (2 versus 1) does not have a systematic effect for all subjects.

If all subjects had been measured and remeasured in the same sessions and if there were anything specific to the session (for example, time of day, temperature, or calibration of the measurement instrument) that could influence measurements on all subjects in a similar way, then subjects and occasions would be *crossed*. We would then include an occasion-specific term ("main effect of occasion") in the model that takes on the same value for all subjects.

The distinction between nested and crossed factors is illustrated in figure 2.8.

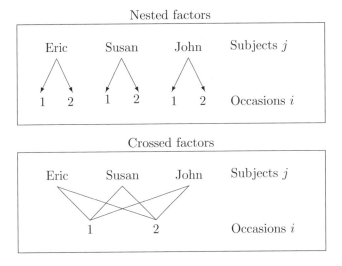

Figure 2.8: Illustration of nested and crossed factors

In the nested case, the effect of occasion 1 (or 2) is different for every subject; in the crossed case, there is a main effect of occasion that is the same for each subject (and possibly an occasion by subject interaction).

In the crossed case, the model can be described as a two-way ANOVA model. A subject-by-occasion interaction could in this case be included in addition to the main effects of each factor. However, because there are no replications for each subject–occasion combination in the peak-expiratory-flow application, an interaction term would be confounded with the error term ϵ_{ij}. If a random effect is specified for subjects and a fixed effect for occasion, we obtain a so-called *mixed-effects two-way* ANOVA *model*. Such a model can be fit by introducing a dummy variable for the second occasion in the fixed part of the model by using the commands

```
. generate occ2 = occasion==2

. xtmixed wm occ2 || id:, mle
```

```
Mixed-effects ML regression                    Number of obs      =         34
Group variable: id                             Number of groups   =         17

                                               Obs per group: min =          2
                                                              avg =        2.0
                                                              max =          2

                                               Wald chi2(1)       =       0.18
Log likelihood = -184.48885                    Prob > chi2        =     0.6714
```

wm	Coef.	Std. Err.	z	P>|z|	[95% Conf. Interval]	
occ2	2.882353	6.793483	0.42	0.671	-10.43263	16.19734
_cons	452.4706	26.40555	17.14	0.000	400.7167	504.2245

Random-effects Parameters	Estimate	Std. Err.	[95% Conf. Interval]	
id: Identity				
sd(_cons)	107.0561	18.67684	76.0523	150.6991
sd(Residual)	19.80624	3.396741	14.15214	27.71927

```
LR test vs. linear regression: chibar2(01) =    46.44 Prob >= chibar2 = 0.0000
```

We could alternatively have used the commands,

```
xtset id
xtreg wm occ2, mle
```

and instead of creating the dummy variable occ2, we could have used the factor-variable notation i.occasion within the xtmixed or xtreg commands.

We see that there is no evidence for an effect of occasion, which in this example could only be interpreted as a practice effect. If there had been considerably more than two occasions, we could have specified a random effect for occasion. Such models with crossed random effects are discussed in chapter 9.

2.10 Parameter estimation

2.10.1 Model assumptions

We now explicitly state a set of assumptions that are sufficient for everything we want to do in this chapter but are not always necessary. We briefly state which assumptions are needed for properties such as consistency and efficiency of the standard estimators for variance-components models discussed in section 2.10.2.

Mean structure and covariance structure

The total residual $\xi_{ij} = \zeta_j + \epsilon_{ij}$ is assumed to have zero expectation:

$$E(\zeta_j + \epsilon_{ij}) \;=\; 0$$

This assumption implies that the expectation or mean of the response, called the *mean structure*, is

$$E(y_{ij}) = \beta$$

If the mean structure is correctly specified, the point estimator $\widehat{\beta}$ of the parameter β will be *consistent*, meaning that $\widehat{\beta}$ approaches β as the sample size tends to infinity. A consistent estimator need not be unbiased in small samples, meaning that the average of $\widehat{\beta}$, over repeated samples, may not equal β. For $\widehat{\beta}$ to be unbiased, the distribution of $\zeta_j + \epsilon_{ij}$ must in general be symmetric (for instance, a normal distribution).

For the covariance structure, it is assumed that the random intercept ζ_j (with variance ψ) and the level-1 residual ϵ_{ij} (with variance θ) are uncorrelated, $\mathrm{Cor}(\epsilon_{ij}, \zeta_j) = 0$. From this it follows that the variance of the total residual is

$$\mathrm{Var}(\zeta_j + \epsilon_{ij}) \;=\; \psi + \theta$$

The ϵ_{ij} are assumed to be uncorrelated across units i, from which it follows that the covariance between total residuals for two units i and i' in the same cluster is

$$\mathrm{Cov}(\zeta_j + \epsilon_{ij}, \zeta_j + \epsilon_{i'j}) \;=\; \psi$$

Both ζ_j and ϵ_{ij} are also uncorrelated across different clusters so that there are no correlations between the total residuals of units in different clusters.

These assumptions imply that the *covariance structure* of the responses is

$$
\begin{aligned}
\mathrm{Var}(y_{ij}) &= \psi + \theta \\
\mathrm{Cov}(y_{ij}, y_{i'j}) &= \psi \quad \text{if } i \neq i' \\
\mathrm{Cov}(y_{ij}, y_{i'j'}) &= 0 \quad \text{if } j \neq j'
\end{aligned}
$$

If both the mean and covariance structure are correct, then the estimators of all parameters in the variance-components model are consistent and asymptotically efficient, and the model-based standard errors are consistent.

An *efficient* estimator is one that has a smaller standard error than any other estimator. Asymptotically efficient estimators acquire that property only asymptotically, as the sample size goes to infinity. For many estimators, the asymptotic sampling distribution is normal, making it easy to construct large-sample confidence intervals and tests. In variance-components models and other two-level models, "large sample" and "asymptotics" refer to the number of clusters going to infinity, possibly with fixed cluster size.

Consistent estimates for β can be obtained even if the covariance structure is not correct. In this case, model-based standard errors will be inconsistent, but robust standard errors can be used instead.

Distributional assumptions

The maximum likelihood estimator is based on the assumption that both ζ_j and $\epsilon_{ij}|\zeta_j$ are normally distributed. (Under normality, the assumptions above that the mean and variance of ϵ_{ij} do not depend on ζ_j actually imply that ϵ_{ij} and ζ_j are independent.) However, normality of the random intercepts and level-1 residuals is not required for consistent estimation of model parameters and standard errors, or for asymptotic normality of the estimators. The assumption does, however, matter in empirical Bayes prediction of the random effects (see section 2.11.2).

2.10.2   Different estimation methods

A classical method for estimating the parameters of statistical models is maximum likelihood (ML). The likelihood function is just the joint probability density of all the observed responses y_{ij}, $(i = 1, \ldots, n_j)$, $(j = 1, \ldots, J)$, as a function of the model parameters β, ψ, and θ. The likelihood contribution for cluster j can be obtained by integrating the joint distribution of the y_{ij} and ζ_j over the random intercept. The product of the likelihood contributions for all clusters is the likelihood, often called the *marginal likelihood* (averaged over ζ_j). The idea is to find parameter estimates $\widehat{\beta}$, $\widehat{\psi}$, and $\widehat{\theta}$ that maximize the likelihood function, thus making the responses appear as likely as possible.

When the data are balanced with the same number of units $n_j = n$ in each of the J clusters ($n = 2$ occasions in the current application), the ML estimators for the two-level variance-components model have relatively simple expressions. The expressions are in terms of the model sum of squares (MSS) and sum of squared errors (SSE) from a one-way ANOVA, treating subjects as a fixed factor (see section 1.4). Here the MSS is the sum of squared deviations of the cluster means $\overline{y}_{.j}$ from the overall mean $\overline{y}_{..}$,

$$\text{MSS} \;=\; \sum_{j=1}^{J}\sum_{i=1}^{n}(\overline{y}_{.j} - \overline{y}_{..})^2, \qquad \overline{y}_{.j} = \frac{1}{n}\sum_{i=1}^{n} y_{ij}, \qquad \overline{y}_{..} = \frac{1}{Jn}\sum_{j=1}^{J}\sum_{i=1}^{n} y_{ij}$$

and the SSE is the sum of squared deviations of the responses from their cluster means,

$$\text{SSE} \;=\; \sum_{j=1}^{J}\sum_{i=1}^{n}(y_{ij} - \overline{y}_{.j})^2$$

The population mean β is estimated by the sample mean,

$$\widehat{\beta} \;=\; \overline{y}_{..}$$

and the ML estimator of the within-cluster variance θ is

$$\widehat{\theta} \;=\; \frac{1}{J(n-1)}\text{SSE} \;=\; \text{MSE}$$

where MSE is the mean squared error from the one-way ANOVA.

The ML estimator of the between-cluster variance ψ is given by

$$\widehat{\psi} = \left\{ \begin{array}{cc} \frac{\text{MSS}}{Jn} - \frac{\widehat{\theta}}{n} & \text{if positive} \\ 0 & \text{otherwise} \end{array} \right.$$

where the subtraction of the second term is required because the level-1 residuals contribute to the MSS. With a small number of clusters, boundary estimates of 0 can occur frequently. The ML estimators for β and θ are unbiased if the model is true, whereas the estimator for ψ has downward bias.

The unbiased moment estimator or ANOVA estimator of ψ is given by

$$\widehat{\psi}^M = \frac{\text{MSS}}{(J-1)n} - \frac{\widehat{\theta}}{n} = \frac{1}{n}\left(\text{MMS} - \text{MSE}\right)$$

where MMS is the model mean square from the one-way ANOVA. The estimate can be negative, making unbiasedness less attractive than it seems. The between-cluster sum of squares is now divided by n times the model degrees of freedom $J-1$ instead of n times J. The difference between the biased ML estimator $\widehat{\psi}$ and the unbiased moment estimator $\widehat{\psi}^M$ becomes small when the number of clusters J is large. In the example considered in this chapter, there are only $J = 17$ clusters, so the difference between ML and ANOVA estimates will not be negligible (see exercise 2.10).

For balanced data, the ANOVA estimator is also the restricted maximum likelihood (REML) estimator. For unbalanced data, the ANOVA, REML, and ML estimators are all different; the latter two estimators are preferable because they are more efficient. The difference between REML and ML is that REML estimates the random-intercept variance taking into account the loss of 1 degree of freedom resulting from the estimation of the overall mean β. In models considered in the next chapters that include covariates, further degrees of freedom are lost because of estimation of additional regression coefficients. Contrary to common belief, REML is not unbiased for ψ when data are unbalanced. Furthermore, it is not clear which method has the smallest mean squared error (MSE).

Another estimation method, particularly popular in econometrics, is feasible generalized least squares (FGLS). For the simple model and the balanced case considered here, the variance-component estimates from FGLS are identical to the ANOVA and REML estimates.

Section 3.10.1 provides more details on ML, REML, and FGLS estimation of models with covariates.

2.10.3 Inference for β

Estimate and standard error: Balanced case

We first consider the balanced case where $n_j = n$. As mentioned in the previous section, the ML estimator $\widehat{\beta}$ of β in the variance-components model is simply the overall sample mean

$$\widehat{\beta} \;=\; \frac{1}{Jn} \sum_{j=1}^{J} \sum_{i=1}^{n} y_{ij} = \frac{1}{J} \sum_{j=1}^{J} \overline{y}_{\cdot j}$$

an unweighted mean of the cluster means. The estimated standard error is given by

$$\widehat{\mathrm{SE}}(\widehat{\beta}) \;=\; \sqrt{\frac{n\widehat{\psi} + \widehat{\theta}}{Jn}} = \sqrt{\frac{\widehat{\psi} + \widehat{\theta}/n}{J}}$$

Remember that β represents the mean of the cluster means for the population of clusters. When the cluster size n is infinite, the cluster means are known with complete precision and uncertainty about β comes only from having a finite sample of J clusters (rather than the infinite population of clusters). The estimated standard error therefore takes the familiar form $\sqrt{\widehat{\psi}/J}$—because $\widehat{\beta}$ is the sample mean of J (precisely known) cluster means, its estimated standard error is the variance of the cluster means divided by the number of clusters.

In the fixed-effects model (with the random ζ_j replaced with fixed α_j; see section 2.8), the estimator of β is the same, but now the estimated standard error is

$$\widehat{\mathrm{SE}}(\widehat{\beta}^F) \;=\; \sqrt{\frac{\widehat{\theta}}{Jn}}$$

which is smaller than the standard error $\widehat{\mathrm{SE}}(\widehat{\beta})$ in the random-effects model if $\widehat{\psi} > 0$. Because β now represents the *sample* mean of the cluster means for the J clusters in the data, the standard error becomes zero when the cluster size n is infinite.

Now consider the model without cluster-specific random or fixed effects (no ζ_j or α_j) that assumes residuals to be independent. Such a single-level model would be used when the nesting of units in clusters is ignored. We refer to the corresponding estimator of β as the pooled ordinary least-squares (OLS) estimator $\widehat{\beta}^{\mathrm{OLS}}$. The estimator is the same as for the random- and fixed-effects models except the estimated standard error is now approximately

$$\widehat{\mathrm{SE}}(\widehat{\beta}^{\mathrm{OLS}}) \;\approx\; \sqrt{\frac{\widehat{\psi} + \widehat{\theta}}{Jn}}$$

where we have approximated the OLS estimate of the residual variance $\widehat{\sigma^2}$ by the sum of the estimated variance components $\widehat{\psi} + \widehat{\theta}$ (the approximation is better for larger n).

We see that

$$\widehat{\mathrm{SE}}(\widehat{\beta}^F) \leq \widehat{\mathrm{SE}}(\widehat{\beta}^{\mathrm{OLS}}) \leq \widehat{\mathrm{SE}}(\widehat{\beta})$$

This relationship is best understood by remembering that the standard error is the standard deviation of the estimates over repeated samples (repeated random draws of y_{ij} for all units). In the fixed-effects case, only the ϵ_{ij} change from sample to sample (with variance θ). In the pooled OLS case, the total residuals change (with variance $\psi+\theta$), but they are drawn independently (because they are assumed to be independent). In contrast, the total residuals are not drawn independently in the random-effects case; they result from drawing ζ_j for all units in the cluster and drawing ϵ_{ij} for each unit. As a consequence, the total residuals $\zeta_j + \epsilon_{ij}$ for a cluster tend to change in the same direction, leading to larger variability in the resulting $\widehat{\beta}$. The difference between the pooled OLS and the random-effects standard error is particularly pronounced if the ζ_j vary considerably (large ψ), and if a change in a ζ_j affects a large number of units (large n).

For the peak-expiratory-flow application, we see from the output of `xtreg` with the `mle` option on page 85 that $\widehat{SE}(\widehat{\beta}) = 26.19$ and from the output of `xtreg` with the `fe` option on page 92 that $\widehat{SE}(\widehat{\beta}^{F}) = 3.41$. We obtain the estimated model-based standard error for the OLS estimator by using the `regress` command,

```
. regress wm
```

Source	SS	df	MS		Number of obs =	34
					F(0, 33) =	0.00
Model	0	0	.		Prob > F =	.
Residual	403082.735	33	12214.6283		R-squared =	0.0000
					Adj R-squared =	0.0000
Total	403082.735	33	12214.6283		Root MSE =	110.52

wm	Coef.	Std. Err.	t	P>|t|	[95% Conf. Interval]	
_cons	453.9118	18.95399	23.95	0.000	415.3496	492.4739

which gives $\widehat{SE}(\widehat{\beta}^{OLS}) = 18.95$. For the application, we see that $\widehat{SE}(\widehat{\beta}) > \widehat{SE}(\widehat{\beta}^{OLS}) > \widehat{SE}(\widehat{\beta}^{F})$, as expected.

Although the clustered nature of the data is not taken into account in the OLS estimator $\widehat{\beta}^{OLS}$ of the population mean, a *sandwich estimator* can be used to produce robust standard errors for $\widehat{\beta}^{OLS}$, taking the clustering into account. Using the `regress` command with the `vce(cluster id)` option, we obtain the following:

```
. regress wm, vce(cluster id)

Linear regression                                        Number of obs =       34
                                                         F(  0,     16) =     0.00
                                                         Prob > F       =        .
                                                         R-squared      =   0.0000
                                                         Root MSE       =   110.52
                                        (Std. Err. adjusted for 17 clusters in id)
```

		Robust			
wm	Coef.	Std. Err.	t	P>|t|	[95% Conf. Interval]
_cons	453.9118	26.99208	16.82	0.000	396.6911 511.1324

The estimated robust standard error is 26.99, which is close to the estimated model-based standard error of 26.19 from ML estimation of the variance-components model. (We have used a sandwich estimator before to obtain robust standard errors for nonclustered data in section 1.6.) By fitting an ordinary regression model with robust standard errors for clustered data instead of fitting variance-components models, we are taking into account within-cluster dependence but treating it as a nuisance, not as a phenomenon we are interested in. We learn nothing about the between and within-cluster variances or intraclass correlation.

Estimate: Unbalanced case

In the unbalanced case, the ML estimator of β under the variance-components model becomes a weighted mean of the cluster means:

$$\widehat{\beta} = \frac{\sum_{j=1}^{J} w_j \overline{y}_{.j}}{\sum_{j=1}^{J} w_j} \quad \text{where} \quad w_j = \frac{1}{\widehat{\psi} + \widehat{\theta}/n_j}$$

Small clusters have a weight similar to large clusters if $\widehat{\theta}$ is small compared with $\widehat{\psi}$. In contrast, the pooled OLS estimator, which disregards clustering and treats the data as single level, is

$$\widehat{\beta}^{\text{OLS}} = \frac{\sum_{j=1}^{J} n_j \overline{y}_{.j}}{\sum_{j=1}^{J} n_j}$$

with cluster means given weights proportional to the cluster sizes n_j. Thus a variance-components model tends to give more weight to smaller clusters than does an ordinary regression model. In the random-effects case, one new observation for a new cluster adds more information regarding the mean for the *population* of clusters than one new observation for a cluster already included in the sample. In pooled OLS, whether the new observation is from a new or existing cluster is immaterial because clustering is ignored.

2.11   Assigning values to the random intercepts

Remember that the cluster-specific intercepts ζ_j are treated as random variables and not as model parameters in multilevel models. However, having obtained estimates $\widehat{\beta}$, $\widehat{\psi}$, and $\widehat{\theta}$ of the model parameters β, ψ, and θ, we may wish to assign values to the random intercepts ζ_j for individual clusters; this would be analogous to obtaining predicted residuals $\widehat{\epsilon}_i$ in ordinary linear regression.

There are a number of reasons why we may want to obtain values for the random intercepts ζ_j for individual clusters. For instance, we will use such assigned values for model diagnostics (see sections 3.9 and 4.8.4), for interpreting and visualizing models (see section 4.8.3), and for inference regarding individual clusters (see section 4.8.5), such as small area estimation (see exercise 3.9) and disease mapping (see section 13.13). An example of the last type would be to assign values to subjects' true expiratory flow $\beta + \zeta_j$ based on the fallible measurements.

It is easy to assign values to the total residuals because $\widehat{\xi}_{ij} = y_{ij} - \widehat{\beta}$. However, the total residuals are partitioned as $\xi_{ij} = \zeta_j + \epsilon_{ij}$, and different methods have been proposed for assigning values to its constituent components ζ_j and ϵ_{ij}. A common feature of the methods is that a value is first assigned to the ζ_j and then ϵ_{ij} is obtained by using the relation $\epsilon_{ij} = \xi_{ij} - \zeta_j$.

Values are assigned to the random intercepts ζ_j by either *prediction* or *estimation*. We continue treating ζ_j as a random variable when prediction is used, whereas the ζ_j are instead viewed as unknown fixed parameters when estimation is used. The predominant approaches to assigning values to the ζ_j are maximum "likelihood" estimation, described in section 2.11.1, and empirical Bayes prediction, described in section 2.11.2.

2.11.1   Maximum "likelihood" estimation

We first substitute the parameter estimate $\widehat{\beta}$ into the variance-components model (2.1), giving

$$y_{ij} \;=\; \widehat{\beta} + \underbrace{\zeta_j + \epsilon_{ij}}_{\xi_{ij}}$$

The ζ_j are now viewed as the only unknown parameters to be estimated. Specifically, for each subject j, we find the value of ζ_j that maximizes the conditional distribution or "likelihood" of the observed responses y_{1j} and y_{2j}, given the random intercept ζ_j,

$$\text{Likelihood}(y_{1j}, y_{2j} | \zeta_j)$$

treating the model parameters as known. This approach of treating ζ_j as an unknown (and fixed) parameter contradicts the original model specification, where ζ_j was treated as a random effect. We put "likelihood" in quotes because it differs from the marginal likelihood that is used to estimate the model parameters in three ways: 1) the model parameters are treated as known, 2) the random effect is treated as an unknown parameter, and 3) the likelihood is based on the data for just one cluster.

We can rearrange the above model by subtracting $\widehat{\beta}$ from y_{ij} to obtain estimated total residuals $\widehat{\xi}_{ij}$ and regard these as the responses:

$$\widehat{\xi}_{ij} \;=\; y_{ij} - \widehat{\beta} = \zeta_j + \epsilon_{ij} \tag{2.8}$$

The ML estimator of ζ_j is simply the cluster mean of the estimated total residual over the n_j occasions (here $n_j = 2$) for which we have data:

$$\widehat{\zeta}_j^{\mathrm{ML}} \;=\; \frac{1}{n_j}\sum_{i=1}^{n_j}\widehat{\xi}_{ij} = \frac{1}{2}(\widehat{\xi}_{1j} + \widehat{\xi}_{2j})$$

Implementation via OLS regression

The model in (2.8) has a different mean ζ_j for each subject, and we can estimate these means by regressing $\widehat{\xi}_{ij}$ on dummy variables for each of the subjects, excluding the overall intercept because this is now redundant. The OLS estimates of the regression coefficients for the subject dummies are the required ML estimates $\widehat{\zeta}_j^{\mathrm{ML}}$ of the ζ_j.

To obtain these estimates in Stata, we first refit the model by using `xtmixed` (with the `quietly` prefix to suppress the output), and then we subtract the predicted fixed part $\widehat{\beta}$, obtained using `predict` with the `xb` option, from the responses:

```
. quietly xtmixed wm || id:, mle
. predict pred, xb
. generate res = wm - pred
```

Next we regress the variable `res` on dummy variables for the 17 subjects, using the factor-variable notation for categorical variables with no base category, `ibn.id`, and with the `noconstant` option to suppress the overall constant:

```
. regress res ibn.id, noconstant
```

Source	SS	df	MS
Model	396343.235	17	23314.308
Residual	6739.5	17	396.441176
Total	403082.735	34	11855.3746

```
Number of obs =      34
F( 17,     17) =   58.81
Prob > F       =  0.0000
R-squared      =  0.9833
Adj R-squared  =  0.9666
Root MSE       =  19.911
```

res	Coef.	Std. Err.	t	P>|t|	[95% Conf. Interval]	
id						
1	64.58823	14.07908	4.59	0.000	34.88396	94.2925
2	-31.41177	14.07908	-2.23	0.039	-61.11604	-1.707504
3	60.08823	14.07908	4.27	0.001	30.38396	89.7925
4	-17.91177	14.07908	-1.27	0.220	-47.61604	11.7925
5	46.08823	14.07908	3.27	0.004	16.38396	75.7925
6	158.5882	14.07908	11.26	0.000	128.884	188.2925
7	-41.91177	14.07908	-2.98	0.008	-71.61604	-12.2075
8	-68.91177	14.07908	-4.89	0.000	-98.61604	-39.2075
9	196.0882	14.07908	13.93	0.000	166.384	225.7925
10	-15.41177	14.07908	-1.09	0.289	-45.11604	14.2925
11	-27.91177	14.07908	-1.98	0.064	-57.61604	1.792496
12	161.5882	14.07908	11.48	0.000	131.884	191.2925
13	-210.4118	14.07908	-14.94	0.000	-240.116	-180.7075
14	18.08823	14.07908	1.28	0.216	-11.61604	47.7925
15	-190.4118	14.07908	-13.52	0.000	-220.116	-160.7075
16	-93.91177	14.07908	-6.67	0.000	-123.616	-64.2075
17	-6.911774	14.07908	-0.49	0.630	-36.61604	22.7925

From the output, we see that, for instance, $\widehat{\zeta}_1^{\mathrm{ML}} = 64.58823$.

Alternatively, after first using `xtreg` with the `mle` option to estimate the model parameters by ML, we can obtain ML estimates of the random intercepts by using `predict` with the `u` option:

```
. quietly xtset id
. quietly xtreg wm, mle
. predict ml2, u
```

Implementation via the mean total residual

The subject-specific means can be calculated using the `egen` command:

```
. egen ml = mean(res), by(id)
```

For the first subject, we get the same result as before:

```
. sort id
. display ml[1]
64.588226
```

In this subsection, we have used all the terminology usually associated with estimating model parameters. However, it is important to remember that ζ_j is not a parameter in the original model. It is only for the purpose of assigning values to ζ_j that we reformulate the problem by treating the original parameters β, ψ, and θ as known constants and the ζ_j as unknown parameters. The likelihood described here must hence be distinguished from the marginal likelihood used to estimate the model parameters.

2.11.2   Empirical Bayes prediction

Having obtained estimates $\widehat{\beta}$, $\widehat{\psi}$, and $\widehat{\theta}$ of the model parameters and treating them as the true parameter values, we can predict values of the random intercepts ζ_j for individual clusters (subjects in the application). Here we continue to treat ζ_j as a random variable, not as a fixed parameter as in ML estimation.

ML estimation of ζ_j uses the responses y_{ij} for subject j as the only information about ζ_j by maximizing the likelihood of observing these particular values:

$$\text{Likelihood}(y_{1j}, y_{2j}|\zeta_j)$$

In contrast, empirical Bayes prediction also uses the *prior distribution* of ζ_j, summarizing our knowledge about ζ_j before seeing the data for subject j:

$$\text{Prior}(\zeta_j)$$

This prior distribution is just the normal distribution specified for the random intercept with zero mean and estimated variance $\widehat{\psi}$. It represents what we know about ζ_j before we have seen the responses y_{1j} and y_{2j} for subject j. For instance, the most likely value of ζ_j is zero. (Obviously, we have already used all responses to obtain the estimate $\widehat{\psi}$, but we now pretend that ψ is known and not estimated.)

Once we have observed the responses, we can combine the prior distribution with the likelihood to obtain the *posterior distribution* of ζ_j given the observed responses y_{1j} and y_{2j}. According to Bayes theorem,

$$\text{Posterior}(\zeta_j|y_{1j}, y_{2j}) \quad \propto \quad \text{Prior}(\zeta_j) \times \text{Likelihood}(y_{1j}, y_{2j}|\zeta_j)$$

where $\propto$ means "proportional to". The posterior of ζ_j represents our updated knowledge regarding ζ_j after seeing the data y_{1j} and y_{2j} for subject j.

The empirical Bayes prediction is just the mean of the posterior distribution with parameter estimates ($\widehat{\beta}$, $\widehat{\psi}$, and $\widehat{\theta}$) plugged in. In a linear model with normal error terms, the posterior is normal and the mean is thus equal to the mode.

Figure 2.9 shows the prior, likelihood, and posterior for a hypothetical example of a subject with $n_j = 2$ responses. In both panels, the estimated total residuals $\widehat{\xi}_{ij}$ are 3 and 5, and the estimated total variance is $\widehat{\psi} + \widehat{\theta} = 5$. In the top panel, 80% of this variance is due to within-subject variability, whereas in the bottom panel, 80% is due to between-subject variability. In both cases, the likelihood (dotted curve) has its maximum at $\zeta_j = 4$, that is, the mode is 4 (see vertical dotted lines). The ML estimate therefore is $\widehat{\zeta}_j^{\text{ML}} = 4$. In contrast, the mode (and mean) of the posterior depends on the relative sizes of the variance components and is 1.33 in the top panel and 3.56 in the bottom panel (see vertical dashed lines). The mean of the posterior lies between the mean of the prior (zero, vertical solid lines) and the mode of the likelihood.

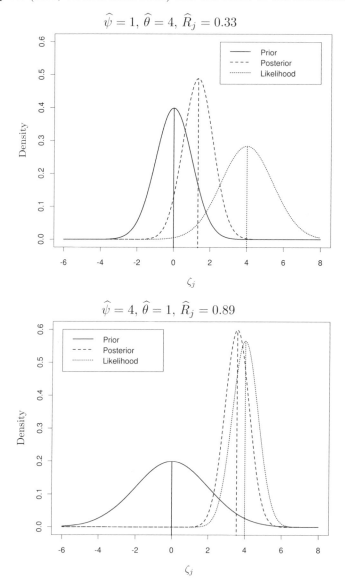

Figure 2.9: Prior distribution, likelihood (normalized), and posterior distribution for a hypothetical subject with $n_j = 2$ responses with total residuals $\widehat{\xi}_{1j} = 3$ and $\widehat{\xi}_{2j} = 5$ [the vertical lines represent modes (and means) of the distributions]

In fact, there is a simple formula relating the empirical Bayes prediction $\widetilde{\zeta}_j^{\mathrm{EB}}$ to the ML estimator $\widehat{\zeta}_j^{\mathrm{ML}}$ in linear random-intercept models:

$$\widetilde{\zeta}_j^{\mathrm{EB}} \;=\; \widehat{R}_j \widehat{\zeta}_j^{\mathrm{ML}}, \qquad \text{where} \qquad \widehat{R}_j \;=\; \frac{\widehat{\psi}}{\widehat{\psi} + \widehat{\theta}/n_j}$$

Here $\widehat{R}_j$ is similar to the estimated intraclass correlation, except that we divide the estimated level-1 variance $\widehat{\theta}$ by the number of responses n_j. $\widehat{R}_j$ can be interpreted as the *reliability* of the ML estimator of ζ_j, defined as the proportion of the variance of the ML estimator that is due to the variance of the random intercept. $\widehat{R}_j$ is also known as the *shrinkage factor* because $0 \leq \widehat{R}_j < 1$ so that the empirical Bayes prediction is shrunken toward 0 (the mean of the prior). There will be more shrinkage (that is, greater influence of the prior) if we have

- a small random-intercept variance $(\widehat{\psi})$ (an informative prior)
- a large level-1 residual variance $(\widehat{\theta})$ (uninformative data)
- a small cluster size (n_j) (uninformative cluster)

A nice feature of empirical Bayes prediction is that the prediction error, defined as the difference $\widetilde{\zeta}_j^{\mathrm{EB}} - \zeta_j$ between the prediction and the truth, has zero mean over repeated samples of ζ_j and ϵ_{ij} (or repeated samples of clusters and units from clusters) when model parameters are treated as fixed and known. Empirical Bayes predictions also have the smallest possible variance (for given model parameters). In linear mixed models, the empirical Bayes predictor is therefore also known as the best linear unbiased predictor (BLUP). When predictions are based on estimated model parameters, the term estimated best linear unbiased predictor (EBLUP) is sometimes used.

Empirical Bayes predictions are conditionally biased in the sense that their mean over repeated samples of ϵ_{ij} for a given ζ_j (or repeated samples of units from the same cluster) will be too close to zero because of shrinkage. In contrast, the ML estimator is conditionally unbiased but has a greater prediction-error variance.

In most applications, shrinkage is desirable because it only affects clusters that provide little information and it effectively downplays their influence, borrowing strength from other clusters. For instance, the empirical Bayes predictor for the cluster mean $\beta + \zeta_j$ becomes

$$\widehat{\beta} + \widetilde{\zeta}_j^{\mathrm{EB}} \;=\; \widehat{\beta} + \widehat{R}_j \widehat{\zeta}_j^{\mathrm{ML}} \;=\; \widehat{\beta} + \widehat{R}_j(\overline{y}_{.j} - \widehat{\beta}) \;=\; (1 - \widehat{R}_j)\widehat{\beta} + \widehat{R}_j \overline{y}_{.j}$$

We see that the empirical Bayes prediction can be expressed as a weighted average of the estimated population mean $\widehat{\beta}$, which is based on data for all clusters, and the cluster mean response $\overline{y}_{.j}$, based only on the data for cluster j. Hence, clusters with low reliabilities borrow more strength from other clusters than do clusters with high reliabilities. Since $\widehat{\beta}$ is based on total pooling of data across clusters, shrinkage estimation is sometimes described as partial pooling.

For the Mini Wright meter measurements, the shrinkage factor can be calculated from the estimates of $\sqrt{\psi}$ and $\sqrt{\theta}$ in table 2.2:

```
display 107 05^2/(107.05^2+(19.91^2)/2)
98299832
```

Instead of typing the rounded estimates, we can access the unrounded counterparts directly after estimation. For regression coefficients, we have already used the syntax _b[*varname*]. To find out how to access the standard deviations, we first refit the model using xtmixed with the estmetric option:

```
xtmixed wm || id:, mle estmetric
```

```
Mixed-effects ML regression                     Number of obs      =        34
Group variable: id                              Number of groups   =        17

                                                Obs per group: min =         2
                                                               avg =       2.0
                                                               max =         2

                                                Wald chi2(0)       =         .
Log likelihood = -184.57839                     Prob > chi2        =         .
```

wm	Coef.	Std. Err.	z	P>|z|	[95% Conf. Interval]	
wm						
_cons	453.9118	26.18617	17.33	0.000	402.5878	505.2357
lns1_1_1						
_cons	4.673263	.1744905	26.78	0.000	4.331268	5.015258
lnsig_e						
_cons	2.991264	.1714985	17.44	0.000	2.655133	3.327395

The estimation metric for each standard deviation is the logarithm of the standard deviation. We can access the estimated logarithms by using the syntax [lns1_1_1]_cons and [lnsig_e]_cons:

```
display exp([lns1_1_1]_cons)
107.04643
display exp([lnsig_e]_cons)
19.910826
```

We can compute the shrinkage factor by using

```
display exp([lns1_1_1]_cons)^2/(exp([lns1_1_1]_cons)^2+ exp([lnsig_e]_cons)^2/2)
98299582
```

We can now obtain empirical Bayes predictions in two ways: either by multiplying the ML estimates obtained in section 2.11.1 by the shrinkage factor,

```
generate eb1 = .98299582*ml
```

or by using the **predict** command with the **reffects** (for "random effects") option after estimation using xtmixed,

```
. predict eb2, reffects
. sort id
. format eb1 eb2 %8.2f
. list id eb1 eb2 if occasion==1, clean noobs
     id        eb1        eb2
      1      63.49      63.49
      2     -30.88     -30.88
      3      59.07      59.07
      4     -17.61     -17.61
      5      45.30      45.30
      6     155.89     155.89
      7     -41.20     -41.20
      8     -67.74     -67.74
      9     192.75     192.75
     10     -15.15     -15.15
     11     -27.44     -27.44
     12     158.84     158.84
     13    -206.83    -206.83
     14      17.78      17.78
     15    -187.17    -187.17
     16     -92.31     -92.31
     17      -6.79      -6.79
```

Both methods give identical results.

It should be kept in mind that empirical Bayes predictions rely on the normality assumptions for the random intercepts unless the cluster sizes are very large.

2.11.3   Empirical Bayes standard errors

There are several different kinds of variances (squared standard errors) for empirical Bayes predictions that can be used to express uncertainty regarding the predictions. These variances do not take into account the uncertainty in the parameter estimates because the parameters are treated as known in empirical Bayes prediction; however, in practice, this only matters in small samples.

Comparative standard errors

The posterior variance is the variance of the random intercept ζ_j given the observed responses, the variance of the posterior distribution shown as a dashed curve in figure 2.9. For the model considered here, the posterior variance is

$$\text{Var}(\zeta_j | y_{1j}, y_{2j}) = \frac{\theta/n_j}{\psi + \theta/n_j}\,\psi = (1 - R_j)\,\psi$$

As expected, the posterior variance is smaller than the prior variance ψ because of the information gained regarding the random intercept by knowing the responses y_{1j} and y_{2j} for cluster (here subject) j.

The posterior variance is also the conditional variance of the prediction errors $\widetilde{\zeta}_j^{\text{EB}} - \zeta_j$, given the responses, $\text{Var}(\widetilde{\zeta}_j^{\text{EB}} - \zeta_j | y_{1j}, y_{2j})$. In the linear variance-components model, the conditional variance equals the unconditional (or marginal) variance of the prediction errors,

$$\mathrm{Var}(\widetilde{\zeta}_j^{\mathrm{EB}} - \zeta_j)$$

over repeated samples of ζ_j and ϵ_{ij} (or repeated samples of clusters j and units i), but with parameters held constant and equal to the estimates. This variance is often referred to as the *mean squared error of prediction* (MSEP). Its square root is referred to as the *comparative standard error* because it can be used for inferences regarding differences between clusters' true random intercepts.

The comparative standard error can be estimated by plugging in estimates for θ and ψ. Such estimated standard errors are produced by the `predict` command after `xtmixed` with the `reses` (for "random effects standard errors") option

```
. predict comp_se, reses
```

These standard errors are identical for all clusters because the clusters have the same size $n_j = 2$ so that $\widehat{R}_j$ is the same for all clusters. We therefore display only the value for the first observation:

```
. display comp_se[1]
13.958865
```

Diagnostic standard errors

For linear variance-components models, the sampling distribution of the empirical Bayes predictions (over repeated samples of ζ_j and ϵ_{ij}, or of clusters and units from clusters) is normal with mean 0 and variance

$$\mathrm{Var}(\widetilde{\zeta}_j^{\mathrm{EB}}) \;=\; \frac{\psi}{\psi + \theta/n_j}\, \psi \;=\; R_j\, \psi$$

This variance is useful for deciding if the empirical Bayes prediction for a given cluster is aberrant. For instance, 95% of predictions should be no larger in absolute value than about two sampling standard deviations. Thus the sampling standard deviation is often called the *diagnostic standard error*.

We can estimate the diagnostic standard error from the estimated prior variance $\widehat{\psi}$ and posterior variance $\mathrm{Var}(\zeta_j|y_{1j}, y_{2j})$ obtained earlier, using the relationship

$$\mathrm{Var}(\widetilde{\zeta}_j^{\mathrm{EB}}) \;=\; R_j\psi \;=\; \psi - (1 - R_j)\,\psi \;=\; \psi - \mathrm{Var}(\zeta_j|y_{1j}, y_{2j})$$

The corresponding estimated standard error can be obtained using

```
. generate diag_se = sqrt(exp([lns1_1_1]_cons)^2 - comp_se^2)
. display diag_se[1]
106.13242
```

If $\widehat{R}_j > 0.5$, as is usually the case in practice, we obtain the following relation among the empirical Bayes variances:

$$\text{Var}(\zeta_j | y_{1j}, y_{2j}) \;=\; \text{Var}(\widetilde{\zeta}_j^{\text{EB}} - \zeta_j) \;<\; \text{Var}(\widetilde{\zeta}_j^{\text{EB}})$$

As we would expect, the relation is satisfied for the Mini Wright data because $\widehat{R}_j = 0.98$.

2.12   Summary and further reading

In this chapter, we introduced the idea of decomposing the total variance of the response variable into variance components, specifically the between-cluster variance ψ and the within-cluster variance θ. This was accomplished by specifying a model that includes corresponding error components; a level-2 random intercept ζ_j for clusters; and a level-1 residual ϵ_{ij} for units within clusters, where ϵ_{ij} is uncorrelated with ζ_j. The random intercept induces correlations among responses for units in the same cluster, known as the *intraclass correlation*.

The random intercept is a random variable and not a model parameter. The realizations of ζ_j change in repeated samples, either because clusters are sampled or because the random intercept is redrawn from the data-generating model for given clusters. An alternative to random intercepts are fixed intercepts. Some guidelines for choosing between random and fixed intercepts were given, and this issue will be revisited in the next chapter.

The concepts discussed in this chapter underlie all multilevel or hierarchical modeling. By considering the simplest case of a multilevel model, we have provided some insight into estimation of model parameters and prediction of random effects. We have also discussed how to conduct hypothesis testing and construct confidence intervals for variance-components models. Although the expressions for estimators and predictors become more complex for the models discussed in later chapters of this volume, the basic ideas remain the same.

For further reading about variance-components models, we recommend Snijders and Bosker (2012, chap. 3), as well as many of the books referred to in later chapters of this volume. Streiner and Norman (2008), Shavelson and Webb (1991), and Dunn (2004) are excellent books on linear measurement models.

The exercises cover a range of applications, such as measurement of peak expiratory flow (exercise 2.1), measurement of psychological distress (exercise 2.2), essay grading (exercise 2.5), neuroticism of twins (exercise 2.3), birthweights of siblings (exercise 2.7), head sizes of brothers (exercise 2.6), and achievement of children nested in neighborhoods and schools (exercise 2.4). Exercise 2.8 is about random-effects meta-analysis, a topic not discussed in this chapter.

2.13   Exercises

2.1 Peak-expiratory-flow data

1. Repeat the analysis of section 2.5 for the Wright peak-flow meter measurements (data are in `pefr.dta`) using `xtmixed`.

2. Compare the estimates with those obtained for the Mini Wright meter (see table 2.2). Does one measurement method appear to be better than the other?

3. Obtain empirical Bayes predictions for both methods, and compare the two sets of predictions graphically.

4. Which method has a smaller prediction-error variance (squared comparative standard error)?

2.2 General-health-questionnaire data

Dunn (1992) reported test–retest data for the 12-item version of Goldberg's (1972) General Health Questionnaire (GHQ) designed to measure psychological distress. Twelve clinical psychology students completed the questionnaire on two occasions, three days apart, giving the scores shown in table 2.3.

Table 2.3: GHQ scores for 12 students tested on two occasions

Student	GHQ1	GHQ2
1	12	12
2	8	7
3	22	24
4	10	14
5	10	8
6	6	4
7	8	5
8	4	6
9	14	14
10	6	5
11	2	5
12	22	16

Source: Dunn (1992).

1. Fit the variance-components model in (2.3) for these data using REML.

2. Obtain ML estimates and empirical Bayes predictions of ζ_j. Produce a scatterplot of empirical Bayes predictions versus ML estimates with a $y = x$ line superimposed. Describe how the graph would change if there were more shrinkage.

3. Extend the model to allow for different means at the two occasions instead of assuming a common mean β as shown in section 2.9. Is there any evidence for a change in mean GHQ scores over time?

2.3 Twin-neuroticism data

Sham (1998) analyzed data on 522 female monozygotic (identical) twin-pairs and 272 female dizygotic (nonidentical or fraternal) twin-pairs.

Specifically, the dataset (from MacDonald 1996) contains scores for the neuroticism dimension of the Eysenck Personality Questionnaire (EPQ). Such twin data are often used to find out to what degree a trait or phenotype (here neuroticism) is due to nature (genes) versus nurture (environment). According to the *equal environment assumption*, monozygotic (MZ) and dizygotic (DZ) twins share the same degree of similarity in their environments, so any excess similarity in neuroticism scores for MZ twins must be due to a greater proportion of shared genes (MZ twins share 100% of their genes, whereas DZ twins only share 50%); see Sham (1998) for a more detailed discussion.

The dataset `twin.dta` has the following variables:

- `twin1`: neuroticism score for twin 1 (the twin with the higher score)
- `twin2`: neuroticism score for twin 2
- `num2`: the number of twin-pairs with a given pair of neuroticism scores
- `dzmz`: a string variable for DZ versus MZ twins (`dz` and `mz`)

1. The data are in collapsed or aggregated form with `num2` representing the number of twin-pairs having a given pair of neuroticism scores. Expand the data by using `expand num2`.
2. Create an identifier for twin-pairs (for example, using `generate pair = _n`) and reshape the data to long form, stacking the neuroticism scores into one variable.
3. Fit the variance-components model in (2.3) separately for MZ and DZ twins by ML.
4. Compare the estimated variance components and total variances between MZ and DZ twins. Do these estimates suggest that there is a genetic contribution to the variability in neuroticism?
5. Obtain the estimated intraclass correlations. Again, do these estimates suggest that there is a genetic contribution to the variability in neuroticism?
6. Why should the Pearson correlation not be used for these data?

See also exercise 8.10 for biometrical genetic modeling of the same data.

2.4 Neighborhood-effects data

Garner and Raudenbush (1991), Raudenbush and Bryk (2002), and Raudenbush et al. (2004) considered neighborhood effects on educational attainment for young people who left school between 1984 and 1986 in one education authority in Scotland.

The dataset `neighborhood.dta` has the following variables:

- `attain`: a measure of end-of-school educational attainment, capturing both attainment and length of schooling (based on the number of O-grades and Higher SCE awards at the A–C levels)
- `neighid`: neighborhood identifier
- `schid`: school identifier

Educational attainment (`attain`) is the response variable.

1. Fit a variance-components model for students nested in schools by ML using `xtmixed`. Obtain the estimated intraclass correlation.
2. Fit a variance-components model for students nested in neighborhoods by ML using `xtmixed`. Obtain the estimated intraclass correlation.
3. Do neighborhoods or schools appear to have a greater influence on educational attainment?

See exercise 3.1 for random-intercept modeling with covariates, and see exercise 9.5 for crossed random-effects modeling using this dataset.

2.5 Essay-grading data

Here we consider a subset of data from Johnson and Albert (1999) on grades assigned to 198 essays by five experts. The grades are on a 10-point scale with 10 being "excellent".

The dataset `grader1.dta` has the following variables:

- `essay`: identifier for essays
- `grade1`: grade from grader 1 on 10-point scale
- `grade4`: grade from grader 4 on 10-point scale

1. Reshape the data to stack the grades from graders 1 and 4 into one variable.
2. Fit a linear variance-components model for the essay grades with variance components within and between graders. Use ML estimation in `xtmixed`.
3. Obtain the estimated intraclass correlation, here interpretable as an inter-rater reliability.
4. Include a dummy variable for grader 4 in the fixed part of the model to allow for bias between the graders, as shown in section 2.9. Does one grader appear to be more generous than the other?
5. Obtain empirical Bayes predictions (using `xtmixed` with the `reffects` option) and plot them using a histogram.

2.6 Head-size data

Frets (1921) analyzed data on the adult head sizes of the first two sons of 25 families. Both the length and breadth of each head were measured in millimeters (mm), and the data are provided by Hand et al. (1994).

The variables in the dataset `headsize.dta` are

- `length1`: length of head of first son (mm)
- `breadth1`: breadth of head of first son (mm)
- `length2`: length of head of second son (mm)
- `breadth2`: breadth of head of second son (mm)

1. We will use a variance-components model to estimate the intraclass correlation between the head lengths of sons nested in families. Why wouldn't it make sense to use this approach for obtaining the intraclass correlation between the head lengths and head breadths nested in heads (or men)?

2. Stack the head lengths of both sons into a variable, `length`, and the breadths into a variable, `breadth`.

3. Fit a linear variance-components model for head length by using `xtreg` and `xtmixed`, both with the `mle` option (the results from both commands should be the same).

4. Interpret the estimated intraclass correlation.

5. Extend the model to allow the mean head length to differ between the first-born and second-born sons (see section 2.9).

    a. Write down the model.
    b. Fit the model by using `xtmixed` with the `mle` option.
    c. Interpret the results.

2.7 Georgian-birthweight data [Solutions]

Adams et al. (1997) analyzed a dataset on all live births that occurred in Georgia, U.S.A, from 1980 to 1992 (regardless of maternal state of residence), or that occurred in other states to Georgia residents and for which birth certificates were sent to Georgia. They linked data on births to the same mother using 27 different variables (maternal social security number was often missing or inaccurately recorded; see the paper for details). Following Neuhaus and Kalbfleisch (1998) we will use a subset of the linked data, including only births to mothers for whom five births were identified.

We will use the following variables in `birthwt.dta`:

- `mother`: mother identifier
- `child`: child identifier
- `birthwt`: child's birthweight (in grams)

1. Fit a variance-components model to the birthweights by using `xtmixed` with the `mle` option, treating children as level 1 and mothers as level 2.

2. At the 5% level, is there significant between-mother variability in birthweights? Fully report the method and result of the test.

3. Obtain the estimated intraclass correlation and interpret it.

4. Obtain empirical Bayes predictions of the random intercept and plot a histogram of the empirical Bayes predictions.

See also exercise 3.5 for further analysis of these data.

2.8 ❖ Teacher expectancy meta-analysis data $\boxed{\text{Solutions}}$

According to Raudenbush and Bryk (2002, 210–211), "the hypothesis that teachers' expectations influence pupils' intellectual development as measured by IQ (intelligence quotient) scores has been the source of sustained and acrimonious controversy for over 20 years."

Raudenbush (1984) found 19 reports of experiments testing this hypothesis. In each study, children were assigned either to an experimental group or to a control group. Teachers were encouraged to have high expectations of the children in the experimental group, whereas no particular expectations were encouraged of the children in the control group.

Raudenbush (1984) analyzed the study results using a meta-analysis. The purpose of a meta-analysis is to pool data from several studies to obtain a more precise estimate of the effect of the intervention (effect size) than provided by any individual study. Typically, the data for a meta-analysis are the estimated effect sizes y_j for the individual studies j (because individual data are rarely published) and the estimated standard errors s_j of the estimated effect sizes. In the teacher expectancy meta-analysis, the effect size is the difference in mean IQ between the experimental and the control groups divided by the pooled within-group standard deviation.

The variables in `expectancy.dta` that we will use here are

- `est`: estimated effect size (y_j) (standardized mean difference)
- `se`: estimated standard error (s_j)

In a random-effects meta-analysis, it is acknowledged that there could be systematic differences between studies, such as target population, implementation of the intervention, or measurement of the outcome. Every study is therefore assumed to have a different true effect size $\beta + \zeta_j$, where β is the population mean effect size and ζ_j is a study-specific random intercept. The estimated effect size for study j differs from its true effect size by a random estimation error e_j with standard deviation estimated by s_j. The model therefore is

$$y_j = \beta + \zeta_j + e_j$$

where ζ_j and e_j are uncorrelated across studies and uncorrelated with each other. ζ_j has zero mean and variance τ^2 (the Greek letter τ is pronounced "tau") to be estimated. e_j has zero mean and variance set equal to the estimated squared standard error, s_j^2. A good book on meta-analysis is Borenstein et al. (2009).

1. Fit the model above by ML using the user-written command `metaan` (Kontopantelis and Reeves 2010). The program can be installed (if your computer is connected to the Internet) using `ssc install metaan`. The syntax is `metaan est se, ml`.

2. Find the estimated model parameters in the output and interpret them.

3. Fit a so-called fixed-effects meta-analysis that simply omits ζ_j from the model and assumes that all true effect sizes are equal to β. This can be accomplished by replacing the `ml` option with the `fe` option in the `metaan` command.

4. Explain how the model differs from what we have referred to as fixed-effects models in this chapter (apart from the fact that the data are in aggregated form and the level-1 variance is assumed known).

5. Compare the width of the confidence intervals for β between the random- and fixed-effects meta-analyses, and explain why they differ the way they do.

2.9 Reliability and empirical Bayes prediction

In a hypothetical test–retest study, the estimates for the measurement model in (2.3) were as shown in table 2.4.

Table 2.4: Estimates for hypothetical test–retest study

	Est
β	30
ψ	9
θ	8

1. What is the estimated test–retest reliability of these measurements?

2. For a person with measurements 34 and 36, obtain the ML estimate and empirical Bayes prediction of ζ_j and of the true score.

3. Two additional measurements of 37 and 33 are made on the same person. Using all four measurements, obtain the ML estimate and empirical Bayes prediction of ζ_j and of the true score. Compare the results with the results for step 2, and explain the reason for any differences.

4. What is the empirical Bayes prediction of ζ_j for a person who has not been measured yet?

2.10 ❖ Maximum likelihood and restricted maximum likelihood

For the Mini Wright meter, use the ML estimates in table 2.2 to calculate the REML or ANOVA estimate of ψ. Also obtain the corresponding REML or ANOVA estimate of the intraclass correlation.

3 Random-intercept models with covariates

3.1 Introduction

In this chapter, we extend the variance-components models introduced in the previous chapter by including observed explanatory variables or covariates x. Seen from another perspective, we extend the linear regression models discussed in chapter 1 by introducing random intercepts ζ_j to handle clustered data. As we will show, ignoring the clustering generally leads to incorrect estimated standard errors and hence incorrect p-values.

Although many of the features of the variance-components models persist, new issues arise in estimating regression coefficients. In particular, we discuss the distinction between within-cluster and between-cluster covariate effects, and the problem of omitted cluster-level covariates and endogeneity. We also discuss coefficients of determination or measures of variance explained by covariates.

3.2 Does smoking during pregnancy affect birthweight?

Abrevaya (2006) investigates the effect of smoking on birth outcomes with the Natality datasets derived from birth certificates by the U.S. National Center for Health Statistics. This is of considerable public health interest because many pregnant women in the U.S. continue to smoke during pregnancy. Indeed, it is estimated that only 18% to 25% of smokers quit smoking once they become pregnant, according to the 2004 Surgeon General's Report on The Health Consequences of Smoking.

Abrevaya identified multiple births from the same mothers in nine datasets from 1990–1998 by matching mothers across the datasets. Unlike, for instance, the Nordic countries, a unique person identifier such as a social security number or name is rarely available in U.S. datasets. Perfect matching is thus precluded, and matching must proceed by identifying mothers who have identical values on a set of variables in all datasets. In this study, matching was accomplished by considering mother's state of birth and child's state of birth, as well as mother's county and city of birth; mother's age, race, education, and marital status; and, if married, father's age and race. For the matching on mother's and child's states of birth to be useful, the data were restricted to combinations of states that occur rarely.

Here we consider the subset of the matches where the observed interval between births was consistent with the interval since the last birth recorded on the birth certificate. The data are restricted to births with complete data for the variables considered by Abrevaya (2006), singleton births (no twins or other multiple births), and births to mothers for whom at least two births between 1990 and 1998 could be matched and whose race was classified as white or black.

The birth outcome we will concentrate on is birthweight. Abrevaya (2006) motivates his study by citing a report from the U.S. Surgeon General:

> "Infants born to women who smoke during pregnancy have a lower average birthweight and are more likely to be small for gestational age than infants born to women who do not smoke . . . "
> (*Women and Smoking: A Report of the Surgeon General*, Centers for Disease Control and Prevention, 2001).

The dataset used by Abrevaya (2006) is available from the Journal of Applied Econometrics Data Archive. Here we took a 10% random sample of the data, yielding 8,604 births from 3,978 mothers. We use the following variables from `smoking.dta`:

- `momid`: mother identifier
- `birwt`: birthweight (in grams)
- `mage`: mother's age at the birth of the child (in years)
- `smoke`: dummy variable for mother smoking during pregnancy (1: smoking; 0: not smoking)
- `male`: dummy variable for baby being male (1: male; 0: female)
- `married`: dummy variable for mother being married (1: married; 0: unmarried)
- `hsgrad`: dummy variable for mother having graduated from high school (1: graduated; 0: did not graduate)
- `somecoll`: dummy variable for mother having some college education, but no degree (1: some college; 0: no college)
- `collgrad`: dummy variable for mother having graduated from college (1: graduated; 0: did not graduate)
- `black`: dummy variable for mother being black (1: black; 0: white)
- `kessner2`: dummy variable for Kessner index = 2, or intermediate prenatal care (1: index=2; 0: otherwise)
- `kessner3`: dummy variable for Kessner index = 3, or inadequate prenatal care (1: index=3; 0: otherwise)
- `novisit`: dummy variable for no prenatal care visit (1: no visit; 0: at least 1 visit)
- `pretri2`: dummy variable for first prenatal care visit having occurred in second trimester (1: yes; 0: no)
- `pretri3`: dummy variable for first prenatal care visit having occurred in third trimester (1: yes; 0: no)

Smoking status was determined from the answer to the question asked on the birth certificate whether there was tobacco use during pregnancy. The dummy variables for mother's education—`hsgrad`, `somecoll`, and `collgrad`—were derived from the years of education given on the birth certificate. The Kessner index is a measure of the adequacy of prenatal care (1: adequate; 2: intermediate; 3: inadequate) based on the timing of the first prenatal visit and the number of prenatal visits, taking into account the gestational age of the fetus.

3.2.1   Data structure and descriptive statistics

The data have a two-level structure with births (or children or pregnancies) as units at level 1 and mothers as clusters at level 2. In multilevel models, the response variable always varies at the lowest level, taking on different values for different level-1 units within the same level-2 cluster. However, covariates can either vary at level 1 (and therefore usually also at level 2) or vary at level 2 only. For instance, while `smoke` can change from one pregnancy to the next, `black` is constant between pregnancies. `smoke` is therefore said to be a level-1 variable, whereas `black` is a level-2 variable. Among the variables listed above, `black` appears to be the only one that cannot in principle change between pregnancies. However, because of the way the matching was done, the education dummy variables (`hsgrad`, `somecoll`, and `collgrad`) and `married` also remain constant across births for the same mother and are thus level-2 variables.

We start by reading the smoking and birthweight data into Stata using the command

```
. use http://www.stata-press.com/data/mlmus3/smoking
```

A useful Stata command for exploring how much variables vary at level 1 and 2 is `xtsum`:

```
. quietly xtset momid
. xtsum birwt smoke black
```

Variable		Mean	Std. Dev.	Min	Max	Observations		
birwt	overall	3469.931	527.1394	284	5642	N =		8604
	between		451.1943	1361	5183.5	n =		3978
	within		276.7966	1528.431	5411.431	T-bar =		2.1629
smoke	overall	.1399349	.3469397	0	1	N =		8604
	between		.3216459	0	1	n =		3978
	within		.1368006	-.5267318	.8066016	T-bar =		2.1629
black	overall	.0717108	.2580235	0	1	N =		8604
	between		.257512	0	1	n =		3978
	within		0	.0717108	.0717108	T-bar =		2.1629

The total number of observations is $N = 8604$; the number of clusters is $J = 3978$ (n in the output); and there are on average about 2.2 births per mother (T-bar in the output) in the dataset.

Three different sample standard deviations are given for each variable. The first is the *overall standard deviation*, s_{xO}, defined as usual as the square root of the mean squared deviation of observations from the overall mean:

$$s_{xO} = \sqrt{\frac{1}{N-1} \sum_{j=1}^{J} \sum_{i=1}^{n_j} (x_{ij} - \overline{x}..)^2}$$

The second is the *between standard deviation*, defined as the square root of the mean squared deviation of the cluster means from the overall mean:

$$s_{xB} = \sqrt{\frac{1}{J-1} \sum_{j=1}^{J} (\overline{x}._j - \overline{x}..)^2}$$

This third is the *within standard deviation*, defined as the square root of the mean squared deviation of observations from the cluster means:

$$s_{xW} = \sqrt{\frac{1}{N-1} \sum_{j=1}^{J} \sum_{i=1}^{n_j} (x_{ij} - \overline{x}._j)^2}$$

We see that birthweight and smoking vary more between mothers than within mothers, whereas being black does not vary at all within mothers, as expected. It is important to be aware of how much level-1 variables vary within clusters because some estimators rely only on the within-cluster variability of covariates.

There are two different ways of expressing the mean or proportion for a level-2 variable: considering the summary either across units (with the level-1 units as unit of analysis) or across clusters (with the clusters as unit of analysis). For instance, the mean for `black` produced by `xtsum` is the mean (or proportion, because `black` is binary) across units, the proportion of children born to black mothers. We could also consider the mean across mothers, or the proportion of mothers who are black. To do so, we define a dummy variable equal to 1 for one child per mother using the `egen` command with the `tag()` function,

```
. egen pickone = tag(momid)
```

and summarize `black` across mothers by specifying `if pickone==1` in the command:

```
. summarize black if pickone==1
```

Variable	Obs	Mean	Std. Dev.	Min	Max
black	3978	.0713927	.257512	0	1

We see that the proportion of mothers who are black is very close to the proportion of children born to black mothers, probably because the number of children per mother does not vary much.

We can calculate the number of children per mother by using `egen` with the `count()` function:

```
. egen num = count(birwt), by(momid)
. tabulate num if pickone==1
```

num	Freq.	Percent	Cum.
2	3,330	83.71	83.71
3	648	16.29	100.00
Total	3,978	100.00	

Most mothers in the data have two children, and about 16% have three.

For categorical variables, including variables with more than two categories, `xttab` is a useful command. For the dichotomous variable `smoke`, the command produces the following table:

```
. quietly xtset momid
. xttab smoke
```

	Overall		Between		Within
smoke	Freq.	Percent	Freq.	Percent	Percent
Nonsmoke	7400	86.01	3565	89.62	95.69
Smoker	1204	13.99	717	18.02	79.03
Total	8604	100.00	4282	107.64	92.90
			(n = 3978)		

In the `Overall` table, we see that mothers smoked during their pregnancies for 14% of the children. According to the `Between` table, 90% of mothers had at least one pregnancy where they did not smoke, and 18% of mothers had at least one pregnancy where they did smoke. The `Within` table shows that the women who were ever nonsmokers during a pregnancy were nonsmokers for an average of 96% of their pregnancies. The women who ever smoked during a pregnancy did so for an average of 79% of their pregnancies.

3.3 The linear random-intercept model with covariates

3.3.1 Model specification

An obvious model to consider for the continuous response variable, birthweight, is a multiple linear regression model (discussed in chapter 1), including smoking status and various other variables as explanatory variables or covariates.

The model for the birthweight y_{ij} of child i of mother j is specified as

$$y_{ij} = \beta_1 + \beta_2 x_{2ij} + \cdots + \beta_p x_{pij} + \xi_{ij} \tag{3.1}$$

where x_{2ij} through x_{pij} are covariates and ξ_{ij} is a residual.

It may be unrealistic to assume that the birthweights of children born to the same mother are uncorrelated given the observed covariates, or in other words that the residuals ξ_{ij} and $\xi_{i'j}$ are uncorrelated. We can therefore use the idea introduced in the previous chapter to split the total residual or error into two error components: ζ_j, which is shared between children of the same mother, and ϵ_{ij}, which is unique for each child:

$$\xi_{ij} \equiv \zeta_j + \epsilon_{ij}$$

Substituting for ξ_{ij} into the multiple-regression model (3.1), we obtain a *linear random-intercept model with covariates*:

$$
\begin{aligned}
y_{ij} &= \beta_1 + \beta_2 x_{2ij} + \cdots + \beta_p x_{pij} + (\zeta_j + \epsilon_{ij}) \\
&= (\beta_1 + \zeta_j) + \beta_2 x_{2ij} + \cdots + \beta_p x_{pij} + \epsilon_{ij}
\end{aligned}
\tag{3.2}
$$

This model can be viewed as a regression model with an added level-2 residual ζ_j, or with a mother-specific intercept $\beta_1 + \zeta_j$. The random intercept ζ_j can be considered a latent variable that is not estimated along with the fixed parameters β_1 through β_p, but whose variance ψ is estimated together with the variance θ of the ϵ_{ij}. The linear random-intercept model with covariates is the simplest example of a *linear mixed (effects) model* where there are both fixed and random effects.

The random intercept or level-2 residual ζ_j is a mother-specific error component, which remains constant across births, whereas the level-1 residual ϵ_{ij} is a child-specific error component, which varies between children i as well as mothers j. The ζ_j are uncorrelated over mothers, the ϵ_{ij} are uncorrelated over mothers and children, and the two error components are uncorrelated with each other.

The mother-specific error component ζ_j represents the combined effects of omitted mother characteristics or unobserved heterogeneity at the mother level. If ζ_j is positive, the total residuals for mother j, ξ_{ij}, will tend to be positive, leading to heavier babies than predicted by the covariates. If ζ_j is negative, the total residuals will tend to be negative. Because ζ_j is shared by all responses for the same mother, it induces within-mother dependence among the total residuals ξ_{ij}.

3.3.2   Model assumptions

We now explicitly state a set of assumptions that are sufficient for everything we want to do in this chapter but are not always necessary. For this purpose, all observed covariates for unit i in cluster j are placed in the vector $\mathbf{x}_{ij}$, and the covariates for all the units in cluster j are placed in the matrix $\mathbf{X}_j$.

It is assumed that the level-1 residual ϵ_{ij} has zero expectation or mean, given the covariates and the random intercept:

$$E(\epsilon_{ij}|\mathbf{X}_j, \zeta_j) = 0 \tag{3.3}$$

This mean-independence assumption implies that $E(\epsilon_{ij}|\mathbf{X}_j) = 0$ and that $\mathrm{Cor}(\epsilon_{ij}, \mathbf{x}_{ij}) = 0$. We call this lack of correlation between covariates and level-1 residual "level-1 exogeneity". (See also section 1.13 on the exogeneity assumption in ordinary linear regression.)

The random intercept ζ_j is assumed to have zero expectation given the covariates,

$$E(\zeta_j|\mathbf{X}_j) \;=\; 0 \tag{3.4}$$

and this mean-independence assumption implies that $\mathrm{Cor}(\zeta_j, \mathbf{x}_{ij}) = 0$. We call the lack of correlation between covariates and random intercept "level-2 exogeneity". Violations of the exogeneity assumptions are called level-1 endogeneity and level-2 endogeneity, respectively.

We assume that the variance of the level-1 residual is homoskedastic for given covariates and random intercept,

$$\mathrm{Var}(\epsilon_{ij}|\mathbf{X}_j, \zeta_j) \;=\; \theta \tag{3.5}$$

which implies that $\mathrm{Var}(\epsilon_{ij}) = \theta$ and that $\mathrm{Cor}(\epsilon_{ij}, \zeta_j) = 0$. It is also assumed that the variance of the random intercept is homoskedastic given the covariates,

$$\mathrm{Var}(\zeta_j|\mathbf{X}_j) \;=\; \psi \tag{3.6}$$

which implies that $\mathrm{Var}(\zeta_j) = \psi$.

It is assumed that the level-1 residuals are uncorrelated for two units i and i' (whether they are nested in the same cluster j or in different clusters j and j') given the covariates and random intercept(s),

$$\mathrm{Cov}(\epsilon_{ij}, \epsilon_{i'j'}|\mathbf{X}_j, \mathbf{X}_{j'}, \zeta_j, \zeta_{j'}) \;=\; 0 \qquad \text{if} \ \ i \neq i' \ \ \text{or} \ \ j \neq j' \tag{3.7}$$

and that random intercepts are uncorrelated for different clusters j and j' given the covariates,

$$\mathrm{Cov}(\zeta_j, \zeta_{j'}|\mathbf{X}_j, \mathbf{X}_{j'}) \;=\; 0 \qquad \text{if} \ \ j \neq j' \tag{3.8}$$

These assumptions imply the mean and residual covariance structure of the responses described in sections 3.3.3 and 3.3.4.

It is sometimes assumed that both $\epsilon_{ij}|\mathbf{X}_j, \zeta_j$ and $\zeta_j|\mathbf{X}_j$ have normal distributions. Together with the assumptions (3.3) and (3.5), this implies that ζ_j and ϵ_{ij} are independent (a stronger property than lack of correlation).

The assumptions necessary for consistency of the standard estimators for regression coefficients are a correct mean structure (correct functional form and correct covariates) and lack of correlation between covariates and the random part of the model (the random intercept and the level-1 residual). Consistency of model-based standard errors relies on the additional assumption that the covariance structure (the variances and covariances) of the total residuals is correctly specified. Likewise, efficient estimation of regression coefficients requires that both mean and covariance structures are correct. For unbiased estimation of regression coefficients, the mean structure must be correct and the distribution of the total residuals must be symmetric (such as normal).

3.3.3   Mean structure

Assumption (3.3) implies that the cluster-specific or conditional regression (averaged over ϵ_{ij} but given ζ_j and $\mathbf{X}_j$) is linear:

$$
\begin{aligned}
E(y_{ij}|\mathbf{X}_j, \zeta_j) &= E(\beta_1 + \beta_2 x_{2ij} + \cdots + \beta_p x_{pij}) + E(\zeta_j|\mathbf{X}_j, \zeta_j) + \underbrace{E(\epsilon_{ij}|\mathbf{X}_j, \zeta_j)}_{0} \\
&= \beta_1 + \beta_2 x_{2ij} + \cdots + \beta_p x_{pij} + \zeta_j \qquad\qquad (3.9)
\end{aligned}
$$

We see that the covariates for other units in the cluster do not affect the mean response for unit i once we control for the covariates $\mathbf{x}_{ij}$ for unit i and the random intercept ζ_j. The covariates for the cluster $\mathbf{X}_j$ are then said to be "strictly exogenous given the random intercept".

It follows from (3.4) that the population-averaged or marginal regression (averaged over ζ_j and ϵ_{ij} but given $\mathbf{X}_j$) is linear:

$$
\begin{aligned}
E(y_{ij}|\mathbf{X}_j) &= E(\beta_1 + \beta_2 x_{2ij} + \cdots + \beta_p x_{pij}) + \underbrace{E(\zeta_j|\mathbf{X}_j)}_{0} + \underbrace{E(\epsilon_{ij}|\mathbf{X}_j)}_{0} \\
&= \beta_1 + \beta_2 x_{2ij} + \cdots + \beta_p x_{pij} \qquad\qquad (3.10)
\end{aligned}
$$

We sometimes refer to this relationship between the mean and covariates as the mean structure.

3.3.4   Residual variance and intraclass correlation

It follows from assumptions (3.5) and (3.6) that total residuals or error terms are homoskedastic (having constant variance) given the covariates $\mathbf{X}_j$,

$$
\mathrm{Var}(\xi_{ij}|\mathbf{X}_j) = \mathrm{Var}(\zeta_j + \epsilon_{ij}|\mathbf{X}_j) = \psi + \theta
$$

or, equivalently, that the responses y_{ij} given the covariates are also homoskedastic,

$$
\mathrm{Var}(y_{ij}|\mathbf{X}_j) = \psi + \theta
$$

The conditional correlation between the total residuals for any two children i and i' of the same mother j given the covariates, also called the residual correlation, is

$$
\rho \equiv \mathrm{Cor}(\xi_{ij}, \xi_{i'j}|\mathbf{X}_j) = \frac{\psi}{\psi + \theta}
$$

where ψ is the corresponding covariance. Thus ρ is also the conditional or residual intraclass correlation of responses y_{ij} and $y_{i'j}$ for mother j given the covariates:

$$
\rho \equiv \mathrm{Cor}(y_{ij}, y_{i'j}|\mathbf{X}_j) = \frac{\psi}{\psi + \theta}
$$

It is important to distinguish between the intraclass correlation in a model not containing any covariates—sometimes called the *unconditional* intraclass correlation—and the *conditional* or *residual* intraclass correlation in a model containing covariates. The residual covariance structure is shown in matrix form in display 3.2.

3.3.5 Graphical illustration of random-intercept model

A graphical illustration of the random-intercept model with a single covariate x_{ij} for a mother j is given in figure 3.1.

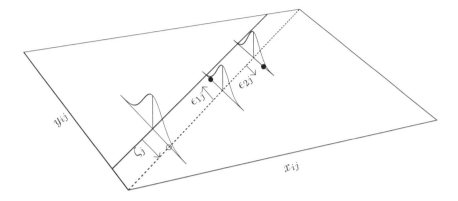

Figure 3.1: Illustration of random-intercept model for one mother

Here the solid line is $E(y_{ij}|x_{ij}) = \beta_1 + \beta_2 x_{ij}$, the population-averaged regression line for the population of all mothers j. The normal density curve centered on this line represents the random-intercept distribution with variance ψ and the hollow circle represents a realization ζ_j from this distribution for mother j (this could have been placed anywhere along the line). This negative random intercept ζ_j produces the dotted mother-specific regression line $E(y_{ij}|x_{ij}, \zeta_j) = (\beta_1 + \zeta_j) + \beta_2 x_{ij}$. This line is parallel to and below the population-averaged regression line. For a mother with a positive ζ_j, the mother-specific regression line would be parallel to and above the population-averaged regression line. Observed responses $y_{ij} = (\beta_1 + \zeta_j) + \beta_2 x_{ij} + \epsilon_{ij}$ are shown for two values of x_{ij}. The responses are sampled from the two normal distributions [with means $(\beta_1 + \zeta_j) + \beta_2 x_{ij}$ and variance θ] shown on the dotted curve.

3.4 Estimation using Stata

We can use `xtreg` or `xtmixed` to fit random-intercept models by maximum likelihood (ML). In addition, `xtreg` can be used to obtain feasible generalized least-squares estimates, and `xtmixed` can be used to obtain restricted ML estimates. Refer to section 3.10.1 for a brief description of these methods.

As discussed in chapter 2, `xtreg` is computationally more efficient than `xtmixed`, but `xtmixed` has a more useful `predict` command. Unless some special feature of `gllamm` or its prediction command `gllapred` are needed, we do not recommend using `gllamm` for linear models. A `gllamm` companion for this book is available from the `gllamm` website.

3.4.1   Using xtreg

The command for fitting the random-intercept model (3.2) by ML using `xtreg` is

```
. quietly xtset momid
. xtreg birwt smoke male mage hsgrad somecoll collgrad married black kessner2
> kessner3 novisit pretri2 pretri3, mle
```

```
Random-effects ML regression              Number of obs      =      8604
Group variable: momid                     Number of groups   =      3978

Random effects u_i ~ Gaussian             Obs per group: min =         2
                                                         avg =       2.2
                                                         max =         3

                                          LR chi2(13)        =    659.47
Log likelihood  = -65145.752              Prob > chi2        =    0.0000
```

birwt	Coef.	Std. Err.	z	P>|z|	[95% Conf. Interval]	
smoke	-218.3289	18.20988	-11.99	0.000	-254.0196	-182.6382
male	120.9375	9.558721	12.65	0.000	102.2027	139.6722
mage	8.100548	1.347266	6.01	0.000	5.459956	10.74114
hsgrad	56.84715	25.03538	2.27	0.023	7.778705	105.9156
somecoll	80.68607	27.30914	2.95	0.003	27.16115	134.211
collgrad	90.83273	27.99598	3.24	0.001	35.96162	145.7038
married	49.9202	25.50319	1.96	0.050	-.0651368	99.90554
black	-211.4138	28.27818	-7.48	0.000	-266.838	-155.9896
kessner2	-92.91883	19.92624	-4.66	0.000	-131.9736	-53.86411
kessner3	-150.8759	40.83414	-3.69	0.000	-230.9093	-70.84246
novisit	-30.03035	65.69213	-0.46	0.648	-158.7846	98.72387
pretri2	92.8579	23.19258	4.00	0.000	47.40127	138.3145
pretri3	178.7295	51.64145	3.46	0.001	77.51416	279.9449
_cons	3117.191	40.97597	76.07	0.000	3036.88	3197.503
/sigma_u	338.7674	6.296444			326.6487	351.3358
/sigma_e	370.6654	3.867707			363.1618	378.324
rho	.4551282	.0119411			.4318152	.4785967

```
Likelihood-ratio test of sigma_u=0: chibar2(01)= 1108.77 Prob>=chibar2 = 0.000
```

The estimated regression coefficients are given next to the corresponding covariate name; for instance, the coefficient β_2 of `smoke` is estimated as -218 grams. This means that, according to the fitted model, the expected birthweight is 218 grams lower for a child of a mother who smoked during the pregnancy compared with a child of a mother who did not smoke, controlling or adjusting for the other covariates. The estimated regression coefficients for the other covariates make sense, although the coefficients for the prenatal care variables (`kessner2`, `kessner3`, `novisit`, `pretri2`, and `pretri3`) are not straightforward to interpret because their definitions are partly overlapping.

The estimate of the random-intercept standard deviation $\sqrt{\psi}$ is given under `/sigma_u` as 339 grams, and the estimate of the level-1 residual standard deviation $\sqrt{\theta}$ is given under `/sigma_e` as 371 grams.

The ML estimates for the random-intercept model are also presented under "Full model" in table 3.1. The estimated regression coefficients are reported under "Fixed part" in the table, and the estimated standard deviations for the random intercept and level-1 residual are given under "Random part".

Table 3.1: Maximum likelihood estimates for smoking data (in grams)

	Full model		Null model		Level-2 covariates	
	Est	(SE)	Est	(SE)	Est	(SE)
Fixed part						
β_1 [_cons]	3,117	(41)	3,468	(7)	3,216	(26)
β_2 [smoke]	−218	(18)				
β_3 [male]	121	(10)				
β_4 [mage]	8	(1)				
β_5 [hsgrad]	57	(25)			131	(25)
β_6 [somecoll]	81	(27)			181	(27)
β_7 [collgrad]	91	(28)			233	(26)
β_8 [married]	50	(26)			115	(25)
β_9 [black]	−211	(28)			−201	(29)
β_{10} [kessner2]	−93	(20)				
β_{11} [kessner3]	−151	(41)				
β_{12} [novisit]	−30	(66)				
β_{13} [pretri2]	93	(23)				
β_{14} [pretri3]	179	(52)				
Random part						
$\sqrt{\psi}$	339		368		348	
$\sqrt{\theta}$	371		378		378	
Derived estimates						
R^2	0.09		0.00		0.05	
ρ	0.46		0.49		0.46	

3.4.2   Using xtmixed

The random-intercept model (3.2) can also be fit by ML using `xtmixed` with the `mle` option:

```
. xtmixed birwt smoke male mage hsgrad somecoll collgrad married black
> kessner2 kessner3 novisit pretri2 pretri3 || momid:, mle
```

Mixed-effects ML regression   Number of obs   =   8604
Group variable: momid   Number of groups   =   3978

                                               Obs per group: min =   2
                                                              avg =   2.2
                                                              max =   3

                                               Wald chi2(13)   =   693.75
Log likelihood = -65145.752   Prob > chi2   =   0.0000

birwt	Coef.	Std. Err.	z	P>|z|	[95% Conf. Interval]	
smoke	-218.3291	18.15944	-12.02	0.000	-253.9209	-182.7372
male	120.9375	9.558013	12.65	0.000	102.2041	139.6708
mage	8.10054	1.344571	6.02	0.000	5.465229	10.73585
hsgrad	56.84715	25.03537	2.27	0.023	7.778723	105.9156
somecoll	80.68608	27.309	2.95	0.003	27.16142	134.2107
collgrad	90.83276	27.99491	3.24	0.001	35.96373	145.7018
married	49.9202	25.50303	1.96	0.050	-.0648328	99.90522
black	-211.4138	28.27757	-7.48	0.000	-266.8368	-155.9908
kessner2	-92.91884	19.92619	-4.66	0.000	-131.9734	-53.86423
kessner3	-150.876	40.83031	-3.70	0.000	-230.9019	-70.85002
novisit	-30.03037	65.69171	-0.46	0.648	-158.7837	98.72301
pretri2	92.85793	23.19069	4.00	0.000	47.40502	138.3108
pretri3	178.7296	51.63681	3.46	0.001	77.52332	279.9359
_cons	3117.192	40.88817	76.24	0.000	3037.052	3197.331

Random-effects Parameters	Estimate	Std. Err.	[95% Conf. Interval]	
momid: Identity				
sd(_cons)	338.7669	6.296455	326.6481	351.3353
sd(Residual)	370.6656	3.867716	363.162	378.3242

LR test vs. linear regression: chibar2(01) = 1108.77 Prob >= chibar2 = 0.0000

The estimates are identical to those reported for `xtreg` in table 3.1.

The `reml` option can be used instead of `mle` to obtain restricted maximum likelihood (REML) estimates (see section 2.10.2 for the basic idea of REML). When there are many level-2 units J, as there are here, the REML and ML estimates will be almost identical. Robust standard errors can be obtained using the `vce(robust)` option.

3.5 Coefficients of determination or variance explained

In section 1.5, we motivated the coefficient of determination, or R-squared, as the proportional reduction in prediction error variance comparing the model without covariates (the null model) with the model of interest.

In ordinary linear regression without covariates, the predictions are $\widehat{y}_i = \overline{y}$, so the estimated prediction error variance is the mean squared error (MSE) for the null model,

$$\text{MSE}_0 \;=\; \frac{1}{N-1} \sum_i (y_i - \overline{y})^2 \;=\; \widehat{\sigma_0^2}$$

where $\widehat{\sigma_0^2}$ is an estimate of the residual variance in the null model. In the ordinary linear regression model including all covariates, the predictions are $\widehat{y}_i = \widehat{\beta}_1 + \widehat{\beta}_2 x_{2i} + \cdots + \widehat{\beta}_p x_{pi}$, and the estimated prediction error variance is the mean squared error in the regression model of interest,

$$\text{MSE}_1 \;=\; \frac{1}{N-p} \sum_i (y_i - \widehat{y}_i)^2 \;=\; \widehat{\sigma_1^2}$$

This is also an estimate of the residual variance σ_1^2 in the model of interest. The coefficient of determination is defined as

$$R^2 \;=\; \frac{\sum_i (y_i - \overline{y})^2 - \sum_i (y_i - \widehat{y}_i)^2}{\sum_i (y_i - \overline{y})^2} \;\approx\; \frac{\frac{\sum_i (y_i - \overline{y})^2}{N-1} - \frac{\sum_i (y_i - \widehat{y}_i)^2}{N-p}}{\frac{\sum_i (y_i - \overline{y})^2}{N-1}} \;=\; \frac{\widehat{\sigma_0^2} - \widehat{\sigma_1^2}}{\widehat{\sigma_0^2}}$$

where the approximation improves as N increases.

In a linear random-intercept model, the total residual variance is given by

$$\text{Var}(\zeta_j + \epsilon_{ij}) \;=\; \psi + \theta$$

An obvious definition of the coefficient of determination for two-level models, discussed by Snijders and Bosker (2012, chap. 7), is therefore the proportional reduction in the estimated total residual variance comparing the null model without covariates with the model of interest,

$$R^2 \;=\; \frac{\widehat{\psi}_0 + \widehat{\theta}_0 - (\widehat{\psi}_1 + \widehat{\theta}_1)}{\widehat{\psi}_0 + \widehat{\theta}_0}$$

where $\widehat{\psi}_0$ and $\widehat{\theta}_0$ are the estimates for the null model, and $\widehat{\psi}_1$ and $\widehat{\theta}_1$ are the estimates for the model of interest.

First, we fit the null model, also often called the *unconditional model*:

```
. quietly xtset momid

. xtreg birwt, mle

Random-effects ML regression              Number of obs      =       8604
Group variable: momid                     Number of groups   =       3978

Random effects u_i ~ Gaussian             Obs per group: min =          2
                                                         avg =        2.2
                                                         max =          3

                                          Wald chi2(0)       =       0.00
Log likelihood  = -65475.486              Prob > chi2        =          .
```

birwt	Coef.	Std. Err.	z	P>|z|	[95% Conf. Interval]
_cons	3467.969	7.137618	485.87	0.000	3453.979	3481.958
/sigma_u	368.2866	6.45442			355.8509	381.1568
/sigma_e	377.6578	3.926794			370.0393	385.4331
rho	.4874391	.0114188			.4650901	.5098276

```
Likelihood-ratio test of sigma_u=0: chibar2(01)= 1315.66 Prob>=chibar2 = 0.000
```

The estimates for this model were also given under "Null model" in table 3.1. The total variance is estimated as

$$\widehat{\psi}_0 + \widehat{\theta}_0 \ = \ 368.2866^2 + 377.6578^2 = 278260.43$$

For the model including all covariates, whose estimates were given under "Full model" in table 3.1, the total residual variance is estimated as

$$\widehat{\psi}_1 + \widehat{\theta}_1 \ = \ 338.7674^2 + 370.6654^2 = 252156.19$$

It follows that

$$R^2 \ = \ \frac{278260.43 - 252156.19}{278260.43} = 0.09$$

so 9% of the variance is explained by the covariates.

Raudenbush and Bryk (2002, chap. 4) suggest considering the proportional reduction in each of the variance components separately. In our example, the proportion of level-2 variance explained by the covariates is

$$R_2^2 \ = \ \frac{\widehat{\psi}_0 - \widehat{\psi}_1}{\widehat{\psi}_0} \ = \ \frac{368.2866^2 - 338.7674^2}{368.2866^2} = 0.15$$

and the proportion of level-1 variance explained is

$$R_1^2 \ = \ \frac{\widehat{\theta}_0 - \widehat{\theta}_1}{\widehat{\theta}_0} \ = \ \frac{377.6578^2 - 370.6654^2}{377.6578^2} = 0.04$$

Let us now fit a random-intercept model that includes only the level-2 covariates:

```
. quietly xtset momid
. xtreg birwt hsgrad somecoll collgrad married black, mle
Random-effects ML regression                   Number of obs      =      8604
Group variable: momid                          Number of groups   =      3978

Random effects u_i ~ Gaussian                  Obs per group: min =         2
                                                              avg =       2.2
                                                              max =         3

                                               LR chi2(5)         =    290.48
Log likelihood  = -65330.247                   Prob > chi2        =    0.0000
```

birwt	Coef.	Std. Err.	z	P>|z|	[95% Conf. Interval]	
hsgrad	131.4395	24.91149	5.28	0.000	82.61384	180.2651
somecoll	180.6879	26.50378	6.82	0.000	128.7414	232.6343
collgrad	232.8944	25.58597	9.10	0.000	182.7468	283.0419
married	114.765	25.45984	4.51	0.000	64.86465	164.6654
black	-201.4773	28.80249	-7.00	0.000	-257.9292	-145.0255
_cons	3216.482	25.82479	124.55	0.000	3165.866	3267.097
/sigma_u	348.1441	6.390242			335.8421	360.8968
/sigma_e	377.7638	3.929694			370.1397	385.5449
rho	.4592642	.0118089			.436201	.4824653

```
Likelihood-ratio test of sigma_u=0: chibar2(01)= 1146.40 Prob>=chibar2 = 0.000
```

In general, and as we can see from comparing the estimates for this model (given under "Level-2 covariates" in table 3.1) with the estimates from the null model, adding level-2 covariates will reduce mostly the level-2 variance. However, adding level-1 covariates can reduce both variances, as we can see by comparing the estimates for the above model with the full model. The reason is that many level-1 covariates vary both within and between clusters and can hence be decomposed as $x_{ij} = (x_{ij} - \overline{x}_{.j}) + \overline{x}_{.j}$, where $x_{ij} - \overline{x}_{.j}$ only varies at level-1 and $\overline{x}_{.j}$ only varies at level-2. Note that the estimated level-2 variance can increase when adding level-1 covariates, potentially producing a negative R_2^2.

Keep in mind that the coefficient of determination expresses to what extent the responses can be predicted from the covariates and not how appropriate the model is for the data. Indeed, a true model could very well have a large total residual variance.

For the intraclass correlation, we see that the unconditional intraclass correlation for the null model without covariates is estimated as 0.487. This reduces to a conditional or residual intraclass correlation of 0.459 when level-2 covariates are added and to 0.455 when all remaining covariates are added. The conditional intraclass correlation can also be larger than the unconditional intraclass correlation if the estimated level-1 variance decreases more than the level-2 variance does when covariates are added.

3.6   Hypothesis tests and confidence intervals

3.6.1   Hypothesis tests for regression coefficients

In ordinary linear regression, we use t tests for testing hypotheses regarding individual regression parameters and F tests for joint hypotheses regarding several regression parameters. Under the null hypothesis, these test statistics have t distributions and F distributions, respectively, with appropriate degrees of freedom in finite samples.

Because finite sample results are not readily available in the multilevel setting, hypothesis testing typically proceeds based on likelihood-ratio or Wald test statistics with asymptotic (large sample) $\chi^2(q)$ null distributions, with the number of restrictions q imposed by the null hypothesis as degrees of freedom. The Wald test and the less commonly used score test and Lagrange multiplier test are approximations of the likelihood-ratio test (see display 2.1). All three tests are asymptotically equivalent to each other but may produce different conclusions in small samples.

Hypothesis tests for individual regression coefficients

The most commonly used hypothesis test concerns an individual regression parameter, say, β_2, with null hypothesis

$$H_0\colon \beta_2 = 0$$

versus the two-sided alternative

$$H_a\colon \beta_2 \neq 0$$

The Wald statistic for testing the null hypothesis is

$$w \;=\; \left(\frac{\widehat{\beta}_2}{\widehat{\mathrm{SE}}(\widehat{\beta}_2)} \right)^2$$

which has an asymptotic $\chi^2(1)$ distribution under the null hypothesis, because the null hypothesis imposes one restriction. In practice, the test statistic

$$z \;=\; \frac{\widehat{\beta}_2}{\widehat{\mathrm{SE}}(\widehat{\beta}_2)}$$

is usually used. It has an asymptotic standard normal null distribution [because its square has a $\chi^2(1)$ distribution].

The z statistic is reported as z in the Stata output. For instance, in the output from xtmixed on page 134, the z statistic for the regression parameter of smoking is -12.02, which gives a two-sided p-value of less than 0.001. If robust standard errors are used in Stata, Wald tests and Wald-based confidence intervals will be based on them.

A likelihood-ratio test (described below) is less commonly used for testing individual regression parameters.

Joint hypothesis tests for several regression coefficients

Consider now the null hypothesis that the regression coefficients of two covariates x_{2ij} and x_{3ij} are both zero,

$$H_0: \beta_2 = \beta_3 = 0$$

versus the alternative hypothesis that at least one of the parameters is nonzero. For example, for the smoking and birthweight application, we may want to test the null hypothesis that the quality of prenatal care (as measured by the Kessner index) makes no difference to birthweight (controlling for the other covariates), where the Kessner index is represented by two dummy variables, `kessner2` and `kessner3`.

Let $\widehat{\beta}_2$ and $\widehat{\beta}_3$ be ML estimates from the model including the covariates x_{2ij} and x_{3ij}. The Wald statistic can be expressed as

$$
\begin{aligned}
w &= (\widehat{\beta}_2, \widehat{\beta}_3) \left\{ \begin{matrix} \widehat{\text{SE}}(\widehat{\beta}_2)^2 & \widehat{\text{Cor}}(\widehat{\beta}_2, \widehat{\beta}_3)\,\widehat{\text{SE}}(\widehat{\beta}_2)\,\widehat{\text{SE}}(\widehat{\beta}_3) \\ \widehat{\text{Cor}}(\widehat{\beta}_2, \widehat{\beta}_3)\,\widehat{\text{SE}}(\widehat{\beta}_2)\,\widehat{\text{SE}}(\widehat{\beta}_3) & \widehat{\text{SE}}(\widehat{\beta}_3)^2 \end{matrix} \right\}^{-1} (\widehat{\beta}_2, \widehat{\beta}_3)' \\
&= \frac{1}{1 - \widehat{\text{Cor}}(\widehat{\beta}_2, \widehat{\beta}_3)^2} \left\{ \left(\frac{\widehat{\beta}_2}{\widehat{\text{SE}}(\widehat{\beta}_2)} \right)^2 + \left(\frac{\widehat{\beta}_3}{\widehat{\text{SE}}(\widehat{\beta}_3)} \right)^2 - 2\,\widehat{\text{Cor}}(\widehat{\beta}_2, \widehat{\beta}_3)\, \frac{\widehat{\beta}_2}{\widehat{\text{SE}}(\widehat{\beta}_2)}\, \frac{\widehat{\beta}_3}{\widehat{\text{SE}}(\widehat{\beta}_3)} \right\}
\end{aligned}
$$

and has an asymptotic $\chi^2(2)$ null distribution because the null hypothesis imposes two restrictions. We see that the Wald statistic for the joint null hypothesis $H_0: \beta_2 = \beta_3 = 0$ decomposes into the sum of the Wald statistics for $H_0: \beta_2 = 0$ and $H_0: \beta_3 = 0$ if $\widehat{\text{Cor}}(\widehat{\beta}_2, \widehat{\beta}_3) = 0$, which would be the case if $\text{Cor}(x_{2i}, x_{3i}) = 0$.

We can also test the simultaneous hypothesis that three or more regression coefficients are all zero, but the expression for the Wald statistic becomes convoluted unless matrix expressions are used.

The Wald test for the null hypothesis that coefficients of the dummy variables `kessner2` and `kessner3` are both zero can be performed by using the `testparm` command:

```
. quietly xtset momid

. quietly xtreg birwt smoke male mage hsgrad somecoll collgrad married
> black kessner2 kessner3 novisit pretri2 pretri3, mle

. testparm kessner2 kessner3
 ( 1)  [birwt]kessner2 = 0
 ( 2)  [birwt]kessner3 = 0
            chi2(  2) =    26.94
          Prob > chi2 =    0.0000
```

We reject the null hypothesis at the 5% level with $w = 26.94$, degrees of freedom (df) = 2, $p < 0.001$. A more robust version of the test is obtained by specifying robust standard errors in the estimation command by using the `vce(robust)` option for `xtmixed, mle` or `xtreg, re`.

The analogous likelihood-ratio test statistic is

$$L = 2(l_1 - l_0)$$

where l_1 and l_0 are now the maximized log likelihoods for the models including and excluding both `kessner2` and `kessner3`, respectively. Under the null hypothesis, the likelihood-ratio statistic also has an asymptotic $\chi^2(2)$ null distribution.

A likelihood-ratio test of the null hypothesis that the coefficients of the dummy variables `kessner2` and `kessner3` are both zero can be performed by using the `lrtest` command:

```
. estimates store full
. quietly xtset momid
. quietly xtreg birwt smoke male mage hsgrad somecoll collgrad married black
> novisit pretri2 pretri3, mle
. lrtest full .
Likelihood-ratio test                           LR chi2(2)  =      26.90
(Assumption: . nested in full)                  Prob > chi2 =     0.0000
```

Note that likelihood-ratio tests for regression coefficients cannot be based on log likelihoods from REML estimation.

Sometimes it is required to test hypotheses regarding linear combinations of coefficients, as demonstrated in section 1.8. In section 3.7.4, we will encounter a special case of this when testing the null hypothesis that two regression coefficients are equal, or in other words that the difference between the coefficients is 0, a simple example of a *contrast*. Wald tests of such hypotheses can be performed in Stata using the `lincom` command.

3.6.2   Predicted means and confidence intervals

We can use the `margins` command (introduced in Stata 11) to obtain predicted means for mothers and pregnancies with particular covariate values, such as education and smoking status. If we evaluate the other covariates at particular values of our choice, we obtain adjusted means, called *adjusted predictions* in Stata. Alternatively, we can obtain what Stata calls *predictive margins*, the mean birthweight we would obtain if the distributions of the other covariates were the same for all combinations of education and smoking status. As mentioned in section 1.7, in linear models, predictive margins can be obtained by evaluating the other covariates at their means.

The `margins` command works only if factor notation is used for the categorical variables for which we want to make predictions (education and smoking status). We therefore define a categorical variable for level of education,

```
. generate education = hsgrad*1 + somecoll*2 + collgrad*3
```

and refit the model, declaring `education` and `smoke` as categorical variables using `i.education` and `i.smoke`:

```
. quietly xtset momid
. quietly xtreg birwt i.smoke male mage i.education married black
> kessner2 kessner3 novisit pretri2 pretri3, mle
```

We then use the `margins` command to obtain predictive margins for all combinations of `smoke` and `education`:

```
. margins i.smoke#i.education
```

```
Predictive margins                              Number of obs    =        8604
Model VCE    : OIM

Expression   : Linear prediction, predict()
```

	Margin	Delta-method Std. Err.	z	P>|z|	[95% Conf.	Interval]
smoke# education						
0 0	3430.916	23.4495	146.31	0.000	3384.956	3476.876
0 1	3487.763	13.22835	263.66	0.000	3461.836	3513.69
0 2	3511.602	14.11543	248.78	0.000	3483.936	3539.268
0 3	3521.749	12.16946	289.39	0.000	3497.897	3545.601
1 0	3212.587	25.18765	127.55	0.000	3163.22	3261.954
1 1	3269.434	19.62569	166.59	0.000	3230.969	3307.9
1 2	3293.273	21.13939	155.79	0.000	3251.841	3334.706
1 3	3303.42	21.2282	155.61	0.000	3261.813	3345.026

Here the interaction syntax `i.smoke#i.education` was used to specify that we want predictions for all combinations of the values of `smoke` and `education`. The standard errors of the predictions are based on the estimated standard errors from the random-intercept model.

We can plot these predictive margins using `marginsplot` (available from Stata 12),

```
. marginsplot, xdimension(education)
```

which gives the graph in figure 3.2.

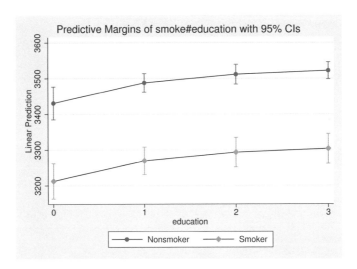

Figure 3.2: Predictive margins and confidence intervals for birthweight data

3.6.3   Hypothesis test for random-intercept variance

Consider testing the null hypothesis that the between-cluster variance is zero:

$$H_0\colon \psi = 0 \quad \text{against} \quad H_a\colon \psi > 0$$

This null hypothesis is equivalent to the hypothesis that $\zeta_j = 0$ or that there is no random intercept in the model. If this is true, a multilevel model is not required.

Likelihood-ratio tests are typically used with the test statistic,

$$L = 2(l_1 - l_0)$$

where l_1 is the maximized log likelihood for the random-intercept model (which includes ζ_j) and l_0 is the maximized log likelihood for an ordinary regression model (without ζ_j). A correct p-value is obtained by dividing the naïve p-value based on the $\chi^2(1)$ by 2, as was discussed in more detail in section 2.6.2. The result for the correct test procedure is provided in the last row of output from `xtreg` and `xtmixed`, giving $L = 1109$ and $p < 0.001$ for the full model.

Alternative tests for the random-intercept variance were described in section 2.6.2.

3.7   Between and within effects of level-1 covariates

We now turn to the estimated regression coefficients for the random-intercept model with covariates. For births where the mother smoked during the pregnancy, the population mean birthweight is estimated to be 218 grams lower than for births where the mother did not smoke, holding all other covariates constant. This estimate represents either a comparison between children of *different* mothers, one of whom smoked during the pregnancy and one of whom did not (holding all other covariates constant), or a comparison between children of the *same* mother, where the mother smoked during one pregnancy and not during the other (holding all other covariates constant). This is neither purely a between-mother comparison (because smoking status can change between pregnancies) nor purely a within-mother comparison (because some mothers either smoke or do not smoke during *all* their pregnancies).

3.7.1 Between-mother effects

If we wanted to obtain purely between-mother effects of the covariates, we could average the response and covariates for each mother j over children i and perform the regression on the resulting means:

$$\frac{1}{n_j}\sum_{i=1}^{n_j} y_{ij} = \frac{1}{n_j}\sum_{i=1}^{n_j}(\beta_1 + \beta_2 x_{2ij} + \cdots + \beta_p x_{pij} + \zeta_j + \epsilon_{ij})$$

or

$$\bar{y}_{\cdot j} = \beta_1 + \beta_2 \bar{x}_{2\cdot j} + \cdots + \beta_p \bar{x}_{p\cdot j} + \zeta_j + \bar{\epsilon}_{\cdot j} \tag{3.11}$$

Here $\bar{y}_{\cdot j}$ is the mean response for mother j, $\bar{x}_{2\cdot j}$ is the mean of the first covariate smoke for mother j, etc., and $\bar{\epsilon}_{\cdot j}$ is the mean of the level-1 residuals in the original regression model (3.2). The error term $\zeta_j + \bar{\epsilon}_{\cdot j}$ has population mean zero, $E(\zeta_j + \bar{\epsilon}_{\cdot j}) = 0$, and heteroskedastic variance $\mathrm{Var}(\zeta_j + \bar{\epsilon}_{\cdot j}) = \psi + \theta/n_j$, unless the data are balanced with $n_j = n$. Any information on the regression coefficients from within-mother variability is eliminated, and the coefficients of covariates that do not vary between mothers are absorbed by the intercept.

Ordinary least-squares (OLS) estimates[1] $\widehat{\boldsymbol{\beta}}^B$ of the parameters $\boldsymbol{\beta}$ in the between regression (3.11) whose corresponding covariates vary between mothers (here all covariates) can be obtained using `xtreg` with the `be` (between) option:

---

1. For unbalanced data $(n_j \neq n)$, *weighted* least-squares (WLS) estimates can be obtained using the `wls` option. Then the cluster weights $1/(\widehat{\psi} + \widehat{\theta}/n_j)$ are used, where $\widehat{\psi}$ and $\widehat{\theta}$ are obtained from OLS. OLS still produces consistent albeit inefficient estimators of the regression coefficients but inconsistent estimators of the corresponding standard errors under heteroskedasticity. We use OLS to estimate the between effects here because this estimator will be used when we spell out the relationships among different estimators later in the chapter.

```
. quietly xtset momid

. xtreg birwt smoke male mage hsgrad somecoll collgrad married black kessner2
> kessner3 novisit pretri2 pretri3, be

Between regression (regression on group means)   Number of obs    =      8604
Group variable: momid                            Number of groups =      3978

R-sq:  within  = 0.0299                           Obs per group: min =         2
       between = 0.1168                                          avg =       2.2
       overall = 0.0949                                          max =         3

                                                  F(13,3964)       =     40.31
sd(u_i + avg(e_i.))=  424.7306                    Prob > F         =    0.0000
```

birwt	Coef.	Std. Err.	t	P>|t|	[95% Conf. Interval]	
smoke	-286.1476	23.22554	-12.32	0.000	-331.6828	-240.6125
male	104.9432	19.49531	5.38	0.000	66.72141	143.165
mage	4.398704	1.505448	2.92	0.003	1.447179	7.35023
hsgrad	58.80977	25.51424	2.30	0.021	8.787497	108.832
somecoll	85.07129	28.1348	3.02	0.003	29.91126	140.2313
collgrad	99.87509	29.35324	3.40	0.001	42.32622	157.424
married	41.91268	26.10719	1.61	0.108	-9.272101	93.09745
black	-218.4045	28.57844	-7.64	0.000	-274.4344	-162.3747
kessner2	-101.4931	37.65605	-2.70	0.007	-175.3202	-27.66607
kessner3	-201.9599	79.28821	-2.55	0.011	-357.4094	-46.51042
novisit	-51.02733	124.2073	-0.41	0.681	-294.5435	192.4889
pretri2	125.4776	44.72006	2.81	0.005	37.80114	213.1541
pretri3	241.1201	100.6567	2.40	0.017	43.77638	438.4637
_cons	3241.45	46.15955	70.22	0.000	3150.951	3331.948

The estimates of the between-mother effects are also shown under "Between effects" in table 3.2. The estimated coefficient $\widehat{\beta}_2^B$ of smoke of -286 grams is considerably larger, in absolute value, than the ML estimate $\widehat{\beta}_2^{\mathrm{ML}}$ for the random-intercept model of -218 grams, shown under "Random effects" in table 3.2. The between-effect can be interpreted as the difference in mean birthweight comparing two different mothers, one of whom smoked during pregnancy while the other did not, given the other covariates.

Table 3.2: Random-, between-, and within-effects estimates for smoking data (in grams); MLE of random-intercept model (3.2), OLS of (3.11), OLS of (3.12), and MLE of random-intercept model including all cluster means

	Random effects $\widehat{\boldsymbol{\beta}}^{\mathrm{ML}}$		Between effects $\widehat{\boldsymbol{\beta}}^{B}$		Within effects $\widehat{\boldsymbol{\beta}}^{W}$		Random effects +clust. mean $\widehat{\boldsymbol{\beta}}^{\mathrm{ML}}$	
	Est	(SE)	Est	(SE)	Est	(SE)	Est	(SE)
Fixed part								
β_1 [_cons]	3,117	(41)	3,241	(46)	2,768	(86)	3,238	(46)
β_2 [smoke]	−218	(18)	−286	(23)	−105	(29)	−105	(29)
β_3 [male]	121	(10)	105	(19)	126	(11)	126	(11)
β_4 [mage]	8	(1)	4	(2)	23	(3)	23	(3)
β_5 [hsgrad]	57	(25)	59	(26)			56	(25)
β_6 [somecoll]	81	(27)	85	(28)			83	(28)
β_7 [collgrad]	91	(28)	100	(29)			98	(29)
β_8 [married]	50	(26)	42	(26)			42	(26)
β_9 [black]	−211	(28)	−218	(29)			−219	(28)
β_{10} [kessner2]	−93	(20)	−101	(38)	−91	(23)	−91	(23)
β_{11} [kessner3]	−151	(41)	−202	(79)	−128	(48)	−128	(48)
β_{12} [novisit]	−30	(66)	−51	(124)	−5	(78)	−5	(78)
β_{13} [pretri2]	93	(23)	125	(45)	81	(27)	81	(27)
β_{14} [pretri3]	179	(52)	241	(101)	153	(60)	153	(60)
β_{15} [m_smok]							−183	(37)
β_{16} [m_male]							−20	(22)
β_{17} [m_mage]							−18	(3)
β_{18} [m_kessner2]							−9	(44)
β_{19} [m_kessner3]							−79	(92)
β_{20} [m_novisit]							−38	(146)
β_{21} [m_pretri2]							45	(52)
β_{21} [m_pretri3]							96	(117)
Random part								
$\sqrt{\psi}$	339				440 [a]		338	
$\sqrt{\theta}$	371				369 [a]		369	

[a] Not parameter estimates, but standard deviations of estimates $\widehat{\epsilon}_{ij}$ and $\widehat{\alpha}_j$.

3.7.2 Within-mother effects

If we wanted to obtain purely within-mother effects, we could subtract the between-mother regression (3.11) from the original model (3.2) to obtain the within model:

$$y_{ij} - \overline{y}_{.j} = \beta_2(x_{2ij} - \overline{x}_{2 \cdot j}) + \cdots + \beta_p(x_{pij} - \overline{x}_{p \cdot j}) + \epsilon_{ij} - \overline{\epsilon}_{.j} \qquad (3.12)$$

Here the response and all covariates have simply been centered around their respective cluster means. The error term $\epsilon_{ij} - \bar{\epsilon}_{.j}$ has population mean zero, $E(\epsilon_{ij} - \bar{\epsilon}_{.j}) = 0$, and heteroskedastic variance $\text{Var}(\epsilon_{ij} - \bar{\epsilon}_{.j}) = \theta(1 - 1/n_j)$, unless the data are balanced. Covariates that do not vary within clusters drop out of the equation because the mean-centered covariate is zero. Importantly, this also includes the random intercept ζ_j.

OLS can be used to estimate the within effects β^W in (3.12). The standard errors of the estimated coefficients of covariates that vary little within clusters will be large because estimation is solely based on the within-cluster variability.

Identical estimates of within-mother effects can be obtained by replacing the random intercept ζ_j for each mother in the original model in (3.2) by a fixed intercept α_j. This could be accomplished by using dummy variables for each mother and omitting the intercept β_1 so that α_j represents the total intercept for mother j, previously represented by $\beta_1 + \zeta_j$. Letting d_{kj} be the dummy variable for the kth mother, ($k = 1, \ldots, 3{,}978$), the fixed-effects model can be written as

$$
\begin{aligned}
y_{ij} &= \beta_2 x_{2ij} + \cdots + \beta_p x_{pij} + \sum_{k=1}^{3{,}978} d_{kj}\alpha_k + \epsilon_{ij} \\
&= \beta_2 x_{2ij} + \cdots + \beta_p x_{pij} + \alpha_j + \epsilon_{ij}
\end{aligned}
\tag{3.13}
$$

and estimated by OLS. In this model, all mother-specific effects are accommodated by α_j, leaving only within-mother variation to be explained by covariates. The coefficients of level-2 covariates can therefore not be estimated, which can also be seen by considering that the set of dummy variables is collinear with any such covariates. In practice, it is more convenient to eliminate the intercepts by mean-centering all covariates, as in (3.12), instead of estimating 3,978 intercepts.

The within estimates $\widehat{\beta}^W$ for the coefficients of covariates that vary within mothers can be obtained using `xtreg` with the `fe` (fixed effects) option:[2]

---

2. Here the overall constant (next to `_cons`) is obtained by adding the overall means $\bar{y}_{..}$, $\bar{x}_{2..}$, etc., and $\bar{\alpha}_{.} + \bar{\epsilon}_{..}$ back onto the corresponding differences:

$$
y_{ij} - \bar{y}_{.j} + \bar{y}_{..} = \beta_1 + \beta_2(x_{2ij} - \bar{x}_{2.j} + \bar{x}_{2..}) + \cdots + \beta_p(x_{pij} - \bar{x}_{p.j} + \bar{x}_{p..}) + \epsilon_{ij} - \bar{\epsilon}_{.j} + \bar{\alpha}_{.} + \bar{\epsilon}_{..}
$$

```
. quietly xtset momid
. xtreg birwt smoke male mage kessner2 kessner3 novisit pretri2 pretri3, fe
Fixed-effects (within) regression          Number of obs      =      8604
Group variable: momid                      Number of groups   =      3978

R-sq:  within  = 0.0465                     Obs per group: min =         2
       between = 0.0557                                    avg =       2.2
       overall = 0.0546                                    max =         3

                                           F(8,4618)          =     28.12
corr(u_i, Xb)  = -0.0733                    Prob > F           =    0.0000
```

birwt	Coef.	Std. Err.	t	P>|t|	[95% Conf. Interval]	
smoke	-104.5494	29.10075	-3.59	0.000	-161.6007	-47.49798
male	125.6355	10.92272	11.50	0.000	104.2217	147.0492
mage	23.15832	3.006667	7.70	0.000	17.26382	29.05282
kessner2	-91.49483	23.48914	-3.90	0.000	-137.5448	-45.4449
kessner3	-128.091	47.79636	-2.68	0.007	-221.7947	-34.38731
novisit	-4.805898	77.7721	-0.06	0.951	-157.2764	147.6646
pretri2	81.29039	27.04974	3.01	0.003	28.25998	134.3208
pretri3	153.059	60.08453	2.55	0.011	35.26462	270.8534
_cons	2767.504	86.23602	32.09	0.000	2598.44	2936.567

sigma_u	440.05052					
sigma_e	368.91787					
rho	.58725545	(fraction of variance due to u_i)				

```
F test that all u_i=0:     F(3977, 4618) =      2.90          Prob > F = 0.0000
```

The estimates of the within-mother effects were also reported under "Within effects" in table 3.2. The estimated coefficient $\widehat{\beta}_2^W$ for `smoke` of -105 grams is dramatically smaller, in absolute value, than the estimate $\widehat{\beta}_2^R$ of -218 grams for the random-intercept model. The within-effect can be interpreted as the difference in mean birthweight between births for a given mother who changes smoking status between pregnancies, given the level-1 covariates. Level-2 covariates, whether observed or unobserved, are implicitly controlled for because mother is held constant in the comparison, along with all her characteristics. Therefore each mother truly serves as her own control.

In the output, `sigma_u` and `sigma_e` are standard deviations of estimated level-2 and level-1 residuals, $\widehat{\alpha}_j = \bar{y}_{\cdot j} - (\widehat{\beta}_1 + \widehat{\beta}_2 \bar{x}_{2\cdot j} + \cdots + \widehat{\beta}_p \bar{x}_{p\cdot j})$ and $\widehat{\epsilon}_{ij} = y_{ij} - \widehat{\alpha}_j - (\widehat{\beta}_1 + \widehat{\beta}_2 x_{2ij} + \cdots + \widehat{\beta}_p x_{pij})$, the latter standard deviation adjusted for the number of estimated means.

3.7.3 ❖ Relations among within estimator, between estimator, and estimator for random-intercept model

We now show that an estimator for the random-intercept model can be expressed as a weighted average of the within estimator and the between estimator.

The original random-intercept model in (3.2) implicitly assumes that the between and within effects of the set of covariates that vary both between and within mothers are identical because the between-mother model (3.11) and the within-mother model (3.12) derived from the random-intercepts model (3.2) have the same regression coefficients, $\boldsymbol{\beta} = (\beta_1, \beta_2, \ldots, \beta_p)'$. The estimators for the random-intercept model therefore use both within- and between-mother information. This can be seen explicitly for the feasible generalized least-squares (FGLS) estimator described in section 3.10.1 (obtained using `xtreg` with the `re` option), which is equivalent to the ML estimator in large samples. The relations among the estimators are most transparent for a single covariate x_{ij} with regression coefficient β_2 and balanced data $n_j = n$.

It can be shown that the sampling variance of the OLS between estimator $\widehat{\beta}_2^B$ can be consistently estimated as

$$\widehat{\mathrm{SE}}(\widehat{\beta}_2^B)^2 \;=\; \frac{\widehat{\mathrm{Var}}(\zeta_j + \bar{\epsilon}_{\cdot j})}{(J-1)s_{xB}^2} \;=\; \frac{\widehat{\psi} + \widehat{\theta}/n}{(J-1)s_{xB}^2}$$

where $\widehat{\mathrm{Var}}(\zeta_j + \bar{\epsilon}_{\cdot j})$ is the mean squared error from the between regression in (3.11) and $s_{xB}^2 = \frac{1}{J-1}\sum_{j=1}^{J}(\bar{x}_{\cdot j} - \bar{x}_{\cdot\cdot})^2$ is the between variance of x_{ij}, given in section 3.2.1. The sampling variance of the OLS within estimator $\widehat{\beta}_2^W$ can be consistently estimated as

$$\widehat{\mathrm{SE}}(\widehat{\beta}_2^W)^2 \;=\; \frac{\frac{Jn-1}{J(n-1)-1}\widehat{\mathrm{Var}}(\epsilon_{ij} - \bar{\epsilon}_{\cdot j})}{(Jn-1)s_{xW}^2} \;=\; \frac{\widehat{\theta}(1-1/n)}{\{J(n-1)-1\}s_{xW}^2} \tag{3.14}$$

where $\widehat{\mathrm{Var}}(\epsilon_{ij} - \bar{\epsilon}_{\cdot j})$ is the mean squared error from the within regression in (3.12) and $s_{xW}^2 = \frac{1}{Jn-1}\sum_{j=1}^{J}\sum_{i=1}^{n}(x_{ij} - \bar{x}_{\cdot j})^2$ is the within variance of x_{ij}, given in section 3.2.1. The term $\frac{Jn-1}{J(n-1)-1}$ is necessary in (3.14) because of the loss of J degrees of freedom due to mean-centering. The numerator after the first equality is the square of `sigma_e` reported by `xtreg` with the `fe` option.

The FGLS estimator $\widehat{\beta}_2^{\mathrm{FGLS}}$ for the random-intercept model can then be written as

$$\widehat{\beta}_2^{\mathrm{FGLS}} \;=\; (1-\widehat{\omega})\widehat{\beta}_2^B + \widehat{\omega}\widehat{\beta}_2^W$$

where

$$\widehat{\omega} \;=\; \frac{\widehat{\mathrm{SE}}(\widehat{\beta}_2^B)^2}{\widehat{\mathrm{SE}}(\widehat{\beta}_2^B)^2 + \widehat{\mathrm{SE}}(\widehat{\beta}_2^W)^2} \qquad \text{and} \qquad 1-\widehat{\omega} \;=\; \frac{\widehat{\mathrm{SE}}(\widehat{\beta}_2^W)^2}{\widehat{\mathrm{SE}}(\widehat{\beta}_2^B)^2 + \widehat{\mathrm{SE}}(\widehat{\beta}_2^W)^2}$$

We see that the FGLS estimator $\widehat{\beta}_2^{\mathrm{FGLS}}$ for the random-intercept model can be expressed as a weighted average of the between estimator $\widehat{\beta}_2^B$ and the within estimator $\widehat{\beta}_2^W$, where the weight of each estimator decreases as its standard error (imprecision) increases.

$\widehat{\beta}_2^{\mathrm{FGLS}}$ approaches the within estimator $\widehat{\beta}_2^W$ when $\widehat{\omega}$ approaches 1, that is, when the within standard error is much smaller than the between standard error. This happens when n becomes large, or $\widehat{\theta}$ becomes small, or $\widehat{\psi}$ becomes large, or s_{xB} becomes small.

$\widehat{\beta}_2^{\text{FGLS}}$ approaches the between estimator $\widehat{\beta}_2^B$ when $\widehat{\omega}$ approaches 0, that is, when the between standard error is much smaller than the within standard error. This happens when n becomes small, or $\widehat{\theta}$ becomes large, or $\widehat{\psi}$ becomes small, or s_{xW} becomes small. If $\widehat{\psi} = 0$, the FGLS estimator reduces to the OLS estimator, known as the pooled OLS estimator because there are several observations per cluster.

Although $\widehat{\beta}_2^{\text{FGLS}}$, $\widehat{\beta}_2^B$, and $\widehat{\beta}_2^W$ are all estimators of the same parameter β_2, the FGLS estimator $\widehat{\beta}_2^{\text{FGLS}}$ is more efficient (varies less in repeated samples) than the other estimators when the between effects equal the within effects because it exploits both within- and between-mother information.

3.7.4 Level-2 endogeneity and cluster-level confounding

The estimated between effect $\widehat{\beta}_2^B$ based on (3.11) may differ from the estimated within-effect $\widehat{\beta}_2^W$ from (3.12) because of omitted mother-specific explanatory variables that affect both $\overline{x}_{2 \cdot j}$ and the mother-specific residual ζ_j and hence the mean response $\overline{y}_{\cdot j}$, given the included explanatory variables.

As discussed by Abrevaya (2006), mothers who smoke during their pregnancy may also adopt other behaviors such as drinking and poor nutritional intake. They are also likely to have lower socioeconomic status. The between-mother effect of smoking is thus confounded with the effects of these omitted level-2 covariates. Because the confounders adversely affect birthweight and have not been adequately controlled for, the between-effect is likely to be an overestimate of the true effect (in absolute value). We thus have cluster-level confounding or cluster-level omitted-variable bias. In contrast, each mother serves as her own control for the within estimate, so all mother-specific explanatory variables have been held constant. Indeed, this was the reason why Abrevaya (2006) constructed the matched dataset: to get closer to the causal effect of smoking by using within-mother estimates.

To illustrate the idea of different between and within effects, figure 3.3 shows data for hypothetical clusters and a continuous covariate where the between-cluster effect (slope of dashed line) is positive and the within-cluster effect (slope of dotted lines) is negative. Here the hollow circles represent the observed data, and the solid circles represent cluster means.

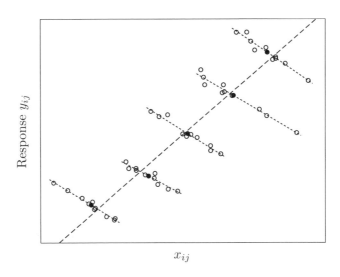

Figure 3.3: Illustration of different within-cluster and between-cluster effects of a covariate

The problem of cluster-level confounding can be described as correlation between the level-1 variable of interest—such as smoking x_{2ij}—and the random intercept ζ_j, which represents the effects of omitted level-2 covariates. In the figure, we see that clusters with larger (mean) values of x_{ij} tend to have larger (mean) values of y_{ij} because they have larger random intercepts. This problem is often referred to as *endogeneity* in econometrics. Specifically, the level-2 exogeneity assumption, discussed in section 3.3.2 is violated.

Cluster-level confounding can also be responsible for the *ecological fallacy*. This fallacy occurs when using aggregated data (cluster means, as in `xtreg, be`) and interpreting the estimated between effects as within effects. The *atomistic fallacy* occurs when ignoring clustering (by using ordinary regression) and interpreting the effects as between effects.

Figure 3.4 illustrates relationships between within and between effects where $\beta_2^W = \beta_2^B$ (left panel) and $\beta_2^W < \beta_2^B$ (right panel). The cluster-specific regression lines are shown for two clusters (solid and dashed lines for clusters 1 and 2, respectively) having the same value of the random intercept ζ_j but nonoverlapping ranges of x_{ij}. β_2^W, the slope of the cluster-specific lines, shown for cluster 1, is the within effect. The bullets represent the cluster means $(\overline{x}_{.1}, \overline{y}_{.1})$ for cluster 1 and $(\overline{x}_{.2}, \overline{y}_{.2})$ for cluster 2. The between effect β_2^B is the increase in the cluster mean of y_{ij} when the cluster mean of x_{ij} increases one unit (here shown for a change from 1.5 to 2.5).

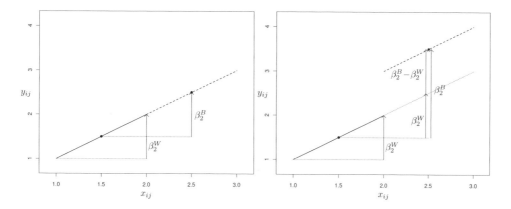

Figure 3.4: Illustration of different within and between effects for two clusters having the same value of ζ_j (β_2^W is the within effect and β_2^B is the between effect); in the left panel, $\beta_2^W = \beta_2^B$, whereas $\beta_2^W < \beta_2^B$ in the right panel

Let the clusters be schools with y_{ij} representing an achievement score for student i in school j and with x_{ij} representing the socioeconomic status (SES) of the student. In the left panel, the difference in school mean achievement is purely due to a *compositional effect*; within the schools, higher SES is associated with greater achievement, which completely explains why the school with greater mean SES has greater mean achievement. In the right panel, the compositional effect does not completely explain the difference in mean achievement between the schools. There is an additional, so-called *contextual effect* $\beta_2^B - \beta_2^W$, an additional increase of the second school's mean $\overline{y}_{\cdot j}$ after allowing for the within (or compositional) effect β_2^W. The contextual effect could be due to nonrandom assignment of high SES students to better schools (confounding), as well as direct peer effects.

It is common to include only the cluster-mean centered covariate in the model but not the cluster mean itself. However, as shown in figure 3.5, setting $\beta_2^B = 0$ makes the unrealistic assumption that the contextual effect equals the negative of the compositional effect. The original model that assumed equal between and within effects (no contextual effects) is no longer a special case if the cluster mean is omitted from the model.

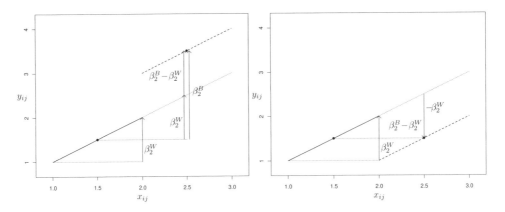

Figure 3.5: Illustration of assuming zero between effect for two clusters having the same value of ζ_j (β_2^W is the within effect and β_2^B is the between effect). The left panel is the same as the right panel of figure 3.4 with $\beta_2^W < \beta_2^B$ whereas in the right panel $\beta_2^B = 0$.

Including the cluster-mean centered covariate only (and not the cluster mean) is likely to lead to cluster-level confounding for other cluster-level covariates. For instance, including school-mean centered SES but not school-mean SES could have the consequence that the compositional effect is attributed to covariates correlated with school mean SES (due to selection effects), such as teacher qualifications. If students with high mean SES tend to have more highly educated teachers, the effect of teacher qualifications would then be overestimated due to confounding with SES. Only purely within-school covariates, such as other school-mean centered covariates and perhaps gender, have thus been controlled for SES.

3.7.5 Allowing for different within and between effects

We can easily relax the assumption that the between and within effects are the same for a particular covariate, say, x_{2ij}, by using the model

$$y_{ij} \;=\; \beta_1 + \beta_2^W(x_{2ij} - \overline{x}_{2\cdot j}) + \beta_2^B \overline{x}_{2\cdot j} + \beta_3 x_{3ij} + \cdots + \beta_p x_{pij} + \zeta_j + \epsilon_{ij} \qquad (3.15)$$

which collapses to the original random-intercept model in (3.2) if $\beta_2^W = \beta_2^B = \beta_2$. The deviation from the cluster mean of smoking $x_{2ij} - \overline{x}_{2\cdot j}$ is uncorrelated with ζ_j because it does not vary between clusters (and ζ_j does not vary within clusters). We can also view the above model as relaxing the assumption that the random intercept is uncorrelated with x_{2ij} if we think of $\beta_2^B \overline{x}_{2\cdot j} + \zeta_j$ as the random intercept.

We do not need to subtract the cluster mean $\overline{x}_{2\cdot j}$ from x_{2ij} as long as we include the cluster mean in the model, because

$$
\begin{aligned}
y_{ij} \;&=\; \beta_1 + \beta_2^W(x_{2ij} - \overline{x}_{2\cdot j}) + \beta_2^B \overline{x}_{2\cdot j} + \beta_3 x_{3ij} + \cdots + \beta_p x_{pij} + \zeta_j + \epsilon_{ij} \\
&=\; \beta_1 + \beta_2^W x_{2ij} + (\beta_2^B - \beta_2^W)\overline{x}_{2\cdot j} + \beta_3 x_{3ij} + \cdots + \beta_p x_{pij} + \zeta_j + \epsilon_{ij} \quad (3.16)
\end{aligned}
$$

Whether x_{2ij} is cluster-mean centered affects only the interpretation of the coefficient of the cluster mean $\overline{x}_{2\cdot j}$. If x_{2ij} is cluster-mean centered, as in the first line of (3.16), the coefficient of the cluster mean represents the between effect. If x_{2ij} is not cluster-mean centered, as in the second line of (3.16), the coefficient represents the difference in between and within effects.

We will now fit (3.15) with the cluster mean of smoke (the proportion of pregnancies in which the mother smokes), as well as the child-specific deviation from the cluster mean of smoke as covariates. These covariates are produced by the commands

```
. egen mn_smok = mean(smoke), by(momid)
. generate dev_smok = smoke - mn_smok
```

We fit the resulting random-intercept model by ML using

```
. quietly xtset momid
. xtreg birwt dev_smok mn_smok male mage hsgrad somecoll collgrad married
> black kessner2 kessner3 novisit pretri2 pretri3, mle
```

```
Random-effects ML regression              Number of obs     =      8604
Group variable: momid                     Number of groups  =      3978

Random effects u_i ~ Gaussian             Obs per group: min =         2
                                                         avg =       2.2
                                                         max =         3

                                          LR chi2(14)       =    684.32
Log likelihood  = -65133.327              Prob > chi2       =    0.0000
```

birwt	Coef.	Std. Err.	z	P>|z|	[95% Conf. Interval]	
dev_smok	-104.2331	29.18164	-3.57	0.000	-161.4281	-47.03815
mn_smok	-289.904	23.12917	-12.53	0.000	-335.2364	-244.5717
male	121.1497	9.543995	12.69	0.000	102.4438	139.8556
mage	8.191991	1.345733	6.09	0.000	5.554404	10.82958
hsgrad	43.10052	25.15802	1.71	0.087	-6.208299	92.40934
somecoll	62.74526	27.51558	2.28	0.023	8.815722	116.6748
collgrad	66.89807	28.37807	2.36	0.018	11.27806	122.5181
married	35.21194	25.6433	1.37	0.170	-15.04801	85.47189
black	-218.9694	28.28467	-7.74	0.000	-274.4064	-163.5325
kessner2	-92.03346	19.89661	-4.63	0.000	-131.0301	-53.03683
kessner3	-149.2771	40.77281	-3.66	0.000	-229.1903	-69.36383
novisit	-23.3923	65.60531	-0.36	0.721	-151.9763	105.1917
pretri2	92.33952	23.15722	3.99	0.000	46.95221	137.7268
pretri3	176.774	51.56358	3.43	0.001	75.7112	277.8367
_cons	3154.8	41.57504	75.88	0.000	3073.314	3236.285
/sigma_u	338.4563	6.27314			326.3818	350.9775
/sigma_e	370.0488	3.856608			362.5666	377.6853
rho	.4554982	.0119054			.4322541	.4788957

```
Likelihood-ratio test of sigma_u=0: chibar2(01)= 1115.22 Prob>=chibar2 = 0.000
```

The estimated between effect of smoking (coefficient of mn_smok) is $\widehat{\beta}_2^B = -290$ grams and is different from $\widehat{\beta}_2^W = -104$ grams, the estimated within effect of smoking (coefficient of dev_smok). Comparing two mothers, one of whom smoked, the expected difference in birthweight is estimated as 290 grams, given the other covariates. Comparing two births of the same mother, where she smoked during one of the pregnancies, the expected difference is estimated as 104 grams, controlling for the other covariates. We can formally test the null hypothesis that the corresponding coefficients are the same, $H_0: \beta_2^W - \beta_2^B = 0$, using the postestimation command `lincom`:

```
. lincom mn_smok - dev_smok
 ( 1)  - [birwt]dev_smok + [birwt]mn_smok = 0
```

birwt	Coef.	Std. Err.	z	P>|z|	[95% Conf. Interval]
(1)	-185.6709	37.21887	-4.99	0.000	-258.6186 -112.7233

There is strong evidence that the within effect of smoking differs from the between effect.

As pointed out previously, $x_{2ij} - \overline{x}_{2\cdot j}$ is correlated with x_{2ij} but uncorrelated with the random intercept ζ_j per construction. However, ζ_j may be correlated with another within-mother covariate x_{3ij} and the inconsistency in estimating the corresponding regression coefficient β_3 can be transmitted to the estimator for β_2^W.

To address this problem, we can follow Mundlak (1978) and include the cluster means of all within-mother covariates. If there is level-1 exogeneity (lack of correlation between covariates and ϵ_{ij}), inclusion of the cluster means ensures consistent estimation of all within effects. The reason is that the deviations from the cluster means are uncorrelated with the cluster means themselves, with any between-mother covariate (such as x_{4j}), and with ζ_j. However, the coefficients of the between-mother covariates and the random-intercept variance ψ are in general not consistently estimated. This is because the cluster means of the within-mother covariates and the between-mother covariates are likely to be correlated with ζ_j. In contrast, the estimator for the level-1 residual variance θ is consistent.

We start by constructing the cluster means (apart from `mn_smok`, which already exists):

```
. egen mn_male = mean(male), by(momid)
. egen mn_mage = mean(mage), by(momid)
. egen mn_kessner2 = mean(kessner2), by(momid)
. egen mn_kessner3 = mean(kessner3), by(momid)
. egen mn_novisit = mean(novisit), by(momid)
. egen mn_pretri2 = mean(pretri2), by(momid)
. egen mn_pretri3 = mean(pretri3), by(momid)
```

A more elegant way of forming cluster means for such a large number of variables is to use a `foreach` loop:

```
foreach var of varlist male mage kessner* novisit pretri* {
    egen mn_'var' = mean('var'), by(momid)
}
```

Here the loop repeats the **egen** command for each of the variables in the variable list **male mage kessner* novisit pretri***. Inside the curly braces, the local macro **var** (which we could have called whatever we like) evaluates to the current variable name (first **male**, then **mage**, etc.). This variable name is accessed by placing the macro name in single quotes (really, a left quote and an apostrophe). We therefore obtain exactly the same commands we previously ran one at a time.

If we include the cluster means of all level-1 covariates but leave the latter variables in the model without cluster-mean centering them, as in (3.16), the coefficients for the cluster means represent the differences in the between and within effects for the set of covariates having both between and within variation. We fit the model using

```
. quietly xtset momid

. xtreg birwt smok mn_smok male mn_male mage mn_mage hsgrad somecoll
> collgrad married black kessner2 mn_kessner2 kessner3 mn_kessner3 novisit
> mn_novisit pretri2 mn_pretri2 pretri3 mn_pretri3, mle
```

Random-effects ML regression				Number of obs	=	8604
Group variable: momid				Number of groups	=	3978
Random effects u_i ~ Gaussian				Obs per group: min =		2
				avg =		2.2
				max =		3
				LR chi2(21)	=	719.28
Log likelihood = -65115.846				Prob > chi2	=	0.0000

birwt	Coef.	Std. Err.	z	P>|z|	[95% Conf.	Interval]
smoke	-104.5494	29.1063	-3.59	0.000	-161.5967	-47.50206
mn_smok	-183.1657	37.19657	-4.92	0.000	-256.0696	-110.2617
male	125.6355	10.9248	11.50	0.000	104.2232	147.0477
mn_male	-20.22363	22.31026	-0.91	0.365	-63.95093	23.50367
mage	23.15832	3.007241	7.70	0.000	17.26424	29.0524
mn_mage	-18.59407	3.360264	-5.53	0.000	-25.18007	-12.00808
hsgrad	56.29698	25.38638	2.22	0.027	6.540583	106.0534
somecoll	83.07017	27.99083	2.97	0.003	28.20914	137.9312
collgrad	98.17599	29.18708	3.36	0.001	40.97037	155.3816
married	42.46127	26.03156	1.63	0.103	-8.559647	93.48219
black	-219.0013	28.41769	-7.71	0.000	-274.699	-163.3037
kessner2	-91.49483	23.49362	-3.89	0.000	-137.5415	-45.44819
mn_kessner2	-9.050791	44.22205	-0.20	0.838	-95.72442	77.62284
kessner3	-128.091	47.80548	-2.68	0.007	-221.788	-34.394
mn_kessner3	-79.42459	92.26946	-0.86	0.389	-260.2694	101.4202
novisit	-4.805899	77.78694	-0.06	0.951	-157.2655	147.6537
mn_novisit	-38.11621	146.3218	-0.26	0.794	-324.9017	248.6693
pretri2	81.29039	27.0549	3.00	0.003	28.26376	134.317
mn_pretri2	44.76713	52.06045	0.86	0.390	-57.26948	146.8037
pretri3	153.059	60.09599	2.55	0.011	35.27303	270.845
mn_pretri3	96.07044	116.707	0.82	0.410	-132.6711	324.8119
_cons	3238.407	45.98903	70.42	0.000	3148.271	3328.544
/sigma_u	338.4422	6.243867			326.423	350.9039
/sigma_e	368.9882	3.840715			361.5368	376.5932
rho	.4569014	.011855			.4337527	.4801973

Likelihood-ratio test of sigma_u=0: chibar2(01)= 1127.34 Prob>=chibar2 = 0.000

The estimates were shown under "Random effects + clust. mean" in table 3.2 on page 145. The estimated coefficients for the covariates varying within mothers are now identical to the within-effects and hence not susceptible to cluster-level confounding.

A Wald test of the joint null hypothesis that all coefficients for the cluster means in the above model are zero can be performed using the `testparm` command:

```
. testparm mn_*
 ( 1)  [birwt]mn_smok = 0
 ( 2)  [birwt]mn_male = 0
 ( 3)  [birwt]mn_mage = 0
 ( 4)  [birwt]mn_kessner2 = 0
 ( 5)  [birwt]mn_kessner3 = 0
 ( 6)  [birwt]mn_novisit = 0
 ( 7)  [birwt]mn_pretri2 = 0
 ( 8)  [birwt]mn_pretri3 = 0

           chi2(  8) =    60.04
         Prob > chi2 =    0.0000
```

The Wald statistic is 60.04 with df $= 8$, so the null hypothesis that the coefficients of the cluster means are all zero is rejected at the 5% level. The above null hypothesis is equivalent to the hypothesis of equal between and within effects (for the covariates having both within and between variation), which is thus also rejected.

A great advantage of clustered or multilevel data is that we can investigate and address level-2 endogeneity of level-1 covariates (correlation between ζ_j and x_{ij}). However, the approaches considered in this chapter do not produce consistent estimates of the coefficients of level-2 covariates and the random-intercept variance in this case. Furthermore, the approaches cannot handle level-2 endogeneity of level-2 covariates (correlation between ζ_j and x_j). Both problems are addressed in an approach suggested by Hausman and Taylor (1981), which we describe in section 5.2.

Unfortunately, it is not straightforward to check for level-1 endogeneity, that is, to check whether ϵ_{ij} is correlated with either cluster-level or unit-level covariates. To correct for level-1 endogeneity, external instrumental variables are usually required.

Fortunately, there is no endogeneity problem due to omitted covariates when estimating a treatment or intervention effect in a randomized experiment.

3.7.6 Hausman endogeneity test

The Hausman test (Hausman 1978), more aptly called the Durbin–Wu–Hausman test, can be used to compare two alternative estimators of β, both of which are consistent if the model is true. In its standard form, one of the estimators is asymptotically efficient if the model is true, but is inconsistent when the model is misspecified. The other estimator is consistent also under misspecification but is not asymptotically efficient when the model is true.

For instance, if the random-intercept model is correctly specified, both the fixed-effects estimator $\widehat{\beta}^W$ and the FGLS estimator $\widehat{\beta}^{\text{FGLS}}$ are consistent for coefficients of covariates that vary within clusters whereas only $\widehat{\beta}^{\text{FGLS}}$ is efficient. However, if the random intercept is correlated with any of the covariates (cluster-level endogeneity), the within effects will differ from the between effects and $\widehat{\beta}^{\text{FGLS}}$ becomes inconsistent, whereas $\widehat{\beta}^W$ remains consistent.

Consider first the simple case of a model with a single covariate x_{ij} that varies both between and within clusters. The Hausman test statistic for endogeneity then takes the form

$$h = \frac{(\widehat{\beta}^W - \widehat{\beta}^{\text{FGLS}})^2}{\widehat{\text{SE}}(\widehat{\beta}^W)^2 - \widehat{\text{SE}}(\widehat{\beta}^{\text{FGLS}})^2} \tag{3.17}$$

which has an asymptotic $\chi^2(1)$ null distribution. The denominator of the test statistic would usually take the form $\widehat{\text{SE}}(\widehat{\beta}^W)^2 + \widehat{\text{SE}}(\widehat{\beta}^{\text{FGLS}})^2 - 2\widehat{\text{Cov}}(\widehat{\beta}^W, \widehat{\beta}^{\text{FGLS}})$, where the covariance between the within and FGLS estimators would be hard to obtain. However, it can be shown that the denominator simplifies to the one in (3.17) because the FGLS estimator is efficient when the random-intercept model is true.

Consider now the case where there are several covariates that all vary both between and within clusters. Let the fixed effects and FGLS estimates be denoted $\widehat{\beta}^W$ and $\widehat{\beta}^{\text{FGLS}}$, respectively, and let the corresponding estimated covariance matrices be denoted $\widehat{\text{Cov}}(\widehat{\beta}^W)$ and $\widehat{\text{Cov}}(\widehat{\beta}^{\text{FGLS}})$. The Hausman test statistic then takes the form

$$h = (\widehat{\beta}^W - \widehat{\beta}^{\text{FGLS}}) \left\{ \widehat{\text{Cov}}(\widehat{\beta}^W) - \widehat{\text{Cov}}(\widehat{\beta}^{\text{FGLS}}) \right\}^{-1} (\widehat{\beta}^W - \widehat{\beta}^{\text{FGLS}})'$$

The h statistic has an asymptotic χ^2 null distribution with degrees of freedom given as the number of overlapping estimated regression coefficients from the two approaches, that is, the number of covariates with both between- and within-cluster variation.

We can use the `hausman` command to perform the Hausman test in Stata, following estimation of $\widehat{\beta}^W$ using `xtreg` with the `fe` option and estimation of $\widehat{\beta}^{\text{FGLS}}$ using `xtreg` with the `re` option:

```
. quietly xtset momid
. quietly xtreg birwt smoke male mage hsgrad somecoll collgrad married
> black kessner2 kessner3 novisit pretri2 pretri3, fe
```

```
. estimates store fixed

. quietly xtreg birwt smoke male mage hsgrad somecoll collgrad married
> black kessner2 kessner3 novisit pretri2 pretri3, re

. estimates store random

. hausman fixed random
```

| | ──── Coefficients ──── | | | |
| | (b) | (B) | (b-B) | sqrt(diag(V_b-V_B)) |
	fixed	random	Difference	S.E.
smoke	-104.5494	-217.7488	113.1995	22.71343
male	125.6355	120.9874	4.648084	5.297981
mage	23.15832	8.137158	15.02116	2.687211
kessner2	-91.49483	-92.89604	1.401212	12.44845
kessner3	-128.091	-150.6366	22.54563	24.87574
novisit	-4.805898	-29.9223	25.11641	41.66561
pretri2	81.29039	92.73087	-11.44048	13.94097
pretri3	153.059	178.4334	-25.37443	30.76114

```
                        b = consistent under Ho and Ha; obtained from xtreg
            B = inconsistent under Ha, efficient under Ho; obtained from xtreg
   Test:  Ho:  difference in coefficients not systematic
                chi2(8) = (b-B)'[(V_b-V_B)^(-1)](b-B)
                        =        60.07
                Prob>chi2 =      0.0000
```

There is strong evidence for model misspecification because the Hausman test statistic is 60.07 with df = 8. The Hausman statistic is practically identical to the Wald statistic for the joint null hypothesis that all regression coefficients of the cluster means are zero, shown in the previous section. This is what we would expect because the corresponding tests are asymptotically equivalent. In fact, equivalent variants of the Hausman test could have been constructed based on instead comparing the between and FGLS estimators, or the OLS and FGLS estimators, or the between and within estimators; but this is not implemented in Stata.

A significant Hausman test is often taken to mean that the random-intercept model should be abandoned in favor of a fixed-effects model that only uses within information. However, if there are covariates having the same within and between effects, we obtain more precise estimates of these coefficients by exploiting both within- and between-cluster information. The fixed-effects estimators are particularly imprecise if the covariates have little within-cluster variation. If the (true) between and within effects differ by a small amount, it may still be advisable to use the random-effects estimator because it may have a smaller mean squared error (some bias but considerably smaller variance) than the fixed-effects estimator.

3.8   Fixed versus random effects revisited

In section 2.8, we discussed whether the effects of clusters should be treated as random or fixed in models without covariates. We argued that this depends on whether inferences are for the population of clusters or only for the clusters included in the sample. In table 3.3, we consider these as the main questions and then ask questions related to each main question. The answers to these questions delineate the main differences between fixed-effects and random-effects approaches.

Table 3.3: Overview of distinguishing features of fixed- and random-effects approaches for linear models that include covariates

Questions	Answers	
	Fixed effects	**Random effects**
Inference for population of clusters?	No	Yes
Minimum number of clusters required?	Any number	At least 10 or 20
What assumptions are required for ζ_j or α_j?		Level-2 exogeneity, constant variance ψ
Can estimate effects of cluster-level covariates?	No	Yes
Inference for clusters in particular sample?	Yes	No, not for βs, but yes, for ζ_j by using empirical Bayes
Minimum cluster size required?	At least 2, but large for est. α_j	Any sizes if many ≥ 2
Is the model parsimonious?	No, J parameters α_j	Yes, one variance parameter ψ for all J clusters
Can estimate within-cluster effects of covariates?	Yes	Yes, by including cluster means

Unlike the fixed-effects model, the random-effects model can be used to make inferences regarding the population of clusters, but at the cost of requiring many clusters and making additional assumptions regarding the random-intercept distribution. The additional assumptions include exogeneity of the observed covariates $\mathbf{x}_{ij}$ with respect to ζ_j (level-2 exogeneity) and a constant variance ψ. The standard assumption of a normal distribution for ζ_j is actually not required for consistent estimation of regression coefficients. Both the fixed-effects and random-effects models assume exogeneity of $\mathbf{x}_{ij}$ with respect to the level-1 residual ϵ_{ij} (level-1 exogeneity). An advantage of the random-effects model is that it can be used to estimate the effects of cluster-level covariates, in contrast to the fixed-effects model, although consistent estimation requires both level-1 and level-2 exogeneity.

While the fixed-effects model is designed for making inferences for the clusters in the sample, the random-effects model can also to some extent be used for this purpose by predicting the ζ_j using empirical Bayes (EB). However, inferences regarding the regression coefficients from random-effects models, such as estimated standard errors, are for the population of clusters as discussed in section 2.10.3. Random-effects models can be used if there are clusters with only one unit as long as there are many clusters with at least two units.

In the fixed-effects approach, singleton clusters do not provide any information on any of the parameters except α_j. Furthermore, the fixed-effects approach requires large cluster sizes if we want to consistently estimate the intercepts α_j. The fixed-effects model is much less parsimonious than the random-intercept model because it includes one parameter α_j for each cluster, whereas the random-intercept model has only one parameter ψ for the variance of the random intercepts ζ_j. Eliminating the α_j by mean centering, as shown in section 3.7.2, simplifies the estimation problem but does not make the estimates of the remaining parameters any more efficient. Unlike the random-effects approach, the fixed-effects approach controls for clusters, providing estimates of within-cluster effects of covariates. The random-effects model can provide estimates of within-cluster effects only with extra effort, namely, by including cluster means of those covariates for which the between effect differs from the within effect.

3.9   Assigning values to random effects: Residual diagnostics

As discussed in section 2.11, we often want to assign values to random effects, for inferences regarding clusters, model visualization, or diagnostics. In data on institutions such as schools or hospitals, the predicted random intercepts can be viewed as measures of institutional performance if the covariates represent intake characteristics of individuals or "case-mix" of institutions (see section 4.8.5). Model visualization is demonstrated in section 4.8.3 and throughout the book.

We now consider residual diagnostics for assessing the normality assumptions for ζ_j and ϵ_{ij}. Inference generally does not hinge on normality, but severe skewness and outliers can pose problems. In contrast, EB prediction of random effects relies much more on normality.

In section 2.11.2, we discussed EB prediction of the random intercepts for different clusters j. Such predictions $\widetilde{\zeta}_j$ can be interpreted as predicted level-2 residuals for the mothers. We can obtain corresponding predicted level-1 residuals for birth i of mother j as

$$\widetilde{\epsilon}_{ij} \;=\; \widehat{\xi}_{ij} - \widetilde{\zeta}_j$$

where

$$\widehat{\xi}_{ij} \;=\; y_{ij} - (\widehat{\beta}_1 + \widehat{\beta}_2 x_{2ij} + \cdots + \widehat{\beta}_p x_{pij})$$

In linear mixed models (but not in the generalized linear mixed models discussed in volume 2), these predicted level-2 and level-1 residuals have normal sampling distributions if it is assumed that the random-intercept model has a normally distributed random intercept and level-1 residual. We can therefore use histograms or normal quantile–quantile plots of the predicted residuals to assess the assumptions that ζ_j and ϵ_{ij} are normally distributed. However, the EB predictions are based on normality assumptions and will tend to look more normal than they are when normality is violated.

We can also try to find outliers by using standardized residuals and looking for values that are unlikely under the standard normal distribution, for example, values outside the range ±4. In section 2.11.3, we discussed the diagnostic standard error of the EB predictions. A standardized level-2 residual can therefore be obtained as

$$r_j^{(2)} = \frac{\widetilde{\zeta}_j}{\sqrt{\widehat{\mathrm{Var}}(\widetilde{\zeta}_j)}}$$

For the level-1 residuals, we simply divide by the estimated level-1 standard deviation:

$$r_{ij}^{(1)} = \frac{\widetilde{\epsilon}_{ij}}{\sqrt{\widehat{\theta}}}$$

After using `xtmixed`, we can use the `predict` command with the `reffects` option to obtain $\widetilde{\zeta}_j$ and with the `rstandard` option to obtain $r_{ij}^{(1)}$. The standardized level-2 residual $r_j^{(2)}$ is, at the time of writing this book, not provided by `predict`, but we can calculate it by first estimating the comparative standard errors using the `reses` option and then estimating the required diagnostic standard error, as shown in section 2.11.3.

We begin by refitting the model in `xtmixed` without producing any output:

```
. quietly xtmixed birwt smoke male mage hsgrad somecoll collgrad
> married black kessner2 kessner3 novisit pretri2 pretri3 || momid: , mle
```

The steps necessary to obtain $r_j^{(2)}$ are (see section 2.11.3)

```
. predict lev2, reffects
. predict comp_se, reses
. generate diag_se = sqrt(exp(2*[lns1_1_1]_cons) - comp_se^2)
. replace lev2 = lev2/diag_se
```

and the $r_{ij}^{(1)}$ are obtained using

```
. predict lev1, rstandard
```

Histograms of the standardized level-1 residuals $r_{ij}^{(1)}$ and the standardized level-2 residuals $r_j^{(2)}$ can be plotted as follows:

```
. histogram lev1, normal xtitle(Standardized level-1 residuals)
. histogram lev2 if idx==1, normal xtitle(Standardized level-2 residuals)
```

These commands produce figures 3.6 and 3.7, respectively.

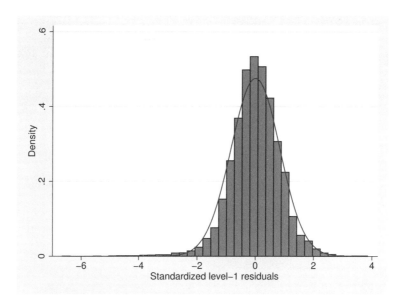

Figure 3.6: Histogram of standardized level-1 residuals

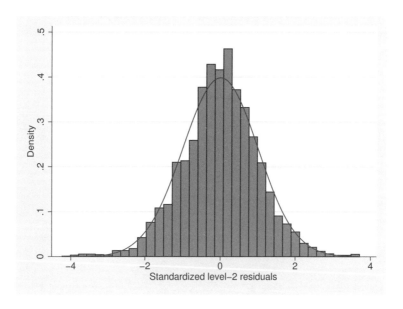

Figure 3.7: Histogram of standardized level-2 residuals

Both histograms look approximately normal, perhaps with thicker tails than the normal distribution. However, there are some extremely small level-1 residuals. The expected number of standardized level-1 residuals less than -4 is approximately 0.27 (=8604*normal(-4)).

It may in this case be judicious to use the sandwich estimator to obtain robust standard errors that do not rely on the model being correctly specified. Robust standard errors can be obtained using xtmixed (from Stata 12) with the vce(robust) option:

```
. quietly xtset momid
. xtreg birwt smoke male mage hsgrad somecoll collgrad married black
> kessner2 kessner3 novisit pretri2 pretri3, vce(robust) re
```

```
Random-effects GLS regression          Number of obs      =      8604
Group variable: momid                  Number of groups   =      3978

R-sq:  within  = 0.0380                Obs per group: min =         2
       between = 0.1128                               avg =       2.2
       overall = 0.0949                               max =         3

Random effects u_i ~ Gaussian          Wald chi2(13)      =    623.99
corr(u_i, X)      = 0 (assumed)        Prob > chi2        =    0.0000
```

(Std. Err. adjusted for 3978 clusters in momid)

birwt	Coef.	Robust Std. Err.	z	P>|z|	[95% Conf. Interval]	
smoke	-217.7488	19.15796	-11.37	0.000	-255.2977	-180.1999
male	120.9874	9.676186	12.50	0.000	102.0224	139.9524
mage	8.137158	1.406836	5.78	0.000	5.37981	10.89451
hsgrad	56.85672	25.83816	2.20	0.028	6.214852	107.4986
somecoll	80.64814	28.40641	2.84	0.005	24.97261	136.3237
collgrad	90.72697	29.14273	3.11	0.002	33.60827	147.8457
married	49.95895	26.63437	1.88	0.061	-2.243452	102.1613
black	-211.3336	29.32198	-7.21	0.000	-268.8037	-153.8636
kessner2	-92.89604	21.66719	-4.29	0.000	-135.363	-50.42913
kessner3	-150.6366	40.99285	-3.67	0.000	-230.9811	-70.29214
novisit	-29.9223	80.94521	-0.37	0.712	-188.572	128.7274
pretri2	92.73087	24.76155	3.74	0.000	44.19912	141.2626
pretri3	178.4334	52.90179	3.37	0.001	74.74785	282.119
_cons	3116.04	42.55804	73.22	0.000	3032.628	3199.452
sigma_u	340.64782					
sigma_e	368.91787					
rho	.46022181	(fraction of variance due to u_i)				

Most of the standard errors are larger than before, but the basic conclusions remain the same.

3.10 More on statistical inference

3.10.1 ❖ Overview of estimation methods

The simplest approach for estimating the fixed part of the model is to use OLS, sometimes referred to as pooled OLS because data on all clusters is combined. The OLS estimator is unbiased and consistent, but the conventional estimated standard errors are invalid if the residuals are correlated (see section 3.10.2). The sandwich estimator for clustered data can be used to obtain standard errors that take the clustering into account without making any assumptions regarding the random part of the model, apart from exogeneity. Such an approach can be viewed as treating clustering as a nuisance because nothing is learned about the residual between-cluster variability or within-cluster dependence.

A disadvantage of the OLS approach is that it is generally not asymptotically efficient if the residuals are correlated. If the residual covariance matrix were known, the efficient estimator would be generalized least squares (GLS), where the inverse covariance matrix of the total residuals is used as a weight matrix.

Display 3.1 shows matrix expressions for the GLS estimator.

Let $\mathbf{y} = (y_{11}, \ldots, y_{n_1 1}, y_{12}, \ldots, y_{n_2 2}, \ldots, y_{1J}, \ldots, y_{n_J J})'$ denote the N-dimensional vector of all responses, $\mathbf{X}$ the $N \times p$ matrix of covariates, $\boldsymbol{\beta}$ the corresponding p-dimensional vector of regression coefficients, and $\boldsymbol{\xi}$ the N-dimensional vector of total residuals. A linear model can then be written as

$$\mathbf{y} = \mathbf{X}\boldsymbol{\beta} + \boldsymbol{\xi}$$

Letting $\mathbf{V}$ denote the $N \times N$ covariance matrix of the vector of residuals $\boldsymbol{\xi}$, the OLS and GLS estimators can be written, respectively, as

$$\widehat{\boldsymbol{\beta}}^{\text{OLS}} = (\mathbf{X}'\mathbf{X})^{-1}\mathbf{X}'\mathbf{y}$$

and

$$\widehat{\boldsymbol{\beta}}^{\text{GLS}} = (\mathbf{X}'\mathbf{V}^{-1}\mathbf{X})^{-1}\mathbf{X}'\mathbf{V}^{-1}\mathbf{y}$$

Display 3.1: Matrix expressions for generalized least-squares estimator

In practice, we know only the structure of the residual covariance matrix, assuming that our model is correct. Namely, according to the random-intercept model, the variance is constant, residuals are correlated within clusters with constant correlation and uncorrelated across clusters. See display 3.2 for the form of the covariance structure.

In FGLS, the regression coefficients are first estimated by OLS, yielding estimated residuals from which the residual covariance matrix is estimated as $\widehat{\mathbf{V}}$. The regression coefficients are then reestimated by substituting $\widehat{\mathbf{V}}$ for $\mathbf{V}$ in the GLS estimator, producing estimates $\widehat{\boldsymbol{\beta}}^{\text{FGLS}}$.

The FGLS estimator is consistent for the regression coefficients even if the covariance structure is incorrectly specified. Under the model assumptions stated in section 3.3.2, the FGLS estimator is asymptotically normal and asymptotically efficient, and the model-based standard errors are valid, if a consistent estimator $\widehat{\mathbf{V}}$ is used for $\mathbf{V}$. Furthermore, the model assumptions imply that the FGLS estimator is unbiased in small samples if the total residuals have a symmetric distribution, such as normal.

In a random intercept model, the covariance matrix $\mathbf{V}$ of all total residuals in the dataset $(\xi_{11}, \ldots, \xi_{n_1 1}, \xi_{12}, \ldots, \xi_{n_2 2}, \ldots, \xi_{1J}, \ldots, \xi_{n_J J})'$ has the block-diagonal form

$$\mathbf{V} = \begin{pmatrix} \mathbf{V}_1 & \mathbf{0} & \cdots & \mathbf{0} \\ \mathbf{0} & \mathbf{V}_2 & \cdots & \mathbf{0} \\ \mathbf{0} & \mathbf{0} & \ddots & \vdots \\ \mathbf{0} & \mathbf{0} & \cdots & \mathbf{V}_J \end{pmatrix}$$

The blocks $\mathbf{V}_1$ to $\mathbf{V}_J$ on the diagonal are the within-cluster covariance matrices, and all other elements are zero. For a cluster with $n_j = 3$ units, the within-cluster covariance matrix has the structure

$$\mathbf{V}_j = \begin{pmatrix} \psi + \theta & \psi & \psi \\ \psi & \psi + \theta & \psi \\ \psi & \psi & \psi + \theta \end{pmatrix}$$

Display 3.2: Residual covariance structure for random-intercept model

The residuals based on FGLS are different from the residuals produced by OLS that were originally used to estimate $\mathbf{V}$. This suggests reestimating $\mathbf{V}$ based on the FGLS residuals and then reestimating the regression coefficients. Iterating this process until convergence yields so-called iterative generalized least squares (IGLS). The IGLS estimator maximizes the likelihood implied by the linear model assuming multivariate normal total residuals.

More commonly, the likelihood is maximized using gradient methods such as the Newton–Raphson algorithm. The basic idea is to find the maximum of the log likelihood iteratively, in each step updating the parameters based on derivatives of the log likelihood at the current parameter values. In Newton–Raphson, the next parameter values are at the maximum of the quadratic function matching the first and second derivatives at the current parameter values. Indeed, if the log likelihood is quadratic, one Newton–Raphson step suffices. It is remarkable how few iterations are often required even for quite complicated models. The negative *Hessian* (the matrix of second derivatives) of the log likelihood at the maximum is called the observed *information matrix*. Its inverse is a model-based estimator of the covariance matrix of the parameter estimates.

Another common approach for ML estimation of multilevel models is the expectation-maximization (EM) algorithm. Here the random effects are treated as missing data. If the random effects (or missing data) were known, or in other words if we had the "complete data", estimation of the parameters would be straightforward.

For instance, in a random-intercept model, the random-intercept variance ψ could be estimated as

$$\widehat{\psi}^{\text{complete data}} = \frac{1}{J} \sum_{j=1}^{J} \zeta_j^2 \tag{3.18}$$

However, the random effects are not known, so in the expectation step, the posterior expectation of the complete data estimators is found. For ψ, this amounts to finding the mean of (3.18) over the posterior distribution of the random effects, given the observed data (where the current parameter estimates are substituted for the unknown parameters; see section 2.11.2):

$$E\left(\frac{1}{J} \sum_{j=1}^{J} \zeta_j^2 \,\middle|\, \mathbf{y}, \mathbf{X}\right)$$

In the maximization step, the parameters are updated by setting them equal to the posterior expectations of the complete data estimators. This yields an updated posterior distribution of the random intercepts, giving updated posterior expectations, etc., until convergence. It has been shown that the EM algorithm converges to a (local) maximum, but the number of iterations required can be large.

An alternative to ML estimation is restricted or residual maximum likelihood (REML) estimation. Here the responses are transformed in such a way that their distribution no longer depends on the regression parameters $\boldsymbol{\beta}$. The likelihood then depends only on the parameters of the random part of the model. Maximizing this residual likelihood produces REML estimates of the variance (and covariance) parameters. Having estimated these parameters by REML, the regression coefficients can be estimated by using the implied residual covariance matrix $\widetilde{\mathbf{V}}$ in GLS. For balanced data, REML gives unbiased estimates of variance (and covariance) parameters (if variances are allowed to be negative), unlike ML. However, it is not clear which estimator has a smaller mean squared error (in balanced or unbalanced data).

Pooled OLS is obtained using the `regress` command, and the sandwich estimator for clustered data is obtained using the `vce(cluster clustvar)` option. FGLS is implemented in `xtreg` with the `re` option and in `xtgls` (see section 6.7.2 and exercise 6.4). The latter command uses IGLS when the `igls` option is specified.

By default, `xtmixed` uses ML (since Stata 12), but the `reml` option can be used for REML estimation. `xtmixed` starts with the EM algorithm and then switches to Newton–Raphson. The options `emiterate()` or `emonly` can be used to increase the number of EM steps or to use EM exclusively. The `technique()` option can be used to replace Newton–Raphson by another gradient method.

3.10.2   Consequences of using standard regression modeling for clustered data

In this section, we presume that the random-intercept model

$$y_{ij} = \beta_1 + \beta_2 x_{ij} + \underbrace{\zeta_j + \epsilon_{ij}}_{\xi_{ij}} \tag{3.19}$$

is the true model, with the assumptions stated in section 3.3.1 satisfied. However, the regression coefficient β_2 in the above random-intercept model is estimated using OLS, based on the assumptions stated in section 1.5, for the linear regression model

$$y_i = \beta_1 + \beta_2 x_i + \epsilon_i$$

which is *misspecified*.

The crucial difference between the two models is that there is a positive within-cluster correlation between the (total) residuals in the random-intercept model if $\psi > 0$, whereas there is no such within-cluster correlation in the linear regression model. For simplicity, we assume that the data are balanced ($n_j = n > 1$) and have large $N = Jn$.

The good news is that the OLS estimator is consistent (approaches the unknown parameter value in large samples) when the random-intercept model is true and $\psi > 0$. Indeed, even for a given dataset, the OLS estimates are usually close to the ML estimates for the correctly specified random-intercept model: $\widehat{\beta}_1^{\mathrm{OLS}} \approx \widehat{\beta}_1^{\mathrm{ML}}$, $\widehat{\beta}_2^{\mathrm{OLS}} \approx \widehat{\beta}_2^{\mathrm{ML}}$, and $\widehat{\sigma^2}^{\mathrm{OLS}} \approx \widehat{\psi}^{\mathrm{ML}} + \widehat{\theta}^{\mathrm{ML}}$.

The results are worse when it comes to fitted model-based standard errors and p-values for hypothesis tests. The estimated standard error of the OLS estimator $\widehat{\beta}_2^{\mathrm{OLS}}$ for the linear regression model is

$$\widehat{\mathrm{SE}}(\widehat{\beta}_2^{\mathrm{OLS}}) = \sqrt{\frac{\widehat{\sigma^2}^{\mathrm{OLS}}}{Jn\, s_{xO}^2}} \approx \sqrt{\frac{\widehat{\psi}^{\mathrm{ML}} + \widehat{\theta}^{\mathrm{ML}}}{Jn\, s_{xO}^2}}$$

where s_{xO}^2 is the overall sample variance of x_{ij} given in section 3.2.1.

For a purely between-cluster covariate (with $x_{ij} = \overline{x}_{\cdot j}$), the within-cluster variance s_{xW}^2 is zero ($s_{xO}^2 = s_{xB}^2$) and the estimated standard error of the ML estimator $\widehat{\beta}_2^{\mathrm{ML}}$ for the random-intercept model is

$$\widehat{\mathrm{SE}}(\widehat{\beta}_2^{\mathrm{ML}}) = \sqrt{\frac{n\widehat{\psi}^{\mathrm{ML}} + \widehat{\theta}^{\mathrm{ML}}}{Jn\, s_{xO}^2}} > \widehat{\mathrm{SE}}(\widehat{\beta}_2^{\mathrm{OLS}}) \qquad \text{if } n > 1,\ \widehat{\psi} > 0$$

Because the OLS standard error is smaller than it should be, the corresponding p-value is too small.

For a purely within-cluster covariate (with $\overline{x}_{\cdot j} = \overline{x}_{\cdot\cdot}$), the between-cluster variance s^2_{xB} is zero ($s^2_{xO} = s^2_{xW}$) and the estimated standard error of $\widehat{\beta}^{\mathrm{ML}}_2$ for the random-intercept model is

$$\widehat{\mathrm{SE}}(\widehat{\beta}^{\mathrm{ML}}_2) = \sqrt{\frac{\widehat{\theta}^{\mathrm{ML}}}{Jn\,s^2_{xO}}} < \widehat{\mathrm{SE}}(\widehat{\beta}^{\mathrm{OLS}}_2) \qquad \text{if } \widehat{\psi} > 0$$

Because the OLS standard error is larger than it should be, the p-value is now too large.

Thus assuming a standard regression model when the random-intercept model is true and $\psi > 0$ will produce too small or too large model-based standard errors and p-values depending on the nature of the covariate. This is contrary to the common belief that the estimated standard errors are always too small.

An important lesson is that robust standard errors for clustered data should be employed when standard regression models are used for clustered data. In Stata, such standard errors can be obtained using the vce(cluster *clustvar*) option, as demonstrated in section 2.10.3.

The above expressions for the fitted model-based standard errors also hold if the model includes other covariates that are uncorrelated with x_{ij}.

3.10.3 ❖ Power and sample-size determination

Here we consider power and sample-size determination for the random-intercept model in (3.19) with a single covariate that either varies only between clusters or varies only within clusters. A typical problem is to determine the sample size to achieve a required power γ at a given significance level α for the two-sided test of the null hypothesis H_0: $\beta_2 = 0$.

From the construction of the z test for the above null hypothesis, it follows that

$$\frac{\beta_2}{\mathrm{SE}(\widehat{\beta}_2)} \approx z_{1-\alpha/2} + z_\gamma \tag{3.20}$$

where $z_{1-\alpha/2}$ is the quantile of a standard normal distribution so that the area under the density curve from $-\infty$ to $z_{1-\alpha/2}$ is $1 - \alpha/2$. Analogously, z_γ is a quantile, making the area equal to γ (see display 3.3 for a derivation).

The above approximation is very accurate if $\gamma > 0.30$. Often the significance level and power are chosen as $\alpha = 0.05$ and $\gamma = 0.80$, respectively, which gives $z_{1-\alpha/2} + z_\gamma = 1.96 + 0.84 = 2.80$.

Under the null hypothesis, H_0: $\beta_2 = 0$, the test statistic z has a standard normal distribution (asymptotically). If the alternative hypothesis is two-sided, the null hypothesis is rejected at significance level $\alpha = 0.05$ if $z > z_{0.975}$ or $z < z_{0.025}$, where $z_{0.975}$ is the 97.5th percentile of the standard normal distribution (equal to 1.96) and $z_{0.025}$ is the 2.5th percentile (equal to -1.96). More generally, the null hypothesis is rejected at level α if $z > z_{1-\alpha/2}$ or $z < z_{\alpha/2}$. The right-tail probability, $P(z > z_{1-\alpha/2}|H_0)$, is shown as a light-shaded area under the standard normal density in the graph below for $\alpha = 0.05$:

Null hypothesis
H_0: $z \sim N(0,1)$

$P(z > z_{1-\alpha/2}|H_0) = \alpha/2$

$P(z > z_{0.975}|H_0) = 0.025$

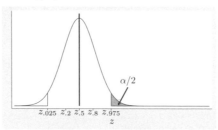

Alternative hypothesis
H_a: $z \sim N(\tau, 1)$

$P(z > z_{1-\alpha/2}|H_a) \approx \gamma$

$P(z < z_{\alpha/2}|H_a) \approx 0$

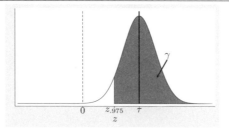

From now on we assume that the alternative hypothesis, H_a: $\beta_2 \neq 0$, is true and that the true coefficient β_2 equals some predetermined positive *effect size*. We let τ be the true coefficient divided by the standard error of its estimate, $\tau = \beta_2/\mathrm{SE}(\widehat{\beta}_2)$. The test statistic $z = \widehat{\beta}_2/\mathrm{SE}(\widehat{\beta}_2)$ now has an asymptotic normal distribution with mean τ and variance 1:

$$z \sim N(\tau, 1) \quad \Longleftrightarrow \quad z - \tau \sim N(0,1)$$

The probability that H_0 is rejected because $z > z_{1-\alpha/2}$ is shown as the dark-shaded area in the graph above. This is approximately the power γ, defined as the probability of rejecting the null hypothesis for an assumed effect size under the alternative hypothesis, because the probability that $z < z_{\alpha/2}$ is negligible if τ is sufficiently large:

$$\text{Power} \equiv \gamma \approx P(z > z_{1-\alpha/2}|H_a) = P(z - \tau > z_{1-\alpha/2} - \tau|H_a)$$

Because $z - \tau$ is standard normal, we know how large $z_{1-\alpha/2} - \tau$ has to be for the probability in the above equation to equal γ, namely,

$$z_{1-\alpha/2} - \tau \approx z_{1-\gamma} \quad \Longleftrightarrow \quad \tau \equiv \frac{\beta_2}{\mathrm{SE}(\widehat{\beta}_2)} \approx z_{1-\alpha/2} - z_{1-\gamma} = z_{1-\alpha/2} + z_\gamma$$

where $z_{1-\gamma}$ is the appropriate percentile of $N(0,1)$.

Display 3.3: Approximate relationship between power, significance level, effect size, and standard error for two-sided test

For a random-intercept model with a purely between-cluster covariate x_j, the standard error of the ML estimate of the coefficient is

$$\mathrm{SE}(\widehat{\beta}_2^B) \;=\; \sqrt{\frac{n\psi + \theta}{Jn\, s_{xO}^2}}$$

We can then substitute this standard error into (3.20) and solve for the total number of clusters J for given cluster size n (or vice versa), with given values of the parameters β_2, ψ, and θ and with a given variance of the covariate s_{xO}^2.

For example, let $n=4$, $\beta_2=1$, $\psi=1$, $\theta=1$, and x_j be a between-cluster treatment dummy variable, equal to 0 for half the clusters and 1 for the other half so that $s_{xO}^2 = 0.5 \times (1-0.5) = 0.25$. We then obtain

$$\mathrm{SE}(\widehat{\beta}_2^B) \;=\; \sqrt{\frac{4 \times 1 + 1}{J \times 4 \times 0.25}} \;=\; \sqrt{5/J}$$

from which it follows that

$$\frac{\beta_2}{\mathrm{SE}(\widehat{\beta}_2^B)} \;=\; \sqrt{\frac{1}{5/J}} \;=\; \sqrt{J/5}$$

Substituting for $\beta_2/\mathrm{SE}(\widehat{\beta}_2^B)$ and $z_{1-\alpha/2}+z_\gamma$ in (3.20), we obtain the equation

$$\sqrt{J/5} \;=\; 2.80$$

which we solve to get $J = 39.2$. We see that about 40 clusters are needed (20 per treatment group) to achieve 80% power to detect the treatment effect at the 5% significance level.

For a random-intercept model with a purely within-cluster covariate x_{ij} that does not vary between clusters, the standard error of the ML estimate of the coefficient is

$$\mathrm{SE}(\widehat{\beta}_2^W) \;=\; \sqrt{\frac{\theta}{Jn\, s_{xO}^2}}$$

Assuming that x_{ij} is a treatment dummy variable equal to 0 for half the units in each cluster and 1 for the other half, and keeping all other assumptions the same as in the example above, we obtain

$$\mathrm{SE}(\widehat{\beta}_2^W) \;=\; \sqrt{\frac{1}{J \times 4 \times 0.25}} \;=\; \sqrt{1/J}$$

and it follows that

$$\frac{\beta_2}{\mathrm{SE}(\widehat{\beta}_2^W)} \;=\; \sqrt{\frac{1}{1/J}} \;=\; \sqrt{J}$$

Substituting in (3.20),

$$\sqrt{J} \;=\; 2.80$$

and solving for J, we see that only about 8 clusters are now needed in total, compared with about 40 if treatment is between clusters.

In randomized experiments, the designs where entire clusters are assigned to treatments are often called cluster-randomized trials whereas studies where units are assigned to treatments, stratified by cluster, are called multisite studies. Multisite studies have more power and thus require smaller sample sizes, particularly if $n\psi$ is large compared with θ as in our example. Sometimes, though, multisite studies are not feasible, for instance, if the clusters are classrooms and the treatment is a new curriculum. Another reason for a cluster-randomized trial is that it avoids "contamination", where units not assigned to the treatment may indirectly benefit from the treatment received by other units in the cluster.

3.11   Summary and further reading

We have discussed linear random-intercept models, which are important for investigating the relationship between a continuous response and a set of covariates when the data have a clustered or hierarchical structure. Topics included hypothesis testing, different kinds of coefficients of determination, the choice between fixed- and random-effects approaches, model diagnostics, consequences of using standard regression for clustered data, and power and sample-size determination.

An important problem in any regression model is the potential for bias due to omitted covariates. In clustered data, we can identify the problem of omitted cluster-level covariates by comparing within-cluster and between-cluster effects of covariates varying both within and between clusters. This problem of cluster-level confounding is the reason for the ecological fallacy and is referred to as an endogeneity problem in econometrics. Although the problem of endogeneity is emphasized in econometrics, where the Hausman test is routinely used, it is not usually considered in other disciplines. However, relatively nontechnical treatments with applications in biostatistics can be found in Neuhaus and Kalbfleisch (1998) and Begg and Parides (2003). Exercise 3.5 uses a dataset considered in the first paper.

Another area where within- and between-cluster effects are contrasted is education (Raudenbush and Bryk 2002, 135–141). Specifically, interest often concerns the difference in within-school and between-school effects of socioeconomic status (SES); see exercise 3.7. As discussed in section 3.7.4, the difference in the within-cluster effect of a covariate (such as students' individual SES) and the between-cluster effect of the aggregated covariate (such as school mean SES) is often called a *contextual effect* in contrast to the *compositional effect* (the within effect). The distinction between compositional and contextual effects on health is discussed by Bingenheimer and Raudenbush (2004) and Duncan, Jones, and Moon (1998). The book by Wooldridge (2010) provides a good but demanding treatment of within and between estimators and endogeneity from an econometric perspective. More accessible discussions of this topic are given in the papers by Palta and Seplaki (2002) and Ebbes, Böckenholt, and Wedel (2004). The endogeneity problem is revisited in chapter 5, where more advanced methods are introduced in section 5.3.2.

An excellent discussion of linear random-intercept models can be found in Snijders and Bosker (2012, chap. 4), which is also a useful reference on model diagnostics and power analysis. A useful overview of power analysis and sample-size determination for multilevel models is provided in the encyclopedia entry by Snijders (2005). Swaminathan and Rogers (2008) give an excellent overview and detailed explanation of estimation methods for random-intercept and random-coefficient models. Many of the references mentioned above are collected in Skrondal and Rabe-Hesketh (2010), the first volume of a recent anthology on multilevel modeling.

In the exercises, random-intercept models are applied to data from different disciplines with clustering due to longitudinal data (exercises 3.2, 3.3, and 3.6), children nested in neighborhoods (exercise 3.1), rat pups nested in litters (exercise 3.4), children nested in mothers (exercise 3.5), and students nested in schools (exercises 3.7 and 3.8). Exercise 3.9 is about small-area estimation, a topic not covered in this chapter. Exercises 3.8 and 3.11 involve power analysis.

3.12   Exercises

3.1 Neighborhood-effects data

Garner and Raudenbush (1991), Raudenbush and Bryk (2002), and Raudenbush et al. (2004) considered neighborhood effects on educational attainment for young people who left school between 1984 and 1986 in one education authority in Scotland.

The dataset `neighborhood.dta` (previously used in exercise 2.4) has the following variables:

- Level 1 (students)
    - `attain`: a measure of end-of-school educational attainment capturing both attainment and length of schooling (based on the number of O-grades and Higher SCE awards at the A–C levels)
    - `p7vrq`: verbal-reasoning quotient (test at age 11–12 in primary school)
    - `p7read`: reading test score (test at age 11–12 in primary school)
    - `dadocc`: father's occupation scaled on the Hope–Goldthorpe scale in conjunction with the Registrar General's social-class index (Willms 1986)
    - `dadunemp`: dummy variable for father being unemployed (1: unemployed; 0: not unemployed)
    - `daded`: dummy variable for father's schooling being past the age of 15
    - `momed`: dummy variable for mother's schooling being past the age of 15
    - `male`: dummy variable for student being male

- Level 2 (neighborhoods)
  - neighid: neighborhood identifier
  - deprive: social-deprivation score derived from poverty concentration, health, and housing stock of local community

1. Fit a random-intercept model with attain as the response variable and without any covariates by ML using xtmixed. What are the estimated variance components between and within neighborhoods? Obtain the estimated intraclass correlation.

2. Include the covariate deprive in the model and interpret the estimates. Discuss the changes in the estimated standard deviations of the random intercept and level-1 residual.

3. Include the student-level covariates and interpret the estimates. Also comment on how the estimated standard deviations have changed.

4. Obtain the overall coefficient of determination R^2 for the model in step 3.

Crossed random-effects models are applied to the same data in exercise 9.5.

3.2 Grade-point-average data

Hox (2010) analyzed simulated longitudinal or panel data on 200 college students whose grade point average (GPA) was recorded over six successive semesters.

The variables in the dataset gpa.dta are

- gpa1–gpa6: grade point average (GPA) for semesters 1–6
- student: student identifier
- highgpa: high school GPA
- job1–job6: amount of time per week spent working for pay (0: not at all; 1: 1 hour; 2: 2 hours; 3: 3 hours; 4: 4 or more hours) in semesters 1–6
- sex: sex (1: male; 2: female)

1. Reshape the data to long form, stacking the time-varying variables gpa1–gpa6 and job1–job6 into two new variables, gpa and job, and generating a new variable, time, taking the values 1–6 for semesters 1–6.

2. Fit a random-intercept model with covariates time, highgpa, and job, and a dummy variable for males.

3. Assess whether the linearity assumptions for the three continuous covariates appear to be reasonable. You could use graphical methods, include quadratic terms, or use dummy variables for the different values of the covariates (if there are not too many).

4. Test whether there are interactions between each of the two student-level covariates and time.

5. For the chosen model, obtain empirical Bayes predictions of the random intercepts and produce graphs to assess their normality.

3.3 Jaw-growth data

In this jaw growth dataset from Potthoff and Roy (1964), eleven boys and sixteen girls had the distance between the center of the pituitary to the pterygomaxillary fissure recorded at ages 8, 10, 12, and 14.

The dataset `growth.dta` has the following variables:

- `idnr`: subject identifier
- `measure`: distance between pituitary and maxillary fissure in millimeters
- `age`: age in years
- `sex`: gender (1: boys; 2: girls)

1. Plot the observed growth trajectories—that is, plot `measure` against `age`—and connect successive observations on the same subject using the option `connect(ascending)`. Use the `by()` option to obtain separate graphs by sex.
2. Fit a linear random-intercept model with `measure` as the response variable and with `age` and a dummy variable for girls as explanatory variables.
3. Test whether there is a significant interaction between sex and age at the 5% level.
4. For the chosen model (with or without the interaction), add the predicted mean trajectories for boys and girls to the graph of the observed growth trajectories. (Hint: use `predict, xb`.)

Random-coefficient models are applied to the same data in exercise 7.3.

3.4 Rat-pups data

Dempster et al. (1984) analyzed data from a reproductive study on rats to assess the effect of an experimental compound on general reproductive performance and pup weights. Thirty dams (rat mothers) were randomized to three groups of 10 dams: control, low dose, and high dose of the compound. In the high-dose group, one female did not conceive, one cannibalized her litter, and one delivered one still birth, so that data on only seven litters were available for that group.

The dataset `pups.dta` has the following variables:

- dam: dam (pup's mother) identifier
- `sex`: sex of pup (0: male; 1: female)
- dose: dose group (0: controls; 1: low dose; 2: high dose)
- `w`: weight of pup in grams

1. Construct a variable, `size`, representing the size of each litter.
2. Construct a variable, `mnw`, representing the mean weight of each litter.
3. Plot `mnw` versus `size`, using different symbols for the three treatment groups. Describe the graph.

4. Fit a random-intercept model for pup weights with `sex`, `dose` (dummy variables for low and high doses), and `size` as covariates and a random intercept for `dam`. Use ML estimation.

5. Obtain level-1 and level-2 residuals, and produce graphs to assess the normality assumptions for ζ_j and ϵ_{ij}.

3.5 Georgian birthweight data

Here we use the data described in exercise 2.7. Following Neuhaus and Kalbfleisch (1998) and Pan (2002), we consider the relationship between the child's birthweight and his or her mother's age at the time of the birth, distinguishing between within-mother and between-mother effects of age.

The variables in `birthwt.dta` are

- `mother`: mother identifier
- `child`: child identifier
- `birthwt`: child's birthweight (in grams)
- `age`: mother's age at the time of the child's birth

1. Fit a random-intercept model with birthweight as the response variable and age as the explanatory variable.
2. Perform a Hausman specification test.
3. Modify the model to estimate both within-mother and between-mother effects of age.
4. Discuss what mother-level omitted covariates might be responsible for the difference between the estimated within-mother and between-mother effects.

We recommend reading either Neuhaus and Kalbfleisch (1998) or Pan (2002), the latter being less technical.

3.6 Wage-panel data

Vella and Verbeek (1998) analyzed panel data for 545 young males taken from the U.S. National Longitudinal Survey (Youth Sample) for the period 1980–1987.

The dataset `wagepan.dta` was supplied by Wooldridge (2010). The subset of variables considered here is

- `nr`: person identifier (j)
- `year`: 1980 to 1987 (i)
- `lwage`: log of hourly wage in U.S. dollars (y_{ij})
- `educ`: years of schooling (x_{2j})
- `black`: dummy variable for being black (x_{3j})
- `hisp`: dummy variable for being Hispanic (x_{4j})
- `exper`: labor-market experience defined as age$-6-$educ (x_{5ij})
- `expersq`: labor-market experience squared (x_{6ij})

- married: dummy variable for being married (x_{7ij})
- union: dummy variable for being a member of a union (that is, wage being set in collective bargaining agreement) (x_{8ij})

1. Ignore the clustered nature of the data, and use the regress command to fit the regression model

$$y_{ij} = \alpha_i + \beta_2 x_{2j} + \beta_3 x_{3j} + \beta_4 x_{4j} + \beta_5 x_{5ij} + \beta_6 x_{6ij} + \beta_7 x_{7ij} + \beta_8 x_{8ij} + \epsilon_{ij}$$

   where α_i are fixed year-specific intercepts and the covariates are defined in the bulleted list above.

2. Refit the above model using the vce(cluster nr) option to get standard errors taking the clustering into account. Compare the estimated standard errors with those from step 1.

3. Fit the random-intercept model

$$y_{ij} = \alpha_i + \beta_2 x_{2j} + \beta_3 x_{3j} + \beta_4 x_{4j} + \beta_5 x_{5ij} + \beta_6 x_{6ij} + \beta_7 x_{7ij} + \beta_8 x_{8ij} + \zeta_j + \epsilon_{ij}$$

   with the usual assumptions. How do the estimates of the βs compare with those from the models ignoring clustering?

4. Modify the random-intercept model to investigate whether the effect of education has increased linearly over time after controlling for the other covariates.

A wide range of longitudinal models are applied to this dataset in chapters 5 and 6.

3.7 High-school-and-beyond data [Solutions]

Raudenbush and Bryk (2002) and Raudenbush et al. (2004) analyzed data from the High School and Beyond Survey.

The variables in the dataset hsb.dta that we will use here are

- schoolid: school identifier (j)
- mathach: a measure of mathematics achievement (y_{ij})
- ses: socioeconomic status (SES) based on parental education, occupation, and income (x_{ij})

1. Use xtreg to fit a model for mathach with a fixed effect for SES and a random intercept for school.

2. Use xtsum to explore the between-school and within-school variability of SES.

3. Produce a variable, mn_ses, equal to the schools' mean SES and another variable, dev_ses, equal to the difference between the students' SES and the mean SES for their school.

4. The model in step 1 assumes that SES has the same effect within and between schools. Check this by using the covariates mn_ses and dev_ses instead of ses and comparing the coefficients using lincom.

5. Interpret the coefficients of `mn_ses` and `dev_ses`.

6. Returning to the model with `ses` as the only covariate, perform a Hausman specification test and comment on the result.

3.8 Cluster-randomized trial of sex education

Wight et al. (2002) reported on a randomized trial of sex education, and the data were also analyzed and provided by Hayes and Moulton (2009).

Twenty-five secondary schools in east Scotland were randomly assigned to a sex education program for adolescents called SHARE (Sexual Health and Relationships: Safe, Happy and Responsible) or to a control group (usual sex education). Trials in which the unit of randomization is a group are often called cluster-randomized trials.

Teachers in schools belonging to the intervention group received five days of training. They delivered 10 sessions in the third year of secondary school (at 13–14 years) and 10 sessions in the fourth year to two successive cohorts of students (students in their third year in 1996 and 1997). Out of 47 non-Catholic, qualifying schools, 25 agreed to participate. Eight thousand four hundred thirty students were recruited to the trial, and 5,854 were successfully followed up after two years.

The variables in the dataset `sex.dta` that we will use are

- `school`: school identifier
- `sex`: gender of the student (1: male; 2: female)
- `arm`: treatment group (0: control; 1: intervention)
- `scpar`: highest social class of mother or father, where social class is defined using the UK Registrar General's classification based on occupation (I is highest). (10: I; 20: II; 31: III nonmanual; 32: III manual; 40: IV; 50: V; 99: not coded)
- `kscore`: knowledge of sexual health at follow-up (score from -8 to $+8$)

1. Discuss whether you expect gender and social class to be approximately balanced between treatment groups (with similar frequency distributions).

2. Produce frequency tables for social class and gender by treatment group (with both absolute frequencies and percentages). Are the treatment groups adequately balanced in these variables?

3. Compare the knowledge of sexual health score at follow-up between students in the intervention and control groups. Include a random intercept for schools to allow for the cluster-randomized design. Interpret the estimated treatment effect.

4. Extend the model in step 3 by including dummy variables for gender and social class (using a dummy variable for the group for whom social class was not coded instead of discarding these data). Comment on any change in the estimated treatment effect.

5. Report the estimated residual intraclass correlation for the model in step 4, and use a likelihood-ratio test to test for zero intraclass correlation.

6. ❖ A similar trial is planned with an improved version of the sex education program. Assume that the estimated effect size from step 4 is a conservative estimate of the effect size for the new study. Also assume that the same covariates will be used and that the residual variances at the school and student levels are equal to the estimates from step 4. The new study will recruit 60 students per school; it will randomize half of the students in each school to the new sex education program and the other half to the control group. How many schools should be recruited to achieve 80% power to detect a treatment effect at the 5% level of significance, allowing for a 30% dropout rate per school?

3.9 ❖ Small-area estimation of crop areas $\boxed{\text{Solutions}}$

Here we consider the problem of estimating quantities for small areas by borrowing strength from other areas using empirical Bayes prediction. The classic reference is Rao (2003), and the classic example is introduced below.

Battese, Harter, and Fuller (1988) analyzed survey and LANDSAT satellite data on 12 Iowa counties. The counties were partitioned into segments of about 250 hectares. In the survey, farm operators reported the number of hectares within 36 of the segments that are devoted to corn. Also available are inaccurate satellite data classifying the crop for all pixels (a pixel is about 0.45 hectares) in the 12 counties as corn, soybean, or neither. The problem is to estimate the number of hectares of corn per segment for an entire county, based on accurate survey information on a small sample of segments within the county, combined with covariate information (satellite measurements) for all segments in the county. This type of problem is called *small-area estimation*.

Battese, Harter, and Fuller (1988) fit a two-level linear random-intercept model, regressing the hectares of corn reported in the survey y_{ij} for the 36 segments i for counties $j = 1, \ldots, 12$ on the number of pixels classified as corn x_{1ij} and soybean x_{2ij} from the satellite data:

$$y_{ij} = \beta_1 + \beta_2 x_{2ij} + \beta_3 x_{3ij} + \zeta_j + \epsilon_{ij}$$

They then predicted the number of hectares devoted to corn per segment for each of the counties by $\widehat{\beta}_1 + \widehat{\beta}_2 \overline{x}_{2 \cdot j} + \widehat{\beta}_3 \overline{x}_{3 \cdot j} + \widetilde{\zeta}_j$, where $\overline{x}_{2 \cdot j}$ and $\overline{x}_{3 \cdot j}$ are the county means of the pixel variables (for all segments in the counties, not just those for which survey data were available) and $\widetilde{\zeta}_j$ is the empirical Bayes prediction of ζ_j.

The dataset `cropareas.dta` contains the following variables:

- `county`: county identifier
- `name`: name of county
- `segment`: segment identifier
- `cornhec`: number of corn hectares in the segment (reported in the survey)
- `soyhec`: number of soybean hectares in the segment (reported in the survey)

- `cornpix`: number of corn pixels in the segment (from satellite data) (x_{2ij})
- `soypix`: number of soybean pixels in the segment (from satellite data) (x_{3ij})
- `mn_cornpix`: mean number of corn pixels per segment, averaged over all segments in the county $(\bar{x}_{2\cdot j})$
- `mn_soypix`: mean number of soybean pixels per segment, averaged over all segments in the county $(\bar{x}_{3\cdot j})$

1. Fit the model above by ML.
2. Obtain predictions following the method of Battese, Harter, and Fuller (1988). (The prediction for Cerro Gordo should be 122.28.)
3. Obtain the estimated comparative standard errors of $\widetilde{\zeta}_j$.
4. Are these standard errors appropriate for expressing the uncertainty in the small-area estimates? Explain.

3.10 ❖ Relations among within, between, and FGLS estimators

For the Georgian birthweight data used in exercise 3.5, the within-mother and between-mother standard deviations of mothers' ages x_{ij} at the birth of the child are $s_{xW} = 2.796$ years and $s_{xB} = 3.693$ years, respectively. Consider a random-intercept model for the child's birthweight (with the usual assumptions):

$$y_{ij} \;=\; \beta_1 + \beta_2 x_{ij} + \zeta_j + \epsilon_{ij}$$

The within-mother and between-mother estimates of the effect of age on birthweight are $\widehat{\beta}_2^W = 11.832$ grams/year and $\widehat{\beta}_2^B = 30.355$ grams/year, respectively. There are 878 mothers with $n = 5$ births each, $\widehat{\mathrm{Var}}(\epsilon_{ij} - \bar{\epsilon}_{\cdot j}) = 150,631$ grams^2 and $\widehat{\mathrm{Var}}(\zeta_j + \bar{\epsilon}_{\cdot j}) = 161,457$ grams^2.

Based solely on the above information, obtain the following:

1. The estimated standard error of $\widehat{\beta}_2^W$.
2. The estimated standard error of $\widehat{\beta}_2^B$.
3. The FGLS estimate of β_2 for the random-intercept model.

3.11 ❖ Power analysis

1. For a random-intercept model with a single between-cluster covariate, calculate the total sample size Jn required to have 80% power to reject the null hypothesis that $\beta_2 = 0$ using a two-sided test at the 5% level. Assume that $n = 20$, $\beta_2 = 1$, $\psi = 1$, $\theta = 5$, and $s_{xO}^2 = 0.25$.
2. Repeat the calculation in step 1 with the same assumptions except that $\psi = 0$ and $\theta = 6$, that is, keeping the total variance as before but setting the intraclass correlation to zero.
3. Now consider the general situation where there are two scenarios, each having the same total residual variance $\psi + \theta$, but one having $\psi > 0$ and the other $\psi = 0$. Obtain an expression for the ratio of the required sample sizes Jn for these two scenarios in terms of the intraclass correlation ρ and the cluster size n. (This factor is sometimes called the "design effect" for clustered data.)

4 Random-coefficient models

4.1 Introduction

In the previous chapter, we considered linear random-intercept models where the overall level of the response was allowed to vary between clusters after controlling for covariates. In this chapter, we include random coefficients or random slopes in addition to random intercepts, thus also allowing the effects of covariates to vary between clusters. Such models involving both random intercepts and random slopes are often called *random-coefficient models*. In longitudinal settings, where the level-1 units are occasions and the clusters are typically subjects, random-coefficient models are also referred to as growth-curve models (see chapter 7).

4.2 How effective are different schools?

We start by analyzing a dataset on inner-London schools that accompanies the ML-wiN software (Rasbash et al. 2009) and is part of the data analyzed by Goldstein et al. (1993).

At age 16, students took their Graduate Certificate of Secondary Education (GCSE) exams in a number of subjects. A score was derived from the individual exam results. Such scores often form the basis for school comparisons, for instance, to allow parents to choose the best school for their child. However, schools can differ considerably in their intake achievement levels. It may be argued that what should be compared is the "value added"; that is, the difference in mean GCSE score between schools after controlling for the students' achievement before entering the school. One such measure of prior achievement is the London Reading Test (LRT) taken by these students at age 11.

The dataset `gcse.dta` has the following variables:

- `school`: school identifier
- `student`: student identifier
- `gcse`: Graduate Certificate of Secondary Education (GCSE) score (z score, multiplied by 10)
- `lrt`: London Reading Test (LRT) score (z score, multiplied by 10)
- `girl`: dummy variable for student being a girl (1: girl; 0: boy)
- `schgend`: type of school (1: mixed gender; 2: boys only; 3: girls only)

One purpose of the analysis is to investigate the relationship between GCSE and LRT and how this relationship varies between schools. The model can then be used to address the question of which schools appear to be most effective, taking prior achievement into account.

We read the data using

```
. use http://www.stata-press.com/data/mlmus3/gcse
```

4.3   Separate linear regressions for each school

Before developing a model for all 65 schools combined, we consider a separate model for each school. For school j, an obvious model for the relationship between GCSE and LRT is a simple regression model,

$$y_{ij} = \beta_{1j} + \beta_{2j}x_{ij} + \epsilon_{ij}$$

where y_{ij} is the GCSE score for the ith student in school j, x_{ij} is the corresponding LRT score, β_{1j} is the school-specific intercept, β_{2j} is the school-specific slope, and ϵ_{ij} is a residual error term with school-specific variance θ_j.

For school 1, OLS estimates of the intercept $\widehat{\beta}_{11}$ and the slope $\widehat{\beta}_{21}$ can be obtained using `regress`,

```
. regress gcse lrt if school==1
```

Source	SS	df	MS		Number of obs =	73
					F(1, 71) =	59.44
Model	4084.89189	1	4084.89189		Prob > F =	0.0000
Residual	4879.35759	71	68.7233463		R-squared =	0.4557
					Adj R-squared =	0.4480
Total	8964.24948	72	124.503465		Root MSE =	8.29

gcse	Coef.	Std. Err.	t	P>|t|	[95% Conf. Interval]	
lrt	.7093406	.0920061	7.71	0.000	.5258856	.8927955
_cons	3.833302	.9822377	3.90	0.000	1.874776	5.791828

where we have selected school 1 by specifying the condition if `school==1`.

To assess whether this is a reasonable model for school 1, we can obtain the predicted (ordinary least squares) regression line for the school,

$$\widehat{y}_{i1} = \widehat{\beta}_{11} + \widehat{\beta}_{21}x_{i1}$$

by using the `predict` command with the `xb` option:

```
. predict p_gcse, xb
```

We superimpose this line on the scatterplot of the data for the school, as shown in figure 4.1.

```
. twoway (scatter gcse lrt) (line p_gcse lrt, sort) if school==1,
> xtitle(LRT) ytitle(GCSE)
```

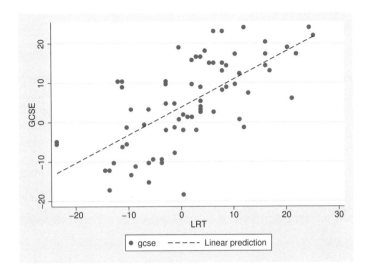

Figure 4.1: Scatterplot of `gcse` versus `lrt` for school 1 with ordinary least-squares regression line

We can also produce a *trellis graph* containing such plots for all 65 schools by using

```
. twoway (scatter gcse lrt) (lfit gcse lrt, sort lpatt(solid)),
> by(school, compact legend(off) cols(5))
> xtitle(LRT) ytitle(GCSE) ysize(3) xsize(2)
```

with the result shown in figure 4.2. The resulting graphs suggest that the model assumptions are reasonably met.

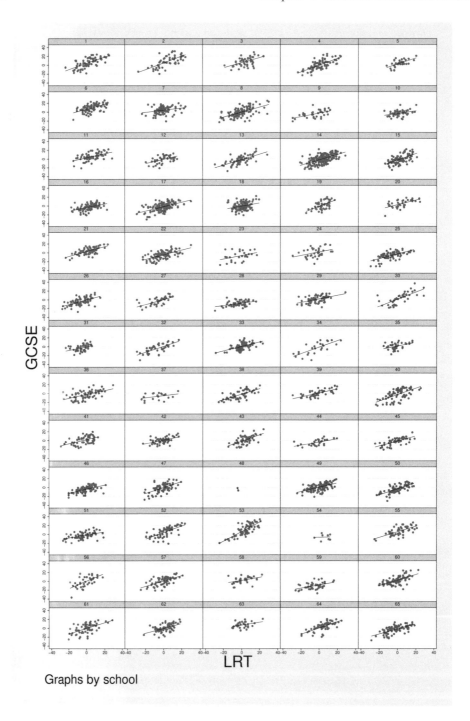

GCSE

LRT

Graphs by school

Figure 4.2: Trellis of scatterplots of `gcse` versus `lrt` with fitted regression lines for all 65 schools

We will now fit a simple linear regression model for each school, which is easily done using Stata's prefix command `statsby`. Then we will examine the variability in the estimated intercepts and slopes.

We first calculate the number of students per school by using `egen` with the `count()` function to preclude fitting lines to schools with fewer than five students:

```
. egen num = count(gcse), by(school)
```

We then use `statsby` to create a new dataset, `ols.dta`, in the local directory with the variables `inter` and `slope` containing OLS estimates of the intercepts (`_b[_cons]`) and slopes (`_b[lrt]`) from the command `regress gcse lrt if num>4` applied to each school (as well as containing the variable `school`):

```
. statsby inter=_b[_cons] slope=_b[lrt], by(school) saving(ols):
> regress gcse lrt if num>4
(running regress on estimation sample)

      command:  regress gcse lrt if num>4
        inter:  _b[_cons]
        slope:  _b[lrt]
           by:  school

Statsby groups
———+—— 1 ——+—— 2 ——+—— 3 ——+—— 4 ——+—— 5
..................................................  50
............
```

We can merge the estimates `inter` and `slope` into the `gcse` dataset by using the `merge` command (after sorting the "master data" that are currently loaded by `school`; the "using data" created by `statsby` are already sorted by `school`):

```
. sort school

. merge m:1 school using ols

    Result                      # of obs.
    ─────────────────────────────────────
    not matched                        2
        from master                    2  (_merge==1)
        from using                     0  (_merge==2)

    matched                        4,057  (_merge==3)
    ─────────────────────────────────────

. drop _merge
```

Here we have specified `m:1` in the `merge` command, which stands for "many-to-one merging" (observations for several students per school in the master data, but only one observation per school in the using data). We see that two of the schools in the master data did not have matches in the using data (because they had fewer than five students per school, so we did not compute OLS estimates for them). We have deleted the variable `_merge` produced by the `merge` command to avoid error messages when we run the `merge` command in the future.

A scatterplot of the OLS estimates of the intercept and slope is produced using the following command and given in figure 4.3:

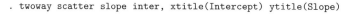

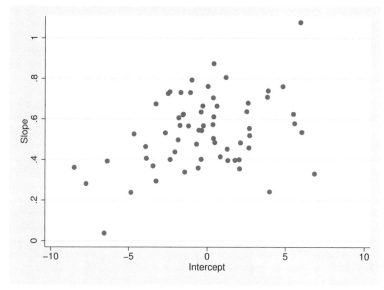

Figure 4.3: Scatterplot of estimated intercepts and slopes for all schools with at least five students

We see that there is considerable variability between the estimated intercepts and slopes of different schools. To investigate this further, we first create a dummy variable to pick out one observation per school,

```
. egen pickone = tag(school)
```

and then we produce summary statistics for the schools by using the `summarize` command:

```
. summarize inter slope if pickone == 1
    Variable │     Obs       Mean    Std. Dev.        Min        Max
─────────────┼─────────────────────────────────────────────────────
       inter │      64  -.1805974    3.291357  -8.519253   6.838716
       slope │      64   .5390514    .1766135   .0380965   1.076979
```

To allow comparison with the parameter estimates obtained from the random-coefficient model considered later on, we also obtain the covariance matrix of the estimated intercepts and slopes:

```
. correlate inter slope if pickone == 1, covariance
(obs=64)
             │    inter      slope
─────────────┼────────────────────
       inter │   10.833
       slope │  .208622    .031192
```

The diagonal elements, 10.83 and 0.03, are the sample variances of the intercepts and slopes, respectively. The off-diagonal element, 0.21, is the sample covariance between the intercepts and slopes, equal to the correlation times the product of the intercept and slope standard deviations.

We can also obtain a *spaghetti plot* of the predicted school-specific regression lines for all schools. We first calculate the fitted values $\widehat{y}_{ij} = \widehat{\beta}_{1j} + \widehat{\beta}_{2j} x_{ij}$,

```
. generate pred = inter + slope*lrt
(2 missing values generated)
```

and sort the data so that `lrt` increases within a given school and then jumps to its lowest value for the next school in the dataset:

```
. sort school lrt
```

We then produce the plot by typing

```
. twoway (line pred lrt, connect(ascending)), xtitle(LRT)
> ytitle(Fitted regression lines)
```

The `connect(ascending)` option is used to connect points only as long as `lrt` is increasing; it ensures that only data for the same school are connected. The resulting graph is shown in figure 4.4.

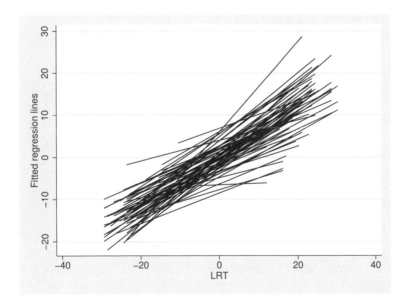

Figure 4.4: Spaghetti plot of ordinary least-squares regression lines for all schools with at least five students

4.4 Specification and interpretation of a random-coefficient model

4.4.1 Specification of a random-coefficient model

How can we develop a joint model for the relationships between `gcse` and `lrt` in all schools?

One way would be to use dummy variables for all schools (omitting the overall constant) to estimate school-specific intercepts and interactions between these dummy variables and `lrt` to estimate school-specific slopes. The only difference between the resulting model and separate regressions is that a common residual error variance $\theta_j = \theta$ is assumed. However, this model has 130 regression coefficients! Furthermore, if the schools are viewed as a (random) sample of schools from a population of schools, we are not interested in the individual coefficients characterizing each school's regression line. Rather, we would like to estimate the mean intercept and slope as well as the (co)variability of the intercepts and slopes in the population of schools.

A parsimonious model for the relationships between `gcse` and `lrt` can be obtained by specifying a school-specific random intercept ζ_{1j} and a school-specific random slope ζ_{2j} for `lrt` (x_{ij}):

$$
\begin{aligned}
y_{ij} &= \beta_1 + \beta_2 x_{ij} + \zeta_{1j} + \zeta_{2j} x_{ij} + \epsilon_{ij} \\
&= (\beta_1 + \zeta_{1j}) + (\beta_2 + \zeta_{2j}) x_{ij} + \epsilon_{ij}
\end{aligned} \tag{4.1}
$$

Here ζ_{1j} represents the deviation of school j's intercept from the mean intercept β_1, and ζ_{2j} represents the deviation of school j's slope from the mean slope β_2.

Given all covariates $\mathbf{X}_j$ in cluster j, it is assumed that the random effects ζ_{1j} and ζ_{2j} have zero expectations:

$$E(\zeta_{1j}|\mathbf{X}_j) = 0$$

$$E(\zeta_{2j}|\mathbf{X}_j) = 0$$

It is also assumed that the level-1 residual ϵ_{ij} has zero expectation, given the covariates and the random effects:

$$E(\epsilon_{ij}|\mathbf{X}_j, \zeta_{1j}, \zeta_{2j}) = 0$$

It follows from these mean-independence assumptions that the random terms ζ_{1j}, ζ_{2j}, and ϵ_{ij} are all uncorrelated with the covariate x_{ij} and that ϵ_{ij} is uncorrelated with both ζ_{1j} and ζ_{2j}. Both the intercepts ζ_{1j} and slopes ζ_{2j} are assumed to be uncorrelated across schools, and the level-1 residuals ϵ_{ij} are assumed to be uncorrelated across schools and students.

An illustration of this random-coefficient model with one covariate x_{ij} for one cluster j is shown in the bottom panel of figure 4.5. A random-intercept model is shown for comparison in the top panel.

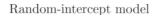

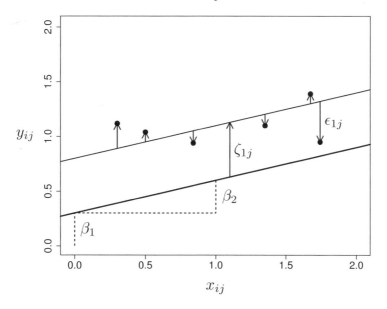

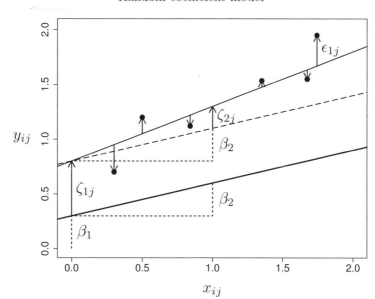

Figure 4.5: Illustration of random-intercept and random-coefficient models

In each panel, the lower bold and solid line represents the population-averaged or marginal regression line

$$E(y_{ij}|x_{ij}) = \beta_1 + \beta_2 x_{ij}$$

across all clusters. The thinner solid line represents the cluster-specific regression line for cluster j. For the random-intercept model, this is

$$E(y_{ij}|x_{ij}, \zeta_{1j}) = (\beta_1 + \zeta_{1j}) + \beta_2 x_{ij}$$

which is parallel to the population-averaged line with vertical displacement given by the random intercept ζ_{1j}. In contrast, in the random-coefficient model, the cluster-specific or conditional regression line

$$E(y_{ij}|x_{ij}, \zeta_{1j}, \zeta_{2j}) = (\beta_1 + \zeta_{1j}) + (\beta_2 + \zeta_{2j})x_{ij}$$

is not parallel to the population-averaged line but has a greater slope because the random slope ζ_{2j} is positive in the illustration. Here the dashed line is parallel to the population-averaged regression line and has the same intercept as cluster j. The vertical deviation between this dashed line and the line for cluster j is $\zeta_{2j}x_{ij}$, as shown in the diagram for $x_{ij}=1$. The bottom panel illustrates that the total intercept for cluster j is $\beta_1 + \zeta_{1j}$ and the total slope is $\beta_2 + \zeta_{2j}$. The arrows from the cluster-specific regression lines to the responses y_{ij} are the within-cluster residual error terms ϵ_{ij} (with variance θ). It is clear that $\zeta_{2j}x_{ij}$ represents an *interaction* between the clusters, treated as random, and the covariate x_{ij}.

Given $\mathbf{X}_j$, the random intercept and random slope have a bivariate distribution assumed to have zero means and covariance matrix $\boldsymbol{\Psi}$:

$$\boldsymbol{\Psi} = \begin{bmatrix} \psi_{11} & \psi_{12} \\ \psi_{21} & \psi_{22} \end{bmatrix} \equiv \begin{bmatrix} \mathrm{Var}(\zeta_{1j}|\mathbf{X}_j) & \mathrm{Cov}(\zeta_{1j},\zeta_{2j}|\mathbf{X}_j) \\ \mathrm{Cov}(\zeta_{2j},\zeta_{1j}|\mathbf{X}_j) & \mathrm{Var}(\zeta_{2j}|\mathbf{X}_j) \end{bmatrix}, \qquad \psi_{21} = \psi_{12}$$

Hence, given the covariates, the variance of the random intercept is ψ_{11}, the variance of the random slope is ψ_{22}, and the covariance between the random intercept and random slope is ψ_{21}. The correlation between the random intercept and random slope given the covariates becomes

$$\rho_{21} \equiv \mathrm{Cor}(\zeta_{1j},\zeta_{2j}|\mathbf{X}_j) = \frac{\psi_{21}}{\sqrt{\psi_{11}\psi_{22}}}$$

It is sometimes assumed that given $\mathbf{X}_j$, the random intercept and random slope have a bivariate normal distribution. An example of a bivariate normal distribution with $\psi_{11} = \psi_{22} = 4$ and $\psi_{21} = \psi_{12} = 1$ is shown as a perspective plot in figure 4.6. Specifying a bivariate normal distribution implies that the (marginal) univariate distributions of the intercept and slope are also normal.

Density

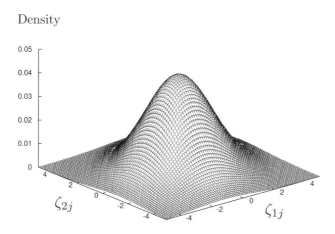

Figure 4.6: Perspective plot of bivariate normal distribution

4.4.2 Interpretation of the random-effects variances and covariances

Interpreting the covariance matrix $\boldsymbol{\Psi}$ of the random effects (given the covariates $\mathbf{X}_j$) is not straightforward.

First, the random-slope variance ψ_{22} and the covariance between random slope and intercept ψ_{21} depend not just on the scale of the response variable but also on the scale of the covariate, here `lrt`. Let the units of the response and explanatory variable be denoted as u_y and u_x, respectively. For instance, in an application considered in chapter 7 on children's increase in weight, u_y is kilograms and u_x is years. The units of ψ_{11} are u_y^2, the units of ψ_{21} are u_y^2/u_x, and the units of ψ_{22} are u_y^2/u_x^2. It therefore does not make sense to compare the magnitude of random-intercept and random-slope variances.

Another issue is that the total residual variance is no longer constant as in random-intercept models. The total residual is now

$$\xi_{ij} \ \equiv \ \zeta_{1j} + \zeta_{2j} x_{ij} + \epsilon_{ij}$$

and the conditional variance of the responses given the covariate, or the conditional variance of the total residual, is

$$\text{Var}(y_{ij}|\mathbf{X}_j) \ = \ \text{Var}(\xi_{ij}|\mathbf{X}_j) \ = \ \psi_{11} + 2\psi_{21}x_{ij} + \psi_{22}x_{ij}^2 + \theta \qquad (4.2)$$

This variance depends on the value of the covariate x_{ij}, and the total residual is therefore *heteroskedastic*. The conditional covariance for two students i and i' with covariate values x_{ij} and $x_{i'j}$ in the same school j is

$$\text{Cov}(y_{ij}, y_{i'j}|\mathbf{X}_j) = \text{Cov}(\xi_{ij}, \xi_{i'j}|\mathbf{X}_j)$$
$$= \psi_{11} + \psi_{21}x_{ij} + \psi_{21}x_{i'j} + \psi_{22}x_{ij}x_{i'j} \qquad (4.3)$$

and the conditional intraclass correlation becomes

$$\text{Cor}(y_{ij}, y_{i'j}|\mathbf{X}_j) = \frac{\text{Cov}(\xi_{ij}, \xi_{i'j}|\mathbf{X}_j)}{\sqrt{\text{Var}(\xi_{ij}|\mathbf{X}_j)\text{Var}(\xi_{i'j}|\mathbf{X}_j)}}$$

When $x_{ij} = x_{i'j} = 0$, the expression for the intraclass correlation is the same as for the random-intercept model and represents the correlation of the residuals (from the overall mean regression line) for two students in the same school who both have lrt scores equal to 0 (the mean). However, for pairs of students in the same school with other values of lrt, the intraclass correlation is a complicated function of lrt (x_{ij} and $x_{i'j}$).

Due to the heteroskedastic total residual variance, it is not straightforward to define coefficients of determination—such as R^2, R_2^2, and R_1^2, discussed in section 3.5—for random-coefficient models. Snijders and Bosker (2012, 114) suggest removing the random coefficient(s) for the purpose of calculating the coefficient of determination because this will usually yield values that are close to the correct version (see their section 7.2.2 for how to obtain the correct version).

Finally, interpreting the parameters ψ_{11} and ψ_{21} can be difficult because their values depend on the translation of the covariate or, in other words, on how much we add or subtract from the covariate. Adding a constant to lrt and refitting the model would result in different estimates of ψ_{11} and ψ_{21} (see also exercise 4.9). This is because the intercept variance is the variability in the vertical positions of school-specific regression lines where lrt=0 (which changes when lrt is translated) and the covariance or correlation is the tendency for regression lines that are higher up where lrt=0 to have higher slopes. This lack of invariance of ψ_{11} and ψ_{21} to translation of the covariate x_{ij} is illustrated in figure 4.7. Here identical cluster-specific regression lines are shown in the two panels, but the covariate $x'_{ij} = x_{ij} - 3.5$ in the lower panel is translated relative to the covariate x_{ij} in the upper panel. The intercepts are the intersections of the regression lines with the vertical lines at zero. Clearly these intercepts vary more in the upper panel than the lower panel, whereas the correlation between intercepts and slopes is negative in the upper panel and positive in the lower panel.

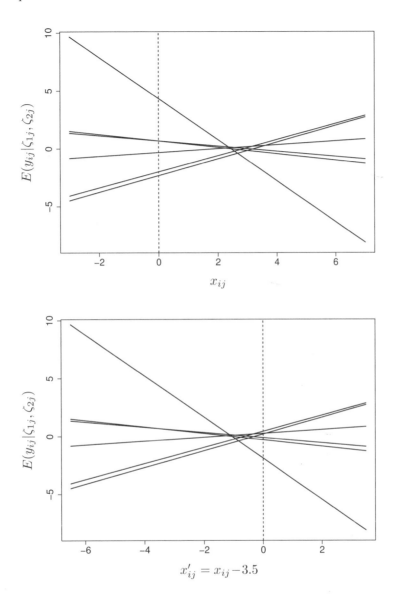

Figure 4.7: Cluster-specific regression lines for random-coefficient model, illustrating lack of invariance under translation of covariate (*Source:* Skrondal and Rabe-Hesketh 2004a)

To make ψ_{11} and ψ_{21} interpretable, it makes sense to translate x_{ij} so that the value $x_{ij} = 0$ is a useful reference point in some way. Typical choices are either mean centering (as for `lrt`) or, if x_{ij} is time, as in growth-curve models, defining 0 to be the initial time in some sense. Because the magnitude and interpretation of ψ_{21} depend on the location (or translation) of x_{ij}, which is often arbitrary, it generally does not make sense to set ψ_{21} to 0 by specifying uncorrelated intercepts and slopes.

A useful way of interpreting the magnitude of the estimated variances $\widehat{\psi}_{11}$ and $\widehat{\psi}_{22}$ is by considering the intervals $\widehat{\beta}_1 \pm 1.96 \sqrt{\widehat{\psi}_{11}}$ and $\widehat{\beta}_2 \pm 1.96 \sqrt{\widehat{\psi}_{22}}$, which contain about 95% of the intercepts and slopes in the population, respectively. To aid interpretation of the random part of the model, it is also useful to produce plots of school-specific regression lines, as discussed in section 4.8.3.

4.5   Estimation using xtmixed

xtmixed can be used to fit linear random-coefficient models by maximum likelihood (ML) estimation or restricted maximum likelihood (REML) estimation. (xtreg can only fit two-level random-intercept models.)

4.5.1   Random-intercept model

We first consider the random-intercept model discussed in the previous chapter:

$$y_{ij} = (\beta_1 + \zeta_{1j}) + \beta_2 x_{ij} + \epsilon_{ij}$$

This model is a special case of the random-coefficient model in (4.1) with $\zeta_{2j} = 0$ or, equivalently, with zero random-slope variance and zero random intercept and slope covariance, $\psi_{22} = \psi_{21} = 0$.

Maximum likelihood estimates for the random-intercept model can be obtained using xtmixed with the mle option (the default):

```
. xtmixed gcse lrt || school:, mle
Mixed-effects ML regression                 Number of obs      =       4059
Group variable: school                      Number of groups   =         65

                                            Obs per group: min =          2
                                                           avg =       62.4
                                                           max =        198

                                            Wald chi2(1)       =    2042.57
Log likelihood = -14024.799                 Prob > chi2        =     0.0000
```

gcse	Coef.	Std. Err.	z	P>|z|	[95% Conf. Interval]	
lrt	.5633697	.0124654	45.19	0.000	.5389381	.5878014
_cons	.0238706	.4002255	0.06	0.952	-.760557	.8082982

Random-effects Parameters	Estimate	Std. Err.	[95% Conf. Interval]	
school: Identity				
sd(_cons)	3.035269	.3052513	2.492261	3.696587
sd(Residual)	7.521481	.0841759	7.358295	7.688285

```
LR test vs. linear regression: chibar2(01) =   403.27 Prob >= chibar2 = 0.0000
```

To allow later comparison with random-coefficient models using likelihood-ratio tests, we store these estimates using

```
. estimates store ri
```

The random-intercept model assumes that the school-specific regression lines are parallel. The common coefficient or slope β_2 of lrt, shared by all schools, is estimated as 0.56 and the mean intercept as 0.02. Schools vary in their intercepts with an estimated standard deviation of 3.04. Within the schools, the estimated residual standard deviation around the school-specific regression lines is 7.52. The within-school correlation, after controlling for lrt, is therefore estimated as

$$\widehat{\rho} = \frac{\widehat{\psi}_{11}}{\widehat{\psi}_{11} + \widehat{\theta}} = \frac{3.035^2}{3.035^2 + 7.521^2} = 0.14$$

The ML estimates for the random-intercept model are also given under "Random intercept" in table 4.1.

Table 4.1: Maximum likelihood estimates for inner-London schools data

Parameter	Random intercept		Random coefficient		Rand. coefficient & level-2 covariates		
	Est	(SE)	Est	(SE)	Est	(SE)	γ_{xx}
Fixed part							
β_1 [_cons]	0.02	(0.40)	-0.12	(0.40)	-1.00	(0.51)	γ_{11}
β_2 [lrt]	0.56	(0.01)	0.56	(0.02)	0.57	(0.03)	γ_{21}
β_3 [boys]					0.85	(1.09)	γ_{12}
β_4 [girls]					2.43	(0.84)	γ_{13}
β_5 [boys_lrt]					-0.02	(0.06)	γ_{22}
β_6 [girls_lrt]					-0.03	(0.04)	γ_{23}
Random part							
$\sqrt{\psi_{11}}$	3.04		3.01		2.80		
$\sqrt{\psi_{22}}$			0.12		0.12		
ρ_{21}			0.50		0.60		
$\sqrt{\theta}$	7.52		7.44		7.44		
Log likelihood	$-14,024.80$		$-14,004.61$		$-13,998.83$		

4.5.2   Random-coefficient model

We now relax the assumption that the school-specific regression lines are parallel by introducing random school-specific slopes $\beta_2 + \zeta_{2j}$ of `lrt`:

$$y_{ij} = (\beta_1 + \zeta_{1j}) + (\beta_2 + \zeta_{2j})x_{ij} + \epsilon_{ij}$$

To introduce a random slope for `lrt` using `xtmixed`, we simply add that variable name in the specification of the random part, replacing `school:` with `school: lrt`. We must also specify the `covariance(unstructured)` option because `xtmixed` will otherwise set the covariance, ψ_{21} (and the corresponding correlation), to zero by default. ML estimates for the random-coefficient model are then obtained using

```
. xtmixed gcse lrt || school: lrt, covariance(unstructured) mle

Mixed-effects ML regression                     Number of obs      =       4059
Group variable: school                          Number of groups   =         65

                                                Obs per group: min =          2
                                                               avg =       62.4
                                                               max =        198

                                                Wald chi2(1)       =     779.79
Log likelihood = -14004.613                     Prob > chi2        =     0.0000
```

gcse	Coef.	Std. Err.	z	P>|z|	[95% Conf. Interval]	
lrt	.556729	.0199368	27.92	0.000	.5176535	.5958044
_cons	-.115085	.3978346	-0.29	0.772	-.8948264	.6646564

Random-effects Parameters	Estimate	Std. Err.	[95% Conf. Interval]	
school: Unstructured				
sd(lrt)	.1205646	.0189827	.0885522	.1641498
sd(_cons)	3.007444	.3044148	2.466258	3.667385
corr(lrt,_cons)	.4975415	.1487427	.1572768	.7322094
sd(Residual)	7.440787	.0839482	7.278058	7.607155

```
LR test vs. linear regression:       chi2(3) =    443.64   Prob > chi2 = 0.0000
Note: LR test is conservative and provided only for reference.
```

Because the `variance` option was not used, the output shows the standard deviations, `sd(lrt)`, of the slope and `sd(_cons)` of the intercept instead of variances. It also shows the correlation between intercepts and slopes, `corr(lrt,_cons)`, instead of the covariance. We can obtain the estimated covariance matrix either by replaying the estimation results with the `variance` option,

```
xtmixed, variance
```

or by using the postestimation command `estat recovariance`:

```
. estat recovariance
Random-effects covariance matrix for level school
                    │      lrt       _cons
          ──────────┼─────────────────────
               lrt  │  .0145358
             _cons  │  .1804042    9.04472
```

The ML estimates for the random-coefficient model were also given under "Random coefficient" in table 4.1 and will be interpreted in section 4.7. We store the estimates under the name `rc` for later use:

```
. estimates store rc
```

Restricted maximum likelihood (REML) estimation is obtained by specifying the `reml` option. Robust standard errors can be obtained using the `vce(robust)` option.

4.6   Testing the slope variance

Before interpreting the parameter estimates, we may want to test whether the random slope is needed in addition to the random intercept. Specifically, we test the null hypothesis

$$H_0\colon \psi_{22} = 0 \quad \text{against} \quad H_a\colon \psi_{22} > 0$$

Note that H_0 is equivalent to the hypothesis that the random slopes ζ_{2j} are all zero. The null hypothesis also implies that $\psi_{21} = 0$, because a variable that does not vary also does not covary with other variables. Setting $\psi_{22} = 0$ and $\psi_{21} = 0$ gives the random-intercept model.

A naïve likelihood-ratio test can be performed using the `lrtest` command in Stata:

```
. lrtest rc ri
Likelihood-ratio test                              LR chi2(2)  =      40.37
(Assumption: ri nested in rc)                      Prob > chi2 =     0.0000

Note: The reported degrees of freedom assumes the null hypothesis is not on the
      boundary of the parameter space.  If this is not true, then the reported
      test is conservative.
```

However, the null hypothesis lies on the boundary of the parameter space because the variance ψ_{22} must be nonnegative, and as discussed in section 2.6.2, the likelihood-ratio statistic L does not have a simple χ^2 distribution under the null hypothesis.

Since Stata 11, the default estimation metric (transformation used during estimation) for the covariance matrix of the random effects is the square root or Cholesky decomposition (which is requested by the `matsqrt` option). This parameterization forces the covariance matrix to be *positive semidefinite* (estimates on the boundary of parameter space, for example, zero variance or perfect correlations, are allowed). It can be shown that the asymptotic null distribution for testing the null hypothesis that the variance of the $q+1$th random effect is zero becomes $0.5\,\chi^2(q) + 0.5\,\chi^2(q+1)$. For our case of testing the random slope variance in a model with a random intercept and a random slope, $q = 1$; it follows that the asymptotic null distribution is $0.5 \, \chi^2(1) + 0.5 \, \chi^2(2)$. The correct p-value can be obtained as

```
. display 0.5*chi2tail(1,40.37) + 0.5*chi2tail(2,40.37)
9.616e-10
```

We see that the conclusion remains the same as for the naïve approach for this application.

If the `matlog` option is used (which was the default in Stata 10), the estimation metric for the covariance matrix of the random effects is matrix logarithms, which forces the covariance matrix to be *positive definite* (estimates on the boundary of the parameter space are not allowed). Consequently, convergence is not achieved if the ML estimates are on the boundary of the parameter space. If this leads to reverting to the model under the null hypothesis, giving a likelihood-ratio statistic equal to zero, then the asymptotic null distribution for testing the null hypothesis that the variance of the $q + 1$th random effect is zero becomes $0.5 \, \chi^2(0) + 0.5 \, \chi^2(q + 1)$, where $\chi^2(0)$ has a probability mass of 1 at 0. For testing the random slope variance in a model with a random intercept and a random slope, the asymptotic null distribution is $0.5 \, \chi^2(0) + 0.5 \, \chi^2(1)$; the correct p-value can simply be obtained by dividing the naïve p-value based on the $\chi^2(2)$ by 2.

Keep in mind that the naïve likelihood-ratio test for testing the slope variance is conservative, as is acknowledged at the bottom of the output from `lrtest`. Hence, if the null hypothesis of a zero slope variance is rejected by the naïve approach, it would also have been rejected if a correct approach had been used.

Unfortunately, there is no straightforward procedure available for testing several variances simultaneously, unless the random effects are independent (see section 8.8), and simulations must be used in this case to obtain the correct null distribution.

4.7 Interpretation of estimates

The population-mean intercept and slope are estimated as -0.12 and 0.56, respectively. These estimates are similar to those for the random-intercept model (see table 4.1) and also similar to the means of the school-specific least-squares regression lines given on page 186.

The estimated random-intercept standard deviation and level-1 residual standard deviation are somewhat lower than for the random-intercept model. The latter is because of a better fit of the school-specific regression lines for the random-coefficient model, which relaxes the restriction of parallel regression lines. The estimated covariance matrix of the intercepts and slopes is similar to the sample covariance matrix of the ordinary least-squares estimates reported on page 186.

As discussed in section 4.4.2, the easiest way to interpret the estimated standard deviations of the random intercept and random slope (conditional on the covariates) is to form intervals within which 95% of the schools' random intercepts and slopes are expected to lie assuming normality. Remember that these intervals represent ranges within which 95% of the realizations of a *random variable* are expected to lie, a concept different from confidence intervals, which are ranges within which an *unknown parameter* is believed to lie.

For the intercepts, we obtain $-0.115 \pm 1.96 \times 3.007$, so 95% of schools have their intercept in the range -6.0 to 5.8. In other words, the school mean GCSE scores for children with average (`lrt=0`) LRT scores vary between -6.0 and 5.8. For the slopes, we obtain $0.557 \pm 1.96 \times 0.121$, giving an interval from 0.32 to 0.80. Thus 95% of schools have slopes between 0.32 and 0.80.

This exercise of forming intervals is particularly important for slopes because it is useful to know whether the slopes have different signs for different schools (which would be odd in the current example). The range from 0.32 to 0.80 is fairly wide and the regression lines for schools may cross: one school could add more value (produce higher mean GCSE scores for given LRT scores) than another school for students with low LRT scores and add less value than the other school for students with high LRT scores.

The estimated correlation $\widehat{\rho}_{21} = 0.50$ between random intercepts and slopes (given the covariates) means that schools with larger mean GCSE scores for students with average LRT scores than other schools also tend to have larger slopes than those other schools. This information, combined with the random-intercept and slope variances and the range of LRT scores, determines how much the lines cross, something that is best explored by plotting the predicted regression lines for the schools, as demonstrated in section 4.8.3.

The variance of the total residual ξ_{ij} (equal to the conditional variance of the responses y_{ij} given the covariates $\mathbf{X}_j$) was given in (4.2). We can estimate the corresponding standard deviation by plugging in the ML estimates:

$$
\begin{aligned}
\sqrt{\widehat{\mathrm{Var}}(\xi_{ij}|\mathbf{X}_j)} &= \sqrt{\widehat{\psi}_{11} + 2\widehat{\psi}_{21}x_{ij} + \widehat{\psi}_{22}x_{ij}^2 + \widehat{\theta}} \\
&= \sqrt{9.0447 + 2 \times 0.1804 \times x_{ij} + 0.0145 \times x_{ij}^2 + 55.3653}
\end{aligned}
$$

A graph of the estimated standard deviation of the total residual against the covariate `lrt` (x_{ij}) can be obtained using the following **twoway function** command, which is graphed in figure 4.8:

```
. twoway function sqrt(9.0447+2*0.1804*x+0.0145*x^2+55.3653), range(-30 30)
> xtitle(LRT) ytitle(Estimated standard deviation of total residual)
```

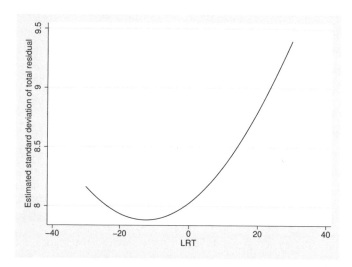

Figure 4.8: Heteroskedasticity of total residual ξ_{ij} as function of `lrt`

The estimated standard deviation of the total residual varies between just under 8 and just under 9.5.

4.8 Assigning values to the random intercepts and slopes

Having obtained estimated model parameters $\widehat{\beta}_1$, $\widehat{\beta}_2$, $\widehat{\psi}_{11}$, $\widehat{\psi}_{22}$, $\widehat{\psi}_{21}$, and $\widehat{\theta}$, we now assign values to the random intercepts and slopes, treating the estimated parameters as known (see also section 2.11). This is useful for model visualization, residual diagnostics, and inference for individual clusters, as will be demonstrated.

4.8.1 Maximum "likelihood" estimation

Maximum likelihood estimates of the random intercepts and slopes can be obtained by first predicting the total residuals $\widehat{\xi}_{ij} = y_{ij} - (\widehat{\beta}_1 + \widehat{\beta}_2 x_{ij})$ and then fitting individual regressions of $\widehat{\xi}_{ij}$ on x_{ij} for each school by OLS. As explained in section 2.11.1, we put "likelihood" in quotes in the section heading because it differs from the marginal likelihood that is used to estimate the model parameters.

We can fit the individual regression models using the `statsby` prefix command. We first retrieve the `xtmixed` estimates stored under `rc`,

```
. estimates restore rc
(results rc are active now)
```

and obtain the predicted total residuals,

```
. predict fixed, xb
. generate totres = gcse - fixed
```

We can then use `statsby` to produce the variables `mli` and `mls`, which contain the ML estimates $\widehat{\zeta}_{1j}$ and $\widehat{\zeta}_{2j}$ of the random intercepts and slopes, respectively:

```
. statsby mli=_b[_cons] mls=_b[lrt], by(school) saving(ols, replace):
> regress totres lrt
(running regress on estimation sample)

      command:  regress totres lrt
          mli:  _b[_cons]
          mls:  _b[lrt]
           by:  school

Statsby groups
──┼── 1 ──┼── 2 ──┼── 3 ──┼── 4 ──┼── 5
..................................................     50
..............

. sort school

. merge m:1 school using ols

    Result                           # of obs.
    ─────────────────────────────────────────
    not matched                              0
    matched                              4,059   (_merge==3)
    ─────────────────────────────────────────

. drop _merge
```

Maximum likelihood estimates will not be available for schools with only one observation or for schools within which x_{ij} does not vary. There are no such schools in the dataset, but school 48 has only two observations, and the ML estimates of the intercept and slope look odd:

```
. list lrt gcse mli mls if school==48, clean noobs
      lrt      gcse      mli        mls
   -4.5541   -1.2908   -32.607   -7.458484
   -3.7276   -6.9951   -32.607   -7.458484
```

Because there are only two students, the fitted line connects the points perfectly. Its intercept and slope are determined by ϵ_{1j} and ϵ_{2j} roughly as much as they are by the true intercept and slope. The intercept and slope are therefore extreme, the so-called "bouncing beta" phenomenon often encountered when using ML estimation of random effects for clusters that provide little information. In general, we therefore do not recommend using this method and suggest using empirical Bayes prediction instead.

4.8.2 Empirical Bayes prediction

As discussed for random-intercept models in section 2.11.2, empirical Bayes (EB) predictions have a smaller prediction error variance than ML estimates because of shrinkage toward the mean (for given model parameters). Furthermore, EB predictions are available for schools with only one observation or only one unique value of x_{ij}.

Empirical Bayes predictions $\widetilde{\zeta}_{1j}$ and $\widetilde{\zeta}_{2j}$ of the random intercepts ζ_{1j} and slopes ζ_{2j}, respectively, can be obtained using the predict command with the reffects option after estimation with xtmixed:

```
. estimates restore rc
. predict ebs ebi, reffects
```

Here we specified the variable names ebs and ebi for the EB predictions $\widetilde{\zeta}_{2j}$ and $\widetilde{\zeta}_{1j}$ of the random slopes and intercepts. The intercept variable comes last because xtmixed treats the intercept as the last random effect, as reflected by the output. This order is consistent with Stata's convention of treating the fixed intercept as the last regression parameter in estimation commands.

To compare the EB predictions with the ML estimates, we list one observation per school for schools 1–9 and school 48:

```
. list school mli ebi mls ebs if pickone==1 & (school<10 | school==48), noobs
```

school	mli	ebi	mls	ebs
1	3.948387	3.749336	.1526116	.1249761
2	4.937838	4.702127	.2045585	.1647271
3	5.69259	4.797687	.0222565	.0808662
4	.1526221	.3502472	.2047174	.1271837
5	2.719525	2.462807	.1232876	.0720581
6	6.147151	5.183819	-.0213858	.0586235
7	4.100312	3.640948	-.314454	-.1488728
8	-.136885	-.1218853	.0106781	.0068856
9	-2.258599	-1.767985	-.1555332	-.0886202
48	-32.607	-.4098203	-7.458484	-.0064854

Most of the time, the EB predictions are closer to zero than the ML estimates because of shrinkage, as discussed for random-intercept models in section 2.11.2. However, for models with several random effects, the relationship between EB predictions and ML estimates is somewhat more complex than for random-intercept models. The benefit of shrinkage is apparent for school 48, where the EB predictions appear more reasonable than the ML estimates.

We can see shrinkage more clearly by plotting the EB predictions against the ML estimates and superimposing a $y = x$ line. For the random intercept, the command is

```
. twoway (scatter ebi mli if pickone==1 & school!=48, mlabel(school))
> (function y=x, range(-10 10)), xtitle(ML estimate)
> ytitle(EB prediction) legend(off) xline(0)
```

and for the random slope, it is

```
. twoway (scatter ebs mls if pickone==1 & school!=48, mlabel(school))
> (function y=x, range(-0.6 0.6)), xtitle(ML estimate)
> ytitle(EB prediction) legend(off) xline(0)
```

These commands produce the graphs in figure 4.9 (we excluded school 48 from the graphs because the ML estimates are so extreme).

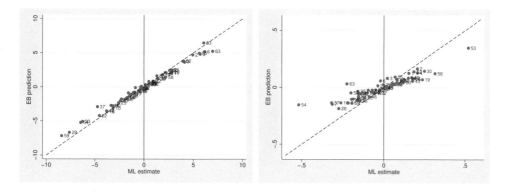

Figure 4.9: Scatterplots of empirical Bayes (EB) predictions versus maximum likelihood (ML) estimates of school-specific intercepts (left) and slopes (right); equality of EB and ML shown as dashed reference lines and ML estimates of 0 shown as solid reference lines

For ML estimates above zero, the EB prediction tends to be smaller than the ML estimate; the reverse is true for ML estimates below zero.

4.8.3   Model visualization

To better understand the random-intercept and random-coefficient models—and in particular, the variability implied by the random part—it is useful to produce graphs of predicted model-implied regression lines for the individual schools.

This can be achieved using the **predict** command with the **fitted** option to obtain school-specific fitted regression lines, with ML estimates substituted for the regression parameters (β_1 and β_2) and EB predictions substituted for the random effects (ζ_{1j} for the random-intercept model, and ζ_{1j} and ζ_{2j} for the random-coefficient model). For instance, for the random-coefficient model, the predicted regression line for school j is

$$\widehat{y}_{ij} = \widehat{\beta}_1 + \widehat{\beta}_2 x_{ij} + \widetilde{\zeta}_{1j} + \widetilde{\zeta}_{2j} x_{ij}$$

These predictions are obtained by typing

```
. predict murc, fitted
```

and a spaghetti plot is produced as follows:

```
. sort school lrt
. twoway (line murc lrt, connect(ascending)), xtitle(LRT)
> ytitle(Empirical Bayes regression lines for model 2)
```

To obtain predictions for the random-intercept model, we must first restore the estimates stored under the name `ri`:

```
. estimates restore ri
(results ri are active now)
. predict muri, fitted
. sort school lrt
. twoway (line muri lrt, connect(ascending)), xtitle(LRT)
> ytitle(Empirical Bayes regression lines for model 1)
```

The resulting spaghetti plots of the school-specific regression lines for both the random-intercept model and the random-coefficient model are given in figure 4.10.

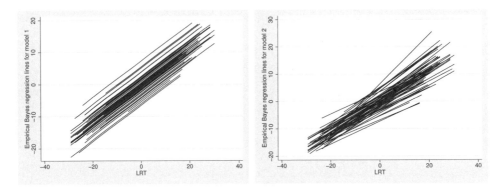

Figure 4.10: Spaghetti plots of empirical Bayes (EB) predictions of school-specific regression lines for the random-intercept model (left) and the random-intercept and random-slope model (right)

The predicted school-specific regression lines are parallel for the random-intercept model (with vertical shifts given by the $\widetilde{\zeta}_{1j}$) but are not parallel for the random-coefficient model, where the slopes $\beta_2 + \widetilde{\zeta}_{2j}$ also vary across schools. Because of shrinkage, the predicted lines vary somewhat less than implied by the estimated variances and covariance.

4.8.4   Residual diagnostics

If normality is assumed for the random intercepts ζ_{1j}, random slopes ζ_{2j}, and level-1 residuals ϵ_{ij}, the corresponding EB predictions should also have normal distributions.

To plot the distributions of the predicted random effects, we must pick one prediction per school, and we can accomplish this using the `pickone` variable created earlier. We can now plot the distributions using

```
. histogram ebi if pickone==1, normal xtitle(Predicted random intercepts)
. histogram ebs if pickone==1, normal xtitle(Predicted random slopes)
```

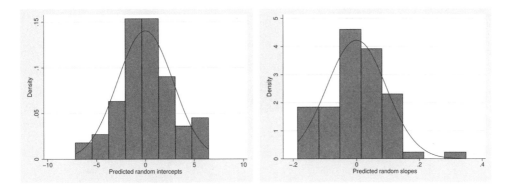

Figure 4.11: Histograms of predicted random intercepts and slopes

The histograms in figure 4.11 look approximately normal although the one for the slopes is perhaps a little positively skewed.

It is also useful to look at the bivariate distribution of the predicted random intercepts and slopes using a scatterplot, or to display such a scatterplot together with the two histograms:

```
. scatter ebs ebi if pickone==1, saving(yx, replace)
> xtitle("Random intercept") ytitle("Random slope") ylabel(, nogrid)
. histogram ebs if pickone==1, freq horizontal saving(hy, replace) normal
> yscale(alt) ytitle(" ") fxsize(35) ylabel(, nogrid)
. histogram ebi if pickone==1, freq saving(hx, replace) normal
> xscale(alt) xtitle(" ") fysize(35) ylabel(, nogrid)
. graph combine hx.gph yx.gph hy.gph, hole(2) imargin(0 0 0 0)
```

Here the scatterplot and histograms are first plotted separately and then combined using the `graph combine` command. In the first `histogram` command, the `horizontal` option is used to produce a rotated histogram of the random slopes. In both histogram commands, the `yscale(alt)` and `xscale(alt)` options are used to put the corresponding axes on the other side, and the `normal` option is used to overlay normal density curves. The `fysize(35)` and `fxsize(35)` options change the aspect ratios of the histograms, making them more flat so that they use up a smaller portion of the combined graph. Finally, in the `graph combine` command, the graphs are listed in lexicographic order, the `hole(2)` option denotes that there should be a hole in the second position—that is, the top-right corner—and the `imargin(0 0 0 0)` option reduces the space between the graphs. The resulting graph is shown in figure 4.12.

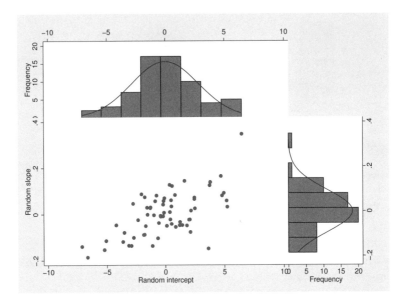

Figure 4.12: Scatterplot and histograms of predicted random intercepts and slopes

After estimation with `xtmixed`, we obtain the predicted level-1 residuals,

$$\widetilde{\epsilon}_{ij} \;=\; y_{ij} - (\widehat{\beta}_1 + \widehat{\beta}_2 x_{ij} + \widetilde{\zeta}_{1j} + \widetilde{\zeta}_{2j} x_{ij})$$

using

```
. predict res1, residuals
```

We plot the residuals using the following command, which produces the graph in figure 4.13:

```
. histogram res1, normal xtitle(Predicted level-1 residuals)
```

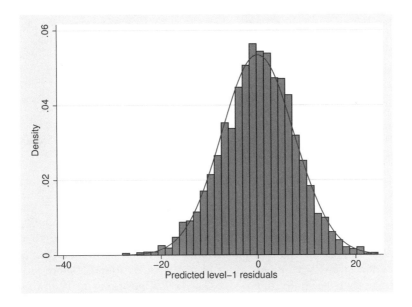

Figure 4.13: Histogram of predicted level-1 residuals

To obtain standardized level-1 residuals, use the `rstandard` option in the `predict` command after estimation using `xtmixed`.

4.8.5   Inferences for individual schools

Random-intercept predictions $\widetilde{\zeta}_{1j}$ are sometimes viewed as measures of institutional performance—in the present context, how much value the schools add for children with LRT scores equal to zero (the mean). However, we may not have adequately controlled for covariates correlated with achievement that are outside the control of the school, such as student SES. Furthermore, the model assumes that the random intercepts are uncorrelated with the LRT scores, so if schools with higher mean LRT scores add more value, their value added would be underestimated. Nevertheless, predicted random intercepts shed some light on the research question: Which schools are most effective for children with LRT = 0?

It does not matter whether we add the predicted fixed part of the model because the ranking of schools is not affected by this.

Returning to the question of comparing the schools' effectiveness for children with LRT scores equal to 0, we can plot the predicted random intercepts with approximate 95% confidence intervals based on the comparative standard errors (see section 2.11.3). These standard errors can be obtained using

```
. estimates restore rc
. predict slope_se inter_se, reses
```

We only need `inter_se`. We first produce ranks for the schools in ascending order of the random intercept predictions `ebi`:

```
. gsort + ebi - pickone
. generate rank = sum(pickone)
```

Here the `gsort` command is used to sort in ascending order of `ebi` (indicated by "+ ebi") and, within `ebi`, in descending order of `pickone` (indicated by "- pickone"). The `sum()` function forms the cumulative sum, so the variable `rank` increases by one every time a new school with higher value of `ebi` is encountered. Before producing the graph, we generate a variable, `labpos`, for the vertical positions in the graph where the school identifiers should go:

```
. generate labpos = ebi + 1.96*inter_se + .5
```

We are now ready to produce a so-called *caterpillar plot*:

```
. serrbar ebi inter_se rank if pickone==1, addplot(scatter labpos rank,
> mlabel(school) msymbol(none) mlabpos(0)) scale(1.96) xtitle(Rank)
> ytitle(Prediction) legend(off)
```

The school labels were added to the graph by superimposing a scatterplot onto the error bar plot with the `addplot()` option, where the vertical positions of the labels are given by the variable `labpos`. The resulting caterpillar plot is shown in figure 4.14.

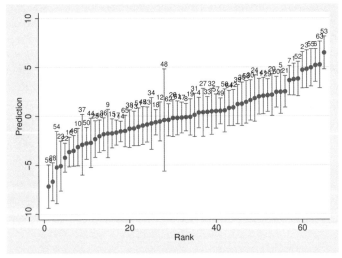

Figure 4.14: Caterpillar plot of random-intercept predictions and approximate 95% confidence intervals versus ranking (school identifiers shown on top of confidence intervals)

The interval for school 48 is particularly wide because there are only two students from this school in the dataset. It is clear from the large confidence intervals that the rankings are not precise and that perhaps only a coarse classification into poor, medium, and good schools can be justified.

An alternative method for producing a caterpillar plot is to first generate the confidence limits `lower` and `upper`,

```
. generate lower = ebi - 1.96*inter_se
. generate upper = ebi + 1.96*inter_se
```

and then use the `rcap` plot type to produce the intervals:

```
. twoway (rcap lower upper rank, blpatt(solid) lcol(black))
> (scatter ebi rank)
> (scatter labpos rank, mlabel(school) msymbol(none) mlabpos(0)
> mlabcol(black) mlabsiz(medium)),
> xtitle(Rank) ytitle(Prediction) legend(off)
> xscale(range(1 65)) xlabel(1/65) ysize(1)
```

Here `scatter` is first used to overlay the point estimates and then the labels. The `ysize()` option is used to change the aspect ratio and obtain the horizontally stretched graph shown in figure 4.15.

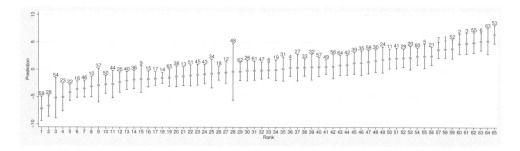

Figure 4.15: Stretched caterpillar plot of random-intercept predictions and approximate 95% confidence intervals versus ranking (school identifiers shown on top of confidence intervals)

We could also produce similar plots for children with different values x^0 of the LRT scores:

$$\widehat{\beta}_1 + \widehat{\beta}_2 x^0 + \widetilde{\zeta}_{1j} + \widetilde{\zeta}_{2j} x^0$$

For instance, in a similar application, Goldstein et al. (2000) substitute the 10th percentile of the intake measure to compare school effectiveness for poorly performing children. (To obtain confidence intervals for different values of x^0 requires posterior correlations that are produced by `gllapred`, the prediction command for `gllamm`.)

4.9   Two-stage model formulation

In this section, we describe an alternative way of specifying random-coefficient models that is popular in some areas such as education (for example, Raudenbush and Bryk 2002). As shown below, models are specified in two stages (for levels 1 and 2), necessitating a distinction between level-1 and level-2 covariates. Many people find this formulation helpful for interpreting and specifying models. Identical models can be formulated using either the approach discussed up to this point or the two-stage formulation.

To express the random-coefficient model using a two-stage formulation, Raudenbush and Bryk (2002) specify a level-1 model:

$$y_{ij} \;=\; \beta_{0j} + \beta_{1j} x_{ij} + r_{ij}$$

where the intercept β_{0j} and slope β_{1j} are school-specific coefficients. Their level-2 models have these coefficients as responses:

$$\begin{aligned}
\beta_{0j} &= \gamma_{00} + u_{0j} \\
\beta_{1j} &= \gamma_{10} + u_{1j}
\end{aligned} \tag{4.4}$$

Sometimes the first of these level-2 models is referred to as a "means as outcomes" or "intercepts as outcomes" model, and the second as a "slopes as outcomes" model. It is typically assumed that given the covariate(s), the residuals or disturbances u_{0j} and u_{1j} in the level-2 model have a bivariate normal distribution with zero mean and covariance matrix

$$\mathbf{T} \;=\; \begin{bmatrix} \tau_{00} & \tau_{01} \\ \tau_{10} & \tau_{11} \end{bmatrix}, \qquad \tau_{10} = \tau_{01}$$

The level-2 models cannot be fit on their own because the random effects β_{0j} and β_{1j} are not observed. Instead, we must substitute the level-2 models into the level-1 model to obtain the *reduced-form* model for the observed responses, y_{ij}:

$$\begin{aligned}
y_{ij} &= \underbrace{\gamma_{00} + u_{0j}}_{\beta_{0j}} + \underbrace{(\gamma_{10} + u_{1j})}_{\beta_{1j}} x_{ij} + r_{ij} \\
&= \underbrace{\gamma_{00} + \gamma_{10} x_{ij}}_{\text{fixed}} + \underbrace{u_{0j} + u_{1j} x_{ij} + r_{ij}}_{\text{random}} \\
&\equiv \beta_1 + \beta_2 x_{ij} \;+\; \zeta_{1j} + \zeta_{2j} x_{ij} + \epsilon_{ij}
\end{aligned}$$

In the reduced form, the fixed part is usually written first, followed by the random part. We can return to our previous notation by defining $\beta_1 \equiv \gamma_{00}$, $\beta_2 \equiv \gamma_{10}$, $\zeta_{1j} \equiv u_{0j}$, $\zeta_{2j} \equiv u_{1j}$, and $\epsilon_{ij} \equiv r_{ij}$. The above model is thus equivalent to the model in (4.1).

Any level-2 covariates (covariates that do not vary at level 1) are included in the level-2 models. For instance, we could include dummy variables for type of school: w_{1j} for boys schools and w_{2j} for girls schools, with mixed schools as the reference category. If we include these dummy variables in the model for the random intercept,

$$\beta_{0j} \;=\; \gamma_{00} + \gamma_{01} w_{1j} + \gamma_{02} w_{2j} + u_{0j}$$

the reduced form becomes

$$
\begin{aligned}
y_{ij} &= \underbrace{\gamma_{00} + \gamma_{01}w_{1j} + \gamma_{02}w_{2j} + u_{0j}}_{\beta_{0j}} + \underbrace{\left(\gamma_{10} + u_{1j}\right)}_{\beta_{1j}} x_{ij} + r_{ij} \\
&= \underbrace{\gamma_{00} + \gamma_{01}w_{1j} + \gamma_{02}w_{2j} + \gamma_{10}x_{ij}}_{\text{fixed}} + \underbrace{u_{0j} + u_{1j}x_{ij} + r_{ij}}_{\text{random}}
\end{aligned}
$$

If we also include the dummy variables for type of school in the model for the random slope,

$$
\beta_{1j} = \gamma_{10} + \gamma_{11}w_{1j} + \gamma_{12}w_{2j} + u_{1j}
$$

we obtain so-called *cross-level interactions* between covariates varying at different levels— w_{1j} and x_{ij} as well as w_{2j} and x_{ij}—in the reduced form

$$
\begin{aligned}
y_{ij} &= \underbrace{\gamma_{00} + \gamma_{01}w_{1j} + \gamma_{02}w_{2j} + u_{0j}}_{\beta_{0j}} + \underbrace{\left(\gamma_{21} + \gamma_{22}w_{2j} + \gamma_{23}w_{3j} + u_{1j}\right)}_{\beta_{1j}} x_{ij} + r_{ij} \\
&= \underbrace{\gamma_{00} + \gamma_{01}w_{1j} + \gamma_{02}w_{2j} + \gamma_{10}x_{ij} + \gamma_{11}w_{1j}x_{ij} + \gamma_{12}w_{2j}x_{ij}}_{\text{fixed}} + \underbrace{u_{0j} + u_{1j}x_{ij} + r_{ij}}_{\text{random}}
\end{aligned}
$$

The effect of `lrt` now depends on the type of school, with γ_{11} representing the additional effect of `lrt` on `gcse` for boys schools compared with mixed schools and γ_{12} representing the additional effect for girls schools compared with mixed schools.

For estimation in `xtmixed`, it is necessary to convert the two-stage formulation to the reduced form because the fixed part of the model is specified first, followed by the random part of the model. Using factor variables in `xtmixed`, the command is

```
. xtmixed gcse i.schgend##c.lrt || school: lrt, covariance(unstructured) mle
Mixed-effects ML regression                    Number of obs      =      4059
Group variable: school                         Number of groups   =        65

                                               Obs per group: min =         2
                                                              avg =      62.4
                                                              max =       198

                                               Wald chi2(5)       =    803.89
Log likelihood = -13998.825                    Prob > chi2        =    0.0000
```

gcse	Coef.	Std. Err.	z	P>|z|	[95% Conf. Interval]	
schgend						
2	.8546715	1.085021	0.79	0.431	-1.271931	2.981274
3	2.43341	.8433413	2.89	0.004	.7804919	4.086329
lrt	.5712361	.0271256	21.06	0.000	.5180708	.6244014
schgend#						
c.lrt						
2	-.0230098	.0573895	-0.40	0.688	-.1354911	.0894716
3	-.029544	.0447032	-0.66	0.509	-.1171606	.0580726
_cons	-.9976073	.506809	-1.97	0.049	-1.990935	-.0042799

Random-effects Parameters	Estimate	Std. Err.	[95% Conf. Interval]	
school: Unstructured				
sd(lrt)	.1199154	.0189129	.0880287	.1633525
sd(_cons)	2.797934	.28868	2.285672	3.425005
corr(lrt,_cons)	.5967727	.1381314	.2614253	.8035676
sd(Residual)	7.441831	.0839662	7.279067	7.608235

```
LR test vs. linear regression:        chi2(3) =    381.45   Prob > chi2 = 0.0000
Note: LR test is conservative and provided only for reference.
```

Here schgend is 1 for mixed schools, 2 for boys schools and 3 for girls schools. We see that students from girls schools perform significantly better at the 5% level than students from mixed schools, whereas students from boys schools do not perform significantly better than students from mixed schools. The effect of lrt does not differ significantly between boys schools and mixed schools or between girls schools and mixed schools. The estimates and the corresponding parameters in the two-stage formulation are given under "Rand. coefficient & level-2 covariates" in the last three columns of table 4.1 on page 195.

Although equivalent models can be specified using either the reduced-form (used by xtmixed) or the two-stage formulation (used in the HLM software of Raudenbush et al. [2004]), in practice, model specification to some extent depends on the approach adopted. For instance, cross-level interactions are easily included using the two-stage specification in the HLM software, whereas same-level interactions must be created outside the program. Papers using HLM tend to include more cross-level interactions and more random coefficients in the models (because the level-2 models look odd without residuals) than papers using, for instance, Stata.

4.10   Some warnings about random-coefficient models

4.10.1   Meaningful specification

It rarely makes sense to include a random slope if there is no random intercept, just like interactions between two covariates usually do not make sense without including the covariates themselves in standard regression models. Similarly, it is seldom sensible to include a random slope without including the corresponding fixed slope because it is usually strange to allow the slope to vary randomly but constrain its population mean to zero.

It is generally not a good idea to include a random coefficient for a covariate that does not vary at a lower level than the random coefficient itself. For example, in the inner-London schools data, it does not make sense to include a random slope for type of school because type of school does not vary within schools. Because we cannot estimate the effect of type of school for individual schools, it also appears impossible to estimate the variability of the effect of type of school between schools. However, level-2 random coefficients of level-2 covariates can be used to construct heteroskedastic random intercepts (see section 7.5.2).

4.10.2   Many random coefficients

It may be tempting to allow many different covariates to have random slopes. However, the number of parameters for the random part of the model increases rapidly with the number of random slopes because there is a variance parameter for each random effect (intercept or slope) and a covariance parameter for each *pair* of random effects. If there are k random slopes (plus one random intercept), then there are $(k + 2)(k + 1)/2 + 1$ parameters in the random part (for example, $k = 3$ gives 11 parameters).

Another problem is that clusters may not provide much information on cluster-specific slopes and hence on the slope variance either if the clusters are small, or if x_{ij} does not vary much within clusters or varies only in a small number of clusters. Perhaps a useful rule is to consider the random part of the model (ignoring the fixed part) and replace the random effects with fixed regression coefficients. It should be possible (even if not very sensible) to fit the resulting model to a good number of clusters (say, 20 or more). Note, however, that it does not matter if some of the clusters have insufficient data as long as there are an adequate number of clusters that do have sufficient data. It is never a good idea to discard clusters merely because they provide little information on some of the parameters of the model.

In general, it makes sense to allow for more flexibility in the fixed part of the model than in the random part. For instance, the fixed part of the model may include a dummy variable for each occasion in longitudinal data, but in the random part of the model it may be sufficient to allow for a random intercept and a random slope of time, keeping in mind that in this case it is only assumed that the *deviation* from the population-average curve is linear in time, not that the relationship itself is linear.

The overall message is that random slopes should be included only if strongly suggested by the subject-matter theory related to the application *and* if the data provide sufficient information.

4.10.3   Convergence problems

Convergence problems can manifest themselves in different ways. Either estimates are never produced, or standard errors are missing, or `xtmixed` produces messages such as "nonconcave", or "backed-up", or "standard error calculation has failed". Sometimes none of these things happen, but the confidence intervals for some of the correlations cover the full permissible range from -1 to 1 (see sections 7.3 and 8.13.2 for examples).

Convergence problems can occur because the estimated covariance matrix "tries" to become negative definite, meaning, for instance, that variances try to become negative or correlations try to be greater than 1 or less than -1. All the commands in Stata force the covariance matrix to be positive (semi)definite, and when parameters approach nonpermissible values, convergence can be slow or even fail. It may help to translate and rescale x_{ij} because variances and covariances are not invariant to these transformations. Often a better remedy is to simplify the model by removing some random slopes. Convergence problems can also occur because of lack of identification, and again, a remedy is to simplify the model.

However, before giving up on a model, it is worth attempting to achieve convergence by trying both the `mle` and the `reml` options, specifying the `difficult` option, trying the `matlog` option (which parameterizes the random part differently during maximization), or increasing the number of EM iterations using either the `emiterate()` option or even the `emonly` option. It can also be helpful to monitor the iterations more closely by using `trace`, which displays the parameter estimates at the end of each iteration (unfortunately, not for the EM iterations). Lack of identification of a parameter might be recognized by that parameter changing wildly between iterations without much of a change in the log likelihood. Problems with variances approaching zero can be detected by noticing that the log-standard deviation takes on very large negative values.

4.10.4   Lack of identification

Sometimes random-coefficient models are simply not identified (or in other words, underidentified). As an important example, consider balanced data with clusters of size $n_j = 2$ and with a covariate x_{ij} taking the same two values $t_1 = 0$ and $t_2 = 1$ for each cluster (an example would be the peak-expiratory-flow data from chapter 2). A model including a random intercept, a random slope of x_{ij}, and a level-1 residual, all of which are normally distributed, is not identified in this case. This can be seen by considering the two distinct variances (for $i = 1$ and $i = 2$) and one covariance of the total residuals when $t_1 = 0$ and $t_2 = 1$:

$$
\begin{aligned}
\mathrm{Var}(\xi_{1j}) &= \psi_{11} + \theta \\
\mathrm{Var}(\xi_{2j}) &= \psi_{11} + 2\psi_{21} + \psi_{22} + \theta \\
\mathrm{Cov}(\xi_{1j}, \xi_{2j}) &= \psi_{11} + \psi_{21}
\end{aligned}
$$

The marginal distribution of y_{ij} given the covariates is normal and therefore completely characterized by the fixed part of the model and these three model-implied moments (two variances and a covariance). However, the three moments are determined by four parameters of the random part (ψ_{11}, ψ_{22}, ψ_{21}, and θ), so fitting the model-implied moments to the data would effectively involve solving three equations for four unknowns. The model is therefore not identified. We could identify the model by setting $\theta = 0$, which does not impose any restrictions on the covariance matrix (however, such a constraint is not allowed in `xtmixed`). The original model becomes identified if the covariate x_{ij}, which has a random slope, varies also between clusters because the model-implied covariance matrix of the total residuals then differs between clusters, yielding more equations to solve for the four parameters.

Still assuming that the random effects and level-1 residual are normally distributed, consider now the case of balanced data with clusters of size $n_j = 3$ and with a covariate x_{ij} taking the same three values t_1, t_2, and t_3 for each cluster. An example would be longitudinal data with three occasions at times t_1, t_2, and t_3. Instead of including a random intercept and a random slope of time, it may be tempting to specify a random-coefficient model with a random-intercept and two random coefficients for the dummy variables for occasions two and three. In total, such a model would contain seven (co)variance parameters: six for the three random effects and one for the level-1 residual variance. Because the covariance matrix of the responses for the three occasions given the covariates only has six elements, it is impossible to solve for all unknowns. The same problem could occur when attempting to fit this kind of model for more than three occasions.

4.11 Summary and further reading

In this chapter, we have introduced the notion of slopes or regression coefficients varying randomly between clusters in linear models. Linear random-coefficient models are parsimonious representations of situations where each cluster has a separate regression model with its own intercept and slope. The linear random-coefficient model was applied to a cross-sectional study of school effectiveness. Here students were nested in schools, and we considered school-specific regressions.

An important consideration when using random-coefficient models is that the interpretation of the covariance matrix of the random effects depends on the scale and location of the covariates having random slopes. One should thus be careful when interpreting the variance and covariance estimates. We briefly demonstrated a two-stage formulation of random-coefficient models that is popular in some fields. This formulation can be used to specify models that are equivalent to models specified using the reduced-form formulation used in this book.

The utility of empirical Bayes prediction was demonstrated for visualizing the model, making inferences for individual clusters, and for diagnostics. See Skrondal and Rabe-Hesketh (2009) for a detailed discussion of prediction of random effects.

Introductory books discussing random-coefficient models include Snijders and Bosker (2012, chap. 5), Kreft and de Leeuw (1998, chap. 3), and Raudenbush and Bryk (2002, chap. 2, 4). Papers and chapters with good overviews of much of the material we covered in chapters 2–4 include Snijders (2004), Duncan, Jones, and Moon (1998), and Steenbergen and Jones (2002); a useful list of multilevel terminology is provided by Diez Roux (2002). These papers and chapters are among those collected in Skrondal and Rabe-Hesketh (2010).

The first six exercises are on standard random-coefficient models applied to data from different disciplines, whereas exercises 4.7 and 4.8 use random-coefficient models to fit biometrical genetic models to nuclear family data. Random-coefficient models for longitudinal data, often called growth-curve models, are considered in chapter 7. Exercises 7.2, 7.5, 7.6, and 7.7, and parts of the other exercises in that chapter can be viewed as supplementary exercises for the current chapter. Parts of exercises 6.1 and 6.2 are also relevant.

4.12   Exercises

4.1 ❖ Inner-London schools data

1. Fit the random-coefficient model fit on page 212.

2. Write down a model with the same covariates as in step 1 that also allows the mean for mixed schools to differ between boys and girls. (The variable `girl` is a dummy variable for the student being a girl.) Write down null hypotheses in terms of linear combinations of regression coefficients for the following research questions:

   a. Do girls do better in girls schools than in mixed schools (after controlling for the other covariates)?

   b. Do boys do better in boys schools than in mixed schools (after controlling for the other covariates)?

3. Fit the model from step 2, and test the null hypotheses from step 2. Discuss whether there is evidence that children of a given gender do better in single-sex schools.

4.2 High-school-and-beyond data

Raudenbush and Bryk (2002) and Raudenbush et al. (2004) analyzed data from the High School and Beyond Survey.

The dataset `hsb.dta` has the following variables:

- Level 1 (student)
  - `mathach`: a measure of mathematics achievement
  - `minority`: dummy variable for student being nonwhite
  - `female`: dummy variable for student being female
  - `ses`: socioeconomic status (SES) based on parental education, occupation, and income
- Level 2 (school)
  - `schoolid`: school identifier
  - `sector`: dummy variable for school being Catholic
  - `pracad`: proportion of students in the academic track
  - `disclim`: scale measuring disciplinary climate
  - `himinty`: dummy variable for more than 40% minority enrollment

Raudenbush et al. (2004) specify a two-level model. We will use their model and notation here. At level 1, math achievement Y_{ij} is regressed on student's SES, centered around the school mean:

$$Y_{ij} = \beta_{0j} + \beta_{1j}(X_{1ij} - \overline{X}_{1.j}) + r_{ij}, \quad r_{ij} \sim N(0, \sigma^2)$$

where X_{1ij} is the student's SES, $\overline{X}_{1.j}$ is the school mean SES, and r_{ij} is a level-1 residual. At level 2, the intercepts and slopes are regressed on the dummy variable W_{1j} for the school being a Catholic school (`sector`) and on the school mean SES

$$\beta_{pj} = \gamma_{p0} + \gamma_{p1}W_{1j} + \gamma_{p2}\overline{X}_{1.j} + u_{pj}, \quad p = 0, 1, \quad (u_{0j}, u_{1j})' \sim N(\mathbf{0}, \mathbf{T})$$

where u_{pj} is a random effect (a random intercept if $p = 0$ and a random slope if $p = 1$). The covariance matrix

$$\mathbf{T} = \begin{bmatrix} \tau_{00} & \tau_{01} \\ \tau_{10} & \tau_{11} \end{bmatrix}$$

has three unique elements with $\tau_{10} = \tau_{01}$.

1. Substitute the level-2 models into the level-1 model, and write down the resulting reduced form using the notation of this book.
2. Construct the variables `meanses`, equal to the school-mean SES ($\overline{X}_{1.j}$), and `devses`, equal to the deviations of the student's SES from their school means ($X_{1ij} - \overline{X}_{1.j}$).
3. Fit the model considered by Raudenbush et al. (2004) using `xtmixed` and interpret the coefficients. In particular, interpret the estimate of γ_{12}.

4. Fit the model that also includes `disclim` in the level-2 models and `minority` in the level-1 model.

4.3 Homework data

Kreft and de Leeuw (1998) consider a subsample of students in eighth grade from the National Education Longitudinal Study of 1988 (NELS–88) collected by the National Center for Educational Statistics of the U.S. Department of Education. The students are viewed as nested in schools.

The data are given in `homework.dta`. In this exercise, we will use the following subset of the variables:

- `schid`: school identifier
- `math`: continuous measure of achievement in mathematics (standardized to have a mean of 50 and a standard deviation of 10)
- `homework`: number of hours of homework done per week
- `white`: student's race (1: white; 0: nonwhite)
- `ratio`: class size as measured by the student–teacher ratio
- `meanses`: school mean socioeconomic status (SES)

1. Write down and state the assumptions of a random-coefficient model with `math` as response variable and `homework`, `white`, and `ratio` as covariates. Let the intercept and the effect of `homework` vary between schools.
2. Fit the model by ML and interpret the estimated parameters.
3. Derive an expression for the estimated variance of math achievement conditional on the covariates.
4. How would you extend the model to investigate whether the effect of homework on math achievement depends on the mean SES of schools? Write down both the two-stage and the reduced-form formulation of your extended model.
5. Fit the model from step 4.

4.4 Wheat and moisture data

Littell et al. (2006) describe data on ten randomly chosen varieties of winter wheat. Each variety was planted on six randomly selected 1-acre plots of land in a 60-acre field. The amount of moisture in the top 36 inches of soil was determined for each plot before planting the wheat. The response variable is the yield in bushels per acre.

The data, `wheat.dta`, contain the following variables:

- `variety`: variety (or type) of wheat (j)
- `plot`: plot (1 acre) on which wheat was planted (i)
- `yield`: yield in bushels per acre (y_{ij})
- `moist`: amount of moisture in top 36 inches of soil prior to planting (x_{ij})

In this exercise, variety of wheat will be treated as the cluster.

1. Write down the model for yield with a fixed and random intercept for variety of wheat and a fixed and variety-specific random slope of moist. State all model assumptions.

2. Fit the random-coefficient model using ML estimation.

3. Use a likelihood-ratio test to test the null hypothesis that the random-coefficient variance is zero.

4. For the chosen model, obtain the predicted yields for each variety (with EB predictions substituted for the random effects).

5. Produce a trellis graph of predicted yield versus moisture, using the by() option to obtain a separate graph for each variety.

6. Produce the same graphs as above but with observed values of yield added as dots.

4.5 Well-being in the U.S. army data   Solutions

Bliese (2009) provides the data analyzed by Bliese and Halverson (1996). The data are on soldiers (with the lowest five enlisted ranks) from 99 U.S. army companies in noncombat environments stationed in the U.S. and Europe.

The variables in the dataset army.dta are the following:

- grp: army company identification number
- wbeing: well-being assessed using the General Well-Being Schedule (Dupuy 1978), an 18-item scale measuring depression, anxiety, somatic complaints, positive well-being, and emotional control
- hrs: answer to the question "How many hours do you usually work in a day?"
- cohes: score on horizontal cohesion scale consisting of eight items, including "My closest relationships are with people I work with"
- lead: score on an 11-item leadership consideration (vertical cohesion) scale with a typical item being "The noncommissioned officer in this company would lead well in combat"

1. Fit a random-intercept model for wbeing with fixed coefficients for hrs, cohes, and lead, and a random intercept for grp. Use ML estimation.

2. Form the cluster means of the three covariates from step 1, and add them as further covariates to the random-intercept model. Which of the cluster means have coefficients that are significant at the 5% level?

3. Refit the model from step 2 after removing the cluster means that are not significant at the 5% level. Interpret the remaining coefficients and obtain the estimated intraclass correlation.

4. We have included soldier-specific covariates x_{ij} in addition to the cluster means $\bar{x}_{\cdot j}$. The coefficient of the cluster means represents the contextual

effects (see section 3.7.5). Use `lincom` to estimate the corresponding between effects.

5. Add a random slope for `lead` to the model in step 3, and compare this model with the model from step 3 using a likelihood-ratio test.

6. Add a random slope for `cohes` to the model chosen in step 5, and compare this model with the model from step 3 using a likelihood-ratio test. Retain the preferred model.

7. Perform residual diagnostics for the level-1 errors, random intercept, and random slope(s). Do the model assumptions appear to be satisfied?

4.6 Dialyzer data

Vonesh and Chinchilli (1997) analyzed data on low-flux dialyzers used to treat patients with end-stage renal disease (kidney disease) to remove excess fluid and waste from their blood. In low-flux hemodialysis, the ultrafiltration rate at which fluid is removed (volume per time) is thought to follow a straight-line relationship with the transmembrane pressure applied across the dialyzer membrane. In a study to investigate this relationship, three centers measured the ultrafiltration rate at several transmembrane pressures for each of several dialyzers, or patients.

The variables in `dialyzer.dta` are as follows:

- `subject`: subject (or dialyzer) identifier
- `tmp`: transmembrane pressure (mmHg)
- `ufr`: ultrafiltration rate (ml/hr)
- `center`: center at which study was conducted

1. For each center, plot a graph of `ufr` versus `tmp` with separate lines for each subject. You may want to use the `by(center)` option.

2. Write down a model that assumes a linear relationship between `ufr` and `tmp` (denoted y_{ij} and x_{ij}, respectively), with mean intercepts and mean slopes differing between the three centers. In the random part of the model, include a random intercept and a random slope of x_{ij}.

3. Fit the model by ML estimation.

4. Test whether the mean slopes differ significantly at the 5% level for each pair of centers.

5. Plot the estimated mean line for each center on one graph, using `twoway function`.

6. For center 1, produce a trellis graph of the data and fitted subject-specific regression lines.

4.7 ❖ Family-birthweight data Solutions

Rabe-Hesketh, Skrondal, and Gjessing (2008) analyzed a random subset of the birthweight data from the Medical Birth Registry of Norway described in Magnus et al. (2001). There are 1,000 nuclear families each comprising mother, father, and one child (not necessarily the only child in the family).

The data are given in `family.dta`. In this exercise, we will use the following variables:

- `family`: family identifier (j)
- `member`: family member (i) (1: mother; 2: father; 3: child)
- `bwt`: birthweight in grams (y_{ij})
- `male`: dummy variable for being male (x_{1ij})
- `first`: dummy variable for being the first child (x_{2ij})
- `midage`: dummy variable for mother of family member being aged 20–35 at time of birth (x_{3ij})
- `highage`: dummy variable for mother of family member being older than 35 at time of birth (x_{4ij})
- `birthyr`: year of birth minus 1967 (1967 was the earliest birth year in the birth registry) (x_{5ij})

In this dataset, family members are nested within families. Because of additive genetic and environmental influences, there will be a particular covariance structure between the members of the same family. Rabe-Hesketh, Skrondal, and Gjessing (2008) show that the following random-coefficient model can be used to induce the required covariance structure (see also exercise 4.8):

$$y_{ij} = \beta_1 + \zeta_{1j}(M_i + K_i/2) + \zeta_{2j}(F_i + K_i/2) + \zeta_{3j}(K_i/\sqrt{2}) + \epsilon_{ij} \qquad (4.5)$$

where M_i is a dummy variable for mothers, F_i is a dummy variable for fathers, and K_i is a dummy variable for children. The random coefficients ζ_{1j}, ζ_{2j}, and ζ_{3j} are constrained to have the same variance ψ and to be uncorrelated with each other. As usual, we assume normality with $\zeta_{1j} \sim N(0, \psi)$, $\zeta_{2j} \sim N(0, \psi)$, $\zeta_{3j} \sim N(0, \psi)$, and $\epsilon_{ij} \sim N(0, \theta)$. The variances ψ and θ can be interpreted as genetic and environmental variances, respectively, and the total residual variance is $\psi + \theta$.

1. Produce the required dummy variables M_i, F_i, and K_i.
2. Generate variables equal to the terms in parentheses in (4.5).
3. Which of the correlation structures available in `xtmixed` should be specified for the random coefficients?
4. Fit the model given in (4.5). Note that the model does not include a random intercept.
5. Obtain the estimated proportion of the total variance that is attributable to additive genetic effects.
6. Now fit the model including all the covariates listed above and having the same random part as the model in step 3.
7. Interpret the estimated coefficients from step 6.

8. Conditional on the covariates, what proportion of the residual variance is estimated to be due to additive genetic effects?

4.8 ❖ Covariance structure for nuclear family data

This exercise concerns family data such as those of exercise 4.7 consisting of a mother, father, and child. Here we consider three types of influences on birth-weight: additive genetic effects (due to shared genes), common environmental effects (due to shared environment), and unique environmental effects. These random effects have variances σ_A^2, σ_C^2, and σ_E^2, respectively.

The additive genetic effects have the following properties:

- The parents share no genes by descent, so their additive genetic effects are uncorrelated.
- The child shares half its genes with each parent by decent, giving a correlation of $1/2$ with each parent.
- The additive genetic variance should be the same for each family member.

For birth outcomes, no two family members share a common environment because they all developed in different wombs. We therefore cannot distinguish between common and unique environmental effects.

Rabe-Hesketh, Skrondal, and Gjessing (2008) show that we can use the following random-coefficient model to produce the required covariance structure:

$$y_{ij} = \beta_1 + \zeta_{1j}(M_i + K_i/2) + \zeta_{2j}(F_i + K_i/2) + \zeta_{3j}(K_i/\sqrt{2}) + \epsilon_{ij} \qquad (4.6)$$

where M_i, F_i, and K_i are dummy variables for mothers, fathers, and children, respectively. The random coefficients ζ_{1j}, ζ_{2j}, and ζ_{3j} produce the required additive genetic correlations and variances. These random coefficients are constrained to have the same variance $\psi = \sigma_A^2$ and to be uncorrelated with each other. As usual, we assume normality with $\zeta_{1j} \sim N(0, \sigma_A^2)$, $\zeta_{2j} \sim N(0, \sigma_A^2)$, $\zeta_{3j} \sim N(0, \sigma_A^2)$, and $\epsilon_{ij} \sim N(0, \theta)$.

1. By substituting the appropriate numerical values for the dummy variables M_i, F_i, and K_i in (4.6), write down three separate models, one for mothers, one for fathers, and one for children. It is useful to substitute $i = 1$ for mothers, $i = 2$ for fathers, and $i = 3$ for children in these equations.

2. Using the equations from step 1, demonstrate that the total variance is the same for mothers, fathers, and children.

3. Using the equations from step 1, demonstrate that the covariance between mothers and fathers from the same families is zero.

4. Using the equations from step 1, demonstrate that the correlation between the additive genetic components (terms involving ζ_{1j}, ζ_{2j}, or ζ_{3j}) of mothers and their children is $1/2$.

5. What is the relationship between θ, σ_C^2, and σ_E^2?

4.9 ❖ Effect of covariate translation on random-effects covariance matrix

Using (4.2) and the estimates for the random-coefficient model without level-2 covariates given in section 4.5.2, calculate what values $\widehat{\psi}_{11}$ and $\widehat{\psi}_{21}$ would take if you were to subtract 5 from the variable lrt and refit the model.

Part III

Models for longitudinal and panel data

Introduction to models for longitudinal and panel data (part III)

In this part, we focus on multilevel models and other methods for longitudinal and panel data. In longitudinal data, subjects have observations at several occasions or time points. Most commonly, longitudinal data are collected prospectively by following a group of subjects over time. Such data are referred to as *panel data*, *repeated measures*, or *cross-sectional time-series data* (the latter term explains the xt prefix in Stata's commands for longitudinal modeling). Longitudinal data can also be collected retrospectively, from archival data or by asking subjects to recall their history. We assume that the data have been collected in a way that makes it reasonable to treat them as prospective longitudinal data. We concentrate on "short panels", where there are many more subjects than occasions per subject. Other types of data that resemble longitudinal data are *time-series* data, where one unit is followed over time (usually at many occasions), and duration or survival data, which are discussed in volume 2 (chapters 14 and 15).

It is useful to distinguish between different types of longitudinal studies. In *panel studies*, all subjects are typically followed up at the same occasions (called "panel waves") leading to balanced or fixed-occasion data, although there may be missing data at some occasions for a subject. Usually, the occasions are also equally spaced with constant time intervals between them. In *cohort studies* (as defined in epidemiology), a group of subjects—sometimes of the same age, as in a birth cohort—may be followed up at subject-specific occasions, which produces unbalanced or variable-occasion data. Intervention studies and clinical trials are special cases of cohort studies with the important difference that subjects are assigned to treatments by researchers. In these studies, the intention is usually to collect balanced data; in practice, however, the data are often unbalanced because it is not feasible to assess all subjects exactly at the intended time points.

Longitudinal data can be viewed as two-level or clustered data with occasions nested in subjects, in which case subjects become clusters. Indeed, we have already applied random-intercept models to longitudinal data, such as the smoking and birthweight data, in previous chapters and exercises. We use the term "occasions" i for level-1 units and the term "subjects" j for level-2 units or clusters. We denote the timing associated with occasion i for subject j by a variable with subscripts i and j, such as t_{ij}, dropping the j subscript for balanced data. Note that the Stata documentation uses the indices t for occasions and i for subjects, and refers to subjects or units as "panels".

A special feature of longitudinal data is that the level-1 units or occasions are ordered in time and not necessarily exchangeable (where permuting units within clusters leaves the multivariate distribution unchanged) unlike, for example, students nested in schools. For this reason, there are a variety of different models designed specifically for longitudinal data. Broadly speaking, these models fall into the following categories:

Random-effects models, where unobserved between-subject heterogeneity is represented by subject-specific effects that are randomly varying.

> Random-effects models are useful for exploring and explaining average trends as well as individual differences by allowing subject-specific relationships to vary randomly around average relationships. Typical application areas are psychological development, physical growth, or learning, where both the nature and reasons for variability are of major interest. Growth-curve models are a special case of random-effects models. In this archetypal multilevel approach to longitudinal data analysis, the focus is on modeling growth (or decline) over time by including random coefficients of time (or functions of time) to represent individual growth trajectories. Random-effects models are discussed in chapter 5. Chapter 7 is devoted to growth-curve models.

Fixed-effects models, where unobserved between-subject heterogeneity is represented by fixed subject-specific effects.

> Fixed-effects models are used to estimate average within-subject relationships between time-varying covariates and the response variable, where every subject acts as its own control. Such models eliminate subject-level confounding and therefore facilitate causal inference. An example was considered in chapter 3, where the effect of smoking on birthweight was investigated by comparing births where the mother smoked with births where the same mother did not smoke. Fixed-effects models are revisited in chapter 5, where several new topics from econometrics are considered, particularly, approaches for handling different kinds of endogeneity or subject-level unmeasured confounding.

Dynamic models, where the response at a given occasion depends on previous or lagged responses.

> Dynamic or lagged-response models allow previous responses to affect future responses directly; for instance, the wages a year ago affect current wages. When combined with random- or fixed-effects approaches, dynamic models allow us to distinguish between two explanations of wage-dependence over time: 1) state dependence, where previous wages cause future wages (either due to fixed percentage salary increases within a job or due to the importance of past wages when negotiating salary in a new job) and 2) individual differences, where ability affected past wages and continues to affect future wages. Dynamic models are discussed in chapter 5.

Marginal models, where within-subject dependence is modeled by direct specification of the residual covariance structure across occasions.

Marginal or population-averaged models focus on average trends while accounting for longitudinal dependence. In these models, a covariance structure is directly specified for (total) residuals, instead of including random effects in the model that imply a certain covariance structure. Such models are often used in randomized controlled clinical trials to estimate average treatment effects. In this case, there are no subject-level confounders by design, and individual differences are of secondary concern. Chapter 6 discusses marginal models.

Which of the approaches to longitudinal modeling outlined above is adopted in practice depends largely on the discipline. In the biomedical sciences, random-effects and marginal models are most common, whereas random-effects models are popular in most of the social sciences. In economics, fixed-effects models and dynamic models are predominant. Repeated measures or split-plot analysis of variance (ANOVA) can in some ways be viewed as a fixed-effects model; this approach is used mostly for experimental designs in areas such as agriculture and psychology but is increasingly being replaced by random-effects models. Growth-curve modeling is particularly popular in education and psychology. We hope that our chapters on longitudinal and panel modeling will make you aware of strengths and weaknesses of methods used in a range of disciplines and encourage you to explore tools not commonly used in your own field.

The rest of this introduction discusses special features and challenges of longitudinal data and provides prerequisite information for all chapters in part III. The ideas are illustrated using a dataset that will be analyzed throughout chapters 5 and 6.

How and why do wages change over time?

Labor economists are interested in research questions such as how hourly wage depends on union membership, labor-market experience, and education, and how hourly wages change over time.

To address these questions, we will use data from the U.S. National Longitudinal Survey of Youth 1979. The original sample is representative of noninstitutionalized civilian youth who were aged 14–21 on December 31, 1978. Here we consider the subsample of the data previously analyzed by Vella and Verbeek (1998) and provided by Wooldridge (2010). The data comprise 545 full-time working males who completed schooling by 1980 and who had complete data for 1980–1987 (note that we would not recommend discarding subjects with incomplete data).

The variables in the dataset, `wagepan.dta`, that we will use here are

- `nr`: person identifier (j)
- `lwage`: log hourly wage in U.S. dollars (y_{ij})
- `black`: dummy variable for being black (x_{2j})

- **hisp**: dummy variable for being Hispanic (x_{3j})
- **union**: dummy variable for being a member of a union (that is, wage being set in collective bargaining agreement) (x_{4ij})
- **married**: dummy variable for being married (x_{5ij})
- **exper**: labor-market experience, defined as age$-6-$educ (L_{ij})
- **year**: calendar year 1980–1987 (P_i)
- **educ**: years of schooling (E_j)

We start by reading in the wage-panel data:

```
. use http://www.stata-press.com/data/mlmus3/wagepan
```

Longitudinal data structure and descriptives

Long and wide form

The wage-panel data are in long form, with one row of data per occasion for each subject. Table III.1 shows data for some of these variables (and two additional variables, C_j and A_{ij}) for two subjects. Longitudinal data are often in wide form, with a separate variable for each occasion and only one row of data per subject. For all analyses discussed in this part, data should be in long form. Stata's **reshape** command is convenient for converting data from long form to wide form and vice versa.

Table III.1: Illustration of longitudinal data in long form

Subject j	Occ. i	Cohort C_j	Age A_{ij}	Period P_i	Black x_{2j}	Hispanic x_{3j}	Union x_{4ij}	Log wage y_{ij}
45	1	1960	20	1980	0	0	1	1.89
45	2	1960	21	1981	0	0	1	1.47
45	3	1960	22	1982	0	0	0	1.47
45	4	1960	23	1983	0	0	0	1.74
45	5	1960	24	1984	0	0	0	1.82
45	6	1960	25	1985	0	0	0	1.91
45	7	1960	26	1986	0	0	0	1.74
45	8	1960	27	1987	0	0	0	2.14
847	1	1959	21	1980	1	0	1	1.56
847	2	1959	22	1981	1	0	1	1.66
847	3	1959	23	1982	1	0	0	1.77
847	4	1959	24	1983	1	0	0	1.79
847	5	1959	25	1984	1	0	0	2.00
847	6	1959	26	1985	1	0	1	1.65
847	7	1959	27	1986	1	0	0	2.13
847	8	1959	28	1987	1	0	1	1.69

To illustrate the **reshape** command, we first delete all unnecessary variables by using the **keep** command:

```
. keep nr lwage black hisp union married exper year educ
```

In the **reshape wide** command, we must list all time-varying variables and give the subject and occasion identifiers in the i() and j() options, respectively:

```
. reshape wide lwage union married exper, i(nr) j(year)
(note: j = 1980 1981 1982 1983 1984 1985 1986 1987)

Data                        long  ->  wide
-----------------------------------------------------------------
Number of obs.              4360  ->     545
Number of variables            9  ->      36
j variable (8 values)       year  ->  (dropped)
xij variables:
                           lwage  ->  lwage1980 lwage1981 ... lwage1987
                           union  ->  union1980 union1981 ... union1987
                         married  ->  married1980 married1981 ... married1987
                           exper  ->  exper1980 exper1981 ... exper1987
-----------------------------------------------------------------
```

We see that the values in the **year** variable have been appended to **lwage** to make eight new variables, **lwage1980** to **lwage1987**, containing the log hourly wages for the eight panel waves, and similarly for the other time-varying variables. Instead of eight rows per subject, a single row is now produced. To see what happened, we list the new log hourly wage variables together with the subject identifier for the first five subjects in the dataset (formatting the variable to make the list fit on the page)

```
. format lwage* %5.3f
. list nr lwage* in 1/5, clean noobs abbreviate(6)
       nr   l~1980   l~1981   l~1982   l~1983   l~1984   l~1985   l~1986   l~1987
       13    1.198    1.853    1.344    1.433    1.568    1.700   -0.720    1.669
       17    1.676    1.518    1.559    1.725    1.622    1.609    1.572    1.820
       18    1.516    1.735    1.632    1.998    2.184    2.267    2.070    2.873
       45    1.894    1.471    1.473    1.741    1.823    1.908    1.742    2.136
      110    1.949    1.962    1.963    2.203    2.135    2.126    1.991    2.112
```

To reshape the data back to long form, we use the `reshape` command again with identical syntax, except for replacing `wide` with `long`:

```
. reshape long lwage union married exper, i(nr) j(year)
(note: j = 1980 1981 1982 1983 1984 1985 1986 1987)
Data                              wide   ->   long

Number of obs.                     545   ->   4360
Number of variables                 36   ->      9
j variable (8 values)                    ->   year
xij variables:
    lwage1980 lwage1981 ... lwage1987    ->   lwage
    union1980 union1981 ... union1987    ->   union
married1980 married1981 ... married1987  ->   married
    exper1980 exper1981 ... exper1987    ->   exper
```

We can then list `lwage` for all eight years for the first subject to have a look at the long form of the data:

```
. list nr year lwage in 1/8, clean noobs
       nr   year    lwage
       13   1980    1.198
       13   1981    1.853
       13   1982    1.344
       13   1983    1.433
       13   1984    1.568
       13   1985    1.700
       13   1986   -0.720
       13   1987    1.669
```

Declaring the data as longitudinal using xtset

As we have already seen in chapters 2 and 3, some Stata commands, such as `xtreg`, require the data to be declared as longitudinal by using the `xtset` command to specify a cluster identifier (called a "panel variable" in Stata) and a time variable

```
. xtset nr year
       panel variable:  nr (strongly balanced)
        time variable:  year, 1980 to 1987
                delta:  1 unit
```

Stata's powerful time-series operators (similar to factor variables) can then be used to refer to lags and other transformations of time-varying variables as will be seen in section 5.7. Note that xtset does not affect the behavior of xtmixed.

For clarity, we will repeat the xtset command (with the quietly prefix to suppress output) whenever we rely on the cluster identifier and the time variable to be defined.

Balance, strong balance, and constant spacing of occasions

When the occasions for all subjects occur at the same sets of points in time so that $t_{ij} = t_i$, the data are called *balanced*. The data are *strongly balanced* if there are no missing data. It also sometimes matters whether the time intervals between occasions are the same across subjects and occasions with $t_{ij} - t_{i-1,j} = \Delta$, where the Greek letter Δ is pronounced "delta". We will refer to this property as *constant spacing of occasions*. We see from the output from xtset that the wage-panel data are strongly balanced with constant spacing of occasions; that is, $\Delta = 1$.

If we had specified a different time variable, such as age, in the xtset command, the data would not be balanced. As we will see later, we can consider several time scales simultaneously, but when first exploring the data, it is best to concentrate on a time scale that is at least approximately balanced.

Missing data

If we did not know that the data are strongly balanced, we might wonder whether lwage was missing at any occasion for any of the subjects. We can investigate missingness patterns by using xtdescribe:

```
. xtdescribe if lwage < .
      nr:  13, 17, ..., 12548                              n =        545
    year:  1980, 1981, ..., 1987                           T =          8
           Delta(year) = 1 unit
           Span(year)  = 8 periods
           (nr*year uniquely identifies each observation)
Distribution of T_i:   min      5%    25%      50%     75%     95%     max
                         8       8      8        8       8       8       8

     Freq.  Percent   Cum. |  Pattern
 ----------------------------------------
       545   100.00 100.00 |  11111111
 ----------------------------------------
       545   100.00        |  XXXXXXXX
```

Here Pattern could be any sequence of eight characters consisting of "1" (for not missing) and "." (for missing). The only pattern in these data is "11111111" corresponding to complete data for everyone (see section 7.7.2 for an example with missing data). For such patterns to be interpretable, it is necessary to specify a time variable or occasion identifier in the xtset command that takes on the same values for all subjects when not missing.

The `xtdescribe` command checks whether there is a row of data for a given subject at a given occasion. Generally, we do not want to count rows where the response variable is missing; we should therefore apply the command only to observations where the response is not missing, here by using the condition if `lwage < .` (The dot, ., in Stata is a very large number denoting a missing observation.)

Time-varying and time-constant variables

Whenever a variable changes over time for some subjects, it is called *time varying*. The response variable in longitudinal data is always time-varying (here log wage y_{ij}). Some explanatory variables or covariates are *subject-specific* or *time-constant* (education E_j and the ethnicity dummies x_{2j} and x_{3j}).

Time-varying covariates can be further classified into (a) occasion-specific (and not subject-specific) covariates (here year P_i) or (b) both subject- and occasion-specific covariates (here labor-market experience L_{ij}, union membership x_{4ij}, and marital status x_{5ij}).

For longitudinal data, it is useful to investigate the within-subject, between-subject, and total variability of the variables, shown here for only some of the variables. (For precise definitions of the different kinds of standard deviations, see section 3.2.1.)

```
. quietly xtset nr
. xtsum lwage union educ year
```

Variable		Mean	Std. Dev.	Min	Max		Observations
lwage	overall	1.649147	.5326094	-3.579079	4.05186	N =	4360
	between		.3907468	.3333435	3.174173	n =	545
	within		.3622636	-2.467201	3.204687	T =	8
union	overall	.2440367	.4295639	0	1	N =	4360
	between		.3294467	0	1	n =	545
	within		.2759787	-.6309633	1.119037	T =	8
educ	overall	11.76697	1.746181	3	16	N =	4360
	between		1.747585	3	16	n =	545
	within		0	11.76697	11.76697	T =	8
year	overall	1983.5	2.291551	1980	1987	N =	4360
	between		0	1983.5	1983.5	n =	545
	within		2.291551	1980	1987	T =	8

As expected, `lwage` and `union` vary both within and between subjects. `educ` does not vary within subjects, and `year` does not vary between subjects (that is, it has the same mean for all subjects). It is especially important to examine how much a variable varies within subjects because some estimation methods, such as fixed-effects approaches (see section 5.4), rely exclusively on within-subject variability for estimation of regression coefficients. We see from the fact that `T` is given in the output instead of `T-bar` that the number of occasions n_j is constant over subjects j as we already know.

For time-varying categorical or binary variables, like union, the xttab command also provides useful information:

```
. quietly xtset nr
. xttab union
```

union	Overall Freq.	Overall Percent	Between Freq.	Between Percent	Within Percent
0	3296	75.60	511	93.76	80.63
1	1064	24.40	280	51.38	47.50
Total	4360	100.00	791	145.14	68.90

(n = 545)

We see from the Overall columns that union takes the value one 24.4% of the time across all subjects and occasions. From the Between columns, we see that 93.8% of subjects were nonunion members for at least one occasion, and 51.4% of subjects were union members for at least one occasion. Finally, the Within column shows that among those who were ever nonunion members the average percentage of occasions for which they were nonunion members is 80.6%. Those who were ever union members were union members for an average of 47.5% of occasions. Whenever the total percentage in the Between columns is greater than 100%, the variable changes over time for some subjects.

Graphical displays for longitudinal data

Box plots of the response variable at each occasion are useful for inspecting the distribution of the variable and detecting outliers:

```
. graph box lwage, over(year) intensity(0) medtype(line)
> marker(1,mlabel(nr) mlabsize(vsmall) msym(i) mlabpos(0)
> mlabcol(black)) ytitle(Log hourly wage)
```

Here we used the mlabel() suboption to show the subject identifiers for extreme log hourly wages (smaller than the lower quartile minus 1.5 interquartile ranges, or greater than the upper quartile plus 1.5 interquartile ranges). We see in figure III.1 that subject 813 had a very low reported log hourly wage in 1984, corresponding to just a few cents. Subject 813 may therefore merit special attention because he may have had a very bad year, the reported wage may be wrong, or an error may have been committed in the coding of the data. (Using box plots requires that there are sufficiently many observations at each occasion.)

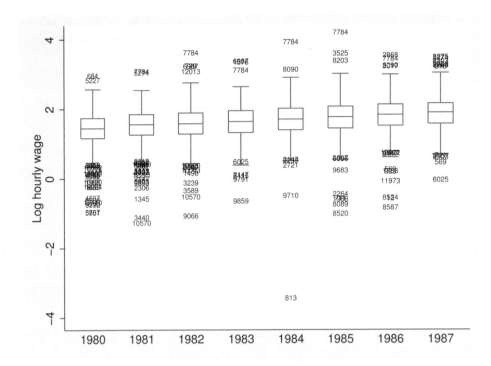

Figure III.1: Box plots of log hourly wages at each occasion

It is also important to get an idea of how log hourly wages change over time for individual subjects. Plotting observed trajectories for all subjects becomes messy, so we will draw a random sample of subjects. We do this by generating a random number, r, for each subject for 1980, with values of r for all other years missing:

```
. format lwage* %9.0g
. sort nr year
. set seed 132144
. generate r = runiform() if year==1980
```

A random sample of 12 subjects can be obtained by choosing the subjects with the 12 largest random numbers r, and similarly for any other required number of subjects. We therefore rank order the random numbers by using egen with the rank() function:

```
. egen num = rank(r) if r<.
```

We then generate a variable, number, that contains the nonmissing value of the rank order, num, for each subject:

```
. egen number = mean(num), by(nr)
```

(The `mean()` function finds the mean of all nonmissing values; here there is only one per subject, so that number is placed in all rows for the subject.) It is now easy to plot the trajectories of log hourly wage for the 12 randomly chosen subjects in a trellis graph:

```
. twoway line lwage year if number<=12, by(nr, compact)
> ytitle(Log hourly wage) xtitle(Year) xlabel(,angle(45))
```

The resulting graph is shown in figure III.2. In the first row, we clearly see the common phenomenon of *tracking*, whereby some individuals consistently remain higher or lower than other individuals. This pattern is consistent with subject-specific random or fixed intercepts, where individual curves are vertically shifted by subject-specific constants. Column 2 of the figure illustrates that the slope of `year` varies between subjects, and column 3 shows that the log hourly wage is volatile for some subjects.

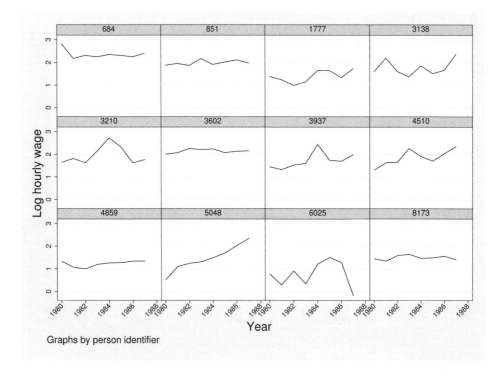

Figure III.2: Trellis graph of trajectories for log hourly wage for 12 randomly chosen subjects

To see these trajectories within the context of the entire sample of subjects, we can add them to a scatterplot for all subjects. We also display the average trajectory on the same graph. First, we find the mean log hourly wage for each year,

```
. egen mn_lwage = mean(lwage), by(year)
```

and then we produce the graph, which is displayed in figure III.3:

```
. sort nr year

. twoway (scatter lwage year, jitter(2) msym(o) msize(tiny))
> (line lwage year if number<=12, connect(ascending)
> lwidth(vthin) lpatt(solid))
> (line mn_lwage year, sort lpatt(longdash)) if lwage>-2,
> ytitle(Log hourly wage) xtitle(Year)
> legend(order(2 "Individual trajectories" 3 "Mean trajectory"))
```

By first sorting the data by `year` within `nr` and then using the `connect(ascending)` option, we ensure that successive observations for a subject are connected, but that the last observation for a subject is not connected to the first observation for the next subject. The outlying observation for subject 813 was discarded by plotting only observations with `lwage>-2` to obtain a better resolution. A small amount of jitter was used for the scatterplot to prevent overlap of data points.

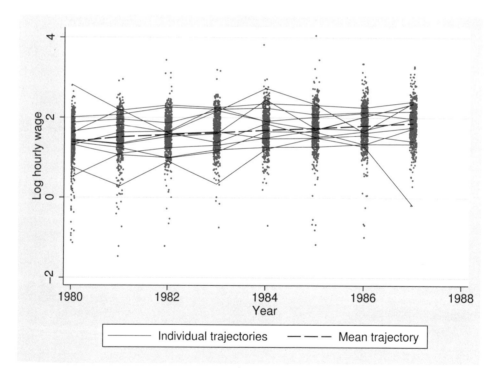

Figure III.3: Scatterplot of log hourly wages versus occasions; individual trajectories for 12 randomly chosen subjects (thin solid lines) and mean trajectory (thick dashed line)

The population-mean trajectory appears to be linear.

Time scales in longitudinal data: Age-period-cohort effects

Consider the longitudinal data on subjects 45 and 847 from the wage-panel data given in table III.1 on page 231. We refer to the birth year as "cohort" and to the calendar year as "period". We have calculated age and cohort from other variables in the data (see below for the relationship among the variables).

When investigating change in the response variable using *age-period-cohort* data, such as those in table III.1, any of three different time scales may be of interest: age, period, or cohort. (The term cohort usually refers to a group of people such as a birth cohort, but here it refers to the birth year associated with the birth cohort.) In our example, the definitions of both cohort and age are in terms of the time of birth; this reference point is important in many investigations. However, for some applications, the reference point could be an occurrence of some other event, such as graduation from university. Then a particular cohort may be referred to as "the class of 2011", and the age-like time scale becomes time since graduation.

The relationship between age, period, and cohort is given by

$$A_{ij} = P_i - C_j$$

as illustrated in figure III.4 for four subjects for the first five waves of the wage-panel data. The top two lines represent subjects 847 and 45 in table III.1. For example, we see that subject 847 was born in 1959 and hence belongs to the 1959 cohort (referred to as C59 in the figure), and his age is 21 years in 1980. In 1984, any member of the 1959 cohort is 25 years old.

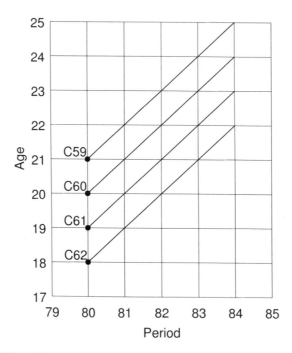

Figure III.4: Illustration of the relationship between age, period, and cohort

A great advantage of this kind of longitudinal study with several cohorts is that we can investigate the effects of more than one time scale simultaneously. In contrast, a cross-sectional study provides data for just one period P, which obviously makes it impossible to estimate the effect of period. Furthermore, we cannot separate the effects of age and cohort in a cross-sectional study because each subject's age A_j is determined by the cohort, $A_j = P - C_j$. For example, there are two competing explanations for older people being more conservative: 1) they are in later stages in life with increased A_j or 2) they were born longer ago (in a different era) with smaller C_j. Cross-sectional data cannot be used to distinguish between these explanations.

A longitudinal study with one cohort C also does not allow us to investigate the effect of more than one time scale. We obviously cannot estimate the effect of cohort, and age is determined by period, $A_i = P_i - C$. For instance, we cannot distinguish between two explanations for salary increases: 1) people get more experience as they get older with increased A_i or 2) there is inflation over calendar time P_i.

Longitudinal studies with several cohorts are sometimes said to have a *cohort-sequential design* or *accelerated longitudinal design*, where the term "accelerated" refers to the range of ages covered exceeding the length of the study. From the relation $A_{ij} = P_i - C_j$, we see that it is possible to estimate the effects of two time scales, but these will be confounded with the third scale. Thus it is necessary to pick the time scales that are believed to be most important. For example, conservatism may be viewed as depending on age and cohort (ignoring period), and salary may be viewed as depending on age and period (ignoring cohort).

To resolve the collinearity among A_{ij}, C_j, and P_i in the wage-panel data, we choose to eliminate C_j. However, other time scales are also of interest because there are two other relevant subject-specific events: entering education (assumed to occur at age 6) and leaving education. The relationship between age A_{ij}, years of education E_j, and number of years in the labor market L_{ij} is

$$A_{ij} = 6 + E_j + L_{ij}$$

We see that we have to eliminate one of the collinear variables, for instance, A_{ij}. We are then left with three time scales: P_i, E_j, and L_{ij}. (We have ignored two other time scales that are collinear with P_i, E_j, and L_{ij}: age at entry into the labor market, $6+E_j$, and calendar year at entry into the labor market, $C_j + 6 + E_j$.)

Several time scales can be relevant in longitudinal studies, and the prospect of disentangling the effects of different time scales depends on the research design. Within the limitations of the chosen research design, the time scales that are deemed most relevant for the research problem should be included in the models. These time scales are not necessarily present in the dataset and may have to be constructed from the data.

Pooled ordinary least-squares estimation

The simplest approach to longitudinal modeling is to ignore the longitudinal structure of the data and proceed as if each row of the data (in long form) corresponds to a different unit of observation. We can then consider a standard linear regression model that includes the three time scales L_{ij}, P_i, and E_j (exper, yeart, and educt) as covariates as well as black, hisp, union, and married:

$$y_{ij} = \beta_1 + \beta_2 x_{2j} + \beta_3 x_{3j} + \beta_4 x_{4ij} + \beta_5 x_{5ij} + \beta_6 L_{ij} + \beta_7 P_i + \beta_8 E_j + \xi_{ij}$$

where ξ_{ij} is a residual. We can estimate the parameters by ordinary least squares (OLS), often called pooled ordinary least squares.

First, we translate the time scales to make the intercept more interpretable here and in later analyses:

```
. generate educt = educ - 12
. generate yeart = year - 1980
```

The variable educt is the number of years of education beyond the usual time (12 years) required to complete high school, and yeart is the number of years since 1980. For simplicity, we will henceforth refer to the translated variables as E_j and P_i, respectively.

We now fit the model by pooled OLS using `regress` with the `vce(cluster nr)` option to obtain appropriate standard errors for clustered data:

```
. regress lwage black hisp union married exper yeart educt, vce(cluster nr)
```

Linear regression	Number of obs	=	4360
	F(7, 544)	=	84.63
	Prob > F	=	0.0000
	R-squared	=	0.1870
	Root MSE	=	.48064

(Std. Err. adjusted for 545 clusters in nr)

lwage	Coef.	Robust Std. Err.	t	P>|t|	[95% Conf. Interval]	
black	-.1371356	.0503591	-2.72	0.007	-.2360577	-.0382134
hisp	.0136874	.0386814	0.35	0.724	-.0622959	.0896707
union	.1863618	.0274214	6.80	0.000	.1324971	.2402265
married	.1119207	.0257939	4.34	0.000	.0612529	.1625885
exper	.0302751	.0111744	2.71	0.007	.0083249	.0522253
yeart	.0267888	.0117137	2.29	0.023	.0037791	.0497985
educt	.0928443	.0110632	8.39	0.000	.0711124	.1145762
_cons	1.298895	.0397477	32.68	0.000	1.220818	1.376973

The estimated coefficients of `exper` and `yeart` suggest that the mere passage of time is associated with a similar effect as each extra year of experience, probably because of inflation. Each additional year of education is associated with an estimated increase in the expectation of log hourly wage of 0.09, controlling for the other covariates. The exponential of this coefficient, 1.10, therefore represents the estimated multiplicative effect of each extra year of education on the expectation of the hourly wage, as shown in display 1.1 on page 64. In other words, there is a 10% $[0.10 = \exp(0.0928443) - 1]$ estimated increase in expected wages per year of education, compared with a 3% $[0.03 = \exp(0.0302751) - 1]$ increase due to experience, when controlling for other covariates. We also estimate that given the other covariates, black men's mean wages are 13% lower than white men's $[-0.13 = \exp(-0.1371356) - 1]$.

Pooled OLS produces consistent estimates for the regression coefficients under the assumption that the mean structure is correctly specified (basically, that the correct covariates are included and that the correct functional form is specified) and that the residuals ξ_{ij} are uncorrelated with the covariates. Furthermore, the sandwich estimator of the standard errors, requested by the `vce(cluster nr)` option, produces consistent estimates of the standard errors even if the residuals are correlated within subjects and have nonconstant variance. However, an important but rarely acknowledged limitation of this approach is the tacit assumption either that there are no missing data or that the probability that a response is missing does not depend on observed or unobserved responses after controlling for the covariates. See section 5.8.1 for a simulation where this assumption is violated.

Correlated residuals

In longitudinal data, the response variable is invariably correlated within subjects even after controlling for covariates. To demonstrate this phenomenon, we now obtain the within-subject correlation matrix of the estimated residuals from pooled OLS using

```
. predict res, residuals
```

To form the within-subject correlation matrix for pairs of occasions, we must reshape the data to wide form, treating the residuals at times 0–7 as variables res0–res7. We first preserve and later restore the data because we will require them in long form when fitting models. (If running the subsequent commands from a do-file, all the commands from preserve to restore must be run in one block, not one command at a time.)

```
. preserve
. keep nr res yeart
. reshape wide res, i(nr) j(year)
(note: j = 0 1 2 3 4 5 6 7)
Data                          long   ->   wide

Number of obs.                4360   ->      545
Number of variables              3   ->        9
j variable (8 values)        yeart   ->   (dropped)
xij variables:
                               res   ->   res0 res1 ... res7
```

The standard deviations and correlations of the residuals are then obtained using

```
. tabstat res*, statistics(sd) format(%3.2f)
    stats |    res0   res1   res2   res3   res4   res5   res6   res7
----------+---------------------------------------------------------
       sd |    0.53   0.50   0.46   0.45   0.50   0.49   0.49   0.43
```

and

```
. correlate res*, wrap
(obs=545)
             res0     res1     res2     res3     res4     res5     res6     res7

    res0   1.0000
    res1   0.3855   1.0000
    res2   0.3673   0.5525   1.0000
    res3   0.3298   0.5220   0.6263   1.0000
    res4   0.2390   0.4451   0.5740   0.6277   1.0000
    res5   0.2673   0.4071   0.5230   0.5690   0.6138   1.0000
    res6   0.2084   0.3220   0.4622   0.4710   0.5020   0.5767   1.0000
    res7   0.2145   0.4019   0.4232   0.4976   0.5349   0.6254   0.6337   1.0000

. restore
```

We see that there are substantial within-subject correlations among the residuals, ranging from 0.21 to 0.63 for pairs of occasions. The correlations tend to decrease down the columns, meaning, for instance, that residuals at occasion 0 (column 1), are more correlated with residuals at occasion 1 than with residuals at occasion 7. As the time interval between occasions increases, the correlation between the corresponding residuals tends to decrease.

Within-subject correlations over time are sometimes referred to as *longitudinal correlations*, *serial correlations*, or *autocorrelations*, and these terms also suggest that correlations may depend on the time interval between occasions. The correlations could be partly due to between-subject heterogeneity in the intercept and possibly in the slopes of covariates, which is not accommodated in the standard linear regression model. Alternatively, the responses at an occasion could depend on previous responses, but lagged responses are mistakenly omitted from the model. Finally, the residuals could be governed by slowly varying processes that induce correlations that decay as the time interval between occasions increases. Within-subject correlations among the residuals could, of course, be due to a combination of these reasons.

Why do we need special models for longitudinal data?

Because pooled OLS with robust standard errors for clustered data gives consistent estimates of regression coefficients and standard errors (assuming a correctly specified mean structure), the question is, why are there three chapters on longitudinal modeling in this part of the book?

In the discussion section of papers based on cross-sectional data, it is often stated that longitudinal data would be required for causal inference. The reason for this is that causal effects are based on comparisons of a subject's hypothetical (potential or counterfactual) responses for different treatments or exposures. It is evident that comparing actual observed responses within subjects is closer to this ideal than comparing observed responses between subjects. Indeed, the great advantage of longitudinal data as compared with cross-sectional data is that each subject can serve as his or her own control. Unfortunately, pooled OLS treats longitudinal data as repeated cross-sectional data (where independent samples of subjects are drawn at each occasion) and conflates within- and between-subject comparisons. Between-subject comparisons are susceptible to omitted variable bias or unmeasured confounding, due to time-constant subject-specific variables that are not included in the model. Within-subject comparisons are free from such bias because subjects truly act as their own controls. In chapter 5, we discuss methods that reap the benefits of longitudinal data for causal inference.

When the causal effects of previous responses are of interest, lagged responses are included as covariates. In this case, pooled OLS is no longer consistent if there are time-constant omitted covariates, which will typically be the case. In chapter 5, we describe methods that address this problem.

Another limitation of pooled OLS is that estimates of regression coefficients are no longer consistent if there are missing data and if missingness depends on observed responses for the same subject, given the covariates. In this common scenario, it becomes necessary to model the within-subject residual covariance matrix to obtain consistent estimates.

If we can model the residual covariance structure appropriately, estimates of regression coefficients that are based on the covariance structure will be more precise or efficient than pooled OLS estimates. Modeling the covariance matrix can also be of interest in its own right because it sheds light on the kinds of processes that lead to within-subject dependence. An estimated residual covariance matrix is also needed, in addition to estimates of regression coefficients, to make forecasts for individual subjects. Modeling the residual covariance matrix is the focus of chapter 6.

Finally, pooled OLS estimates only the population-averaged relationship between the response variable and covariates. Subject-specific effects can be investigated using random-effects and fixed-effects models as discussed in chapter 5. When the nature of changes in the response variable over time is of interest, we can use growth-curve models to investigate how subject-specific growth trajectories vary around the population-averaged trajectory. Such models are discussed in chapter 7.

5 Subject-specific effects and dynamic models

5.1 Introduction

In this chapter, we discuss models where the intercept and possibly some of the coefficients can vary between subjects. In sections 5.2, 5.3, and 5.4, models with subject-specific intercepts are treated, where the intercepts are either random or fixed. As we saw in chapter 3, the random-effects approach treats the intercepts as (unobserved) random variables that can be viewed as residuals, whereas the fixed-effects approach treats the intercepts as model parameters that can be estimated by including dummy variables for subjects. In both cases, the intercepts can be viewed as representing the effects of omitted covariates that are constant over time.

As pointed out in section 3.7.4, an advantage of the fixed-effects approach is that it relaxes the assumption that the covariates are uncorrelated with the subject-specific intercept. For this reason, any estimator that relaxes this exogeneity assumption is called a "fixed-effects estimator" in econometrics, and the assumption is therefore referred to as the random-effects assumption in that literature. However, in modern econometrics, the subject-specific intercept is usually viewed as random even if a fixed-effects approach is used. In fact, some fixed-effects estimators, such as the Hausman–Taylor estimator described in section 5.3, relax the exogeneity assumption for some covariates while explicitly treating the subject-specific intercept as random (by providing an estimate of the random-intercept variance).

Sections 5.5 and 5.6 discuss models where the coefficients of time-varying covariates vary between subjects in addition to the subject-specific intercepts. Again we can distinguish between random-coefficient models where the slopes are treated as random variables and fixed-coefficient models where the slopes are treated as unknown parameters.

Finally, *dynamic models*, where the current response is regressed on previous or lagged responses, are introduced in section 5.7. When these models include random intercepts, they pose special challenges that will be addressed.

Throughout this chapter, we analyze the wage-panel data described in *Introduction to models for longitudinal and panel data (part III)*. We first read in the data

```
. use http://www.stata-press.com/data/mlmus3/wagepan
```

and construct the required variables:

```
. generate educt = educ - 12
. generate yeart = year - 1980
```

5.2   Conventional random-intercept model

Random-intercept models were discussed in detail in chapter 3. For the wage-panel data, a conventional random-intercept model is specified as

$$y_{ij} = (\beta_1 + \zeta_j) + \beta_2 x_{2j} + \beta_3 x_{3j} + \beta_4 x_{4ij} + \beta_5 x_{5ij} + \beta_6 L_{ij} + \beta_7 P_i + \beta_8 E_j + \epsilon_{ij}$$

In econometrics, the error components ζ_j and ϵ_{ij} are sometimes referred to as "permanent" and "transitory" components, respectively. To be poetic, Crowder and Hand (1990) referred to the fixed part of the model as the "immutable constant of the universe", to ζ_j as the "lasting characteristic of the individual", and to ϵ_{ij} as the "fleeting aberration of the moment".

Recall from section 3.7.4 that exogeneity assumptions—such as no correlation between the covariates and either the random intercept ζ_j or the level-1 residuals ϵ_{ij}— are required for consistent estimation of the parameters in the conventional random-intercept model. As we did there, we make the somewhat stronger assumptions that $E(\zeta_j|\mathbf{X}_j) = 0$ and $E(\epsilon_{ij}|\mathbf{X}_j, \zeta_j) = 0$, where $\mathbf{X}_j$ contains the covariates at all occasions for subject j. We also define $\psi \equiv \mathrm{Var}(\zeta_j|\mathbf{X}_j)$ and $\theta \equiv \mathrm{Var}(\epsilon_{ij}|\mathbf{X}_j, \zeta_j)$. It is assumed that the ζ_j are uncorrelated across subjects, that the ϵ_{ij} are uncorrelated across both subjects and occasions, and that the ζ_j and ϵ_{ij} are uncorrelated.

Maximum likelihood (ML) estimation is based on the usual assumptions that given the covariates, the random intercept ζ_j, and the level-1 residual ϵ_{ij} are both normally distributed. However, the normality assumptions are not required for consistent estimation of the parameters and standard errors.

The random-intercept model can be fit by ML using `xtmixed`:

```
. xtmixed lwage black hisp union married exper yeart educt || nr:, mle
Mixed-effects ML regression                     Number of obs     =      4360
Group variable: nr                              Number of groups  =       545

                                                Obs per group: min =         8
                                                               avg =       8.0
                                                               max =         8

                                                Wald chi2(7)      =    894.85
Log likelihood = -2214.3572                     Prob > chi2       =    0.0000
```

lwage	Coef.	Std. Err.	z	P>|z|	[95% Conf. Interval]	
black	-.1338495	.0479549	-2.79	0.005	-.2278395	-.0398595
hisp	.0174169	.0428154	0.41	0.684	-.0664998	.1013336
union	.1105923	.0179007	6.18	0.000	.0755075	.1456771
married	.0753674	.0167345	4.50	0.000	.0425684	.1081664
exper	.0331593	.0112023	2.96	0.003	.0112031	.0551154
yeart	.0259133	.0114064	2.27	0.023	.0035571	.0482695
educt	.0946864	.0107047	8.85	0.000	.0737055	.1156673
_cons	1.317175	.0373979	35.22	0.000	1.243877	1.390474

Random-effects Parameters	Estimate	Std. Err.	[95% Conf. Interval]	
nr: Identity				
sd(_cons)	.3271344	.0114153	.3055088	.3502908
sd(Residual)	.3535088	.0040494	.3456606	.3615351

```
LR test vs. linear regression: chibar2(01) =  1547.76 Prob >= chibar2 = 0.0000
```

The estimates, which are also shown under "Random intercept" in table 5.1 on page 260, are stored for later use:

```
. estimates store ri
```

We can exponentiate the estimated regression coefficients to obtain estimated multiplicative effects on the expected hourly wages (see display 1.1 on page 64). Controlling for the other variables, expected hourly wages increase by about 3% per year of experience and per year of calendar time. Each additional year of education is associated with an estimated 10% increase in expected wages when controlling for other covariates. We also estimate that given the other covariates, black men's mean wages are 13% lower than white men's, although this estimate is likely to be prone to subject-level confounding or bias due to omitted subject-level variables.

From the random part of the model, we see that the residual between-subject standard deviation is estimated as 0.33 compared with an estimate of 0.35 for the residual within-subject standard deviation. The corresponding estimated residual intraclass correlation is

$$\widehat{\rho} \;=\; \frac{\widehat{\psi}}{\widehat{\psi} + \widehat{\theta}} = \frac{0.327^2}{0.327^2 + 0.354^2} = 0.46$$

which is also the within-subject correlation between the residuals. Thus 46% of the variance in log wage that is not explained by the covariates is due to unobserved time-invariant subject-specific characteristics (strictly speaking, the component of the omitted covariates that is uncorrelated with the included covariates—not an issue if the covariates are exogenous).

5.3 Random-intercept models accommodating endogenous covariates

In this section, we discuss methods for accommodating different kinds of endogenous covariates that are correlated with the random intercept ζ_j. It is still assumed that $E(\epsilon_{ij}|\mathbf{X}_j, \zeta_j) = 0$, which implies that all covariates are uncorrelated with the level-1 residual ϵ_{ij}.

5.3.1 Consistent estimation of effects of endogenous time-varying covariates

We now accommodate endogenous time-varying covariates that are correlated with the random-intercept ζ_j by allowing for different within and between effects for time-varying covariates. As discussed in section 3.7.4, subject-mean centered covariates are uncorrelated with ζ_j by construction, and the corresponding coefficients can be consistently estimated. After subtracting the subject mean from `yeart` and `exper`, both deviation variables would be identical because

$$\texttt{yeart}_i = t_j + \texttt{exper}_{ij}$$

where t_j is the time (in years since 1980) when the subject entered the labor market. We therefore omit `yeart` from the model.

Denoting the subject mean of a variable x_{ij} as $\overline{x}_{\cdot j} = \sum_{i=1}^{n_j} x_{ij}/n_j$, the random-intercept model with different within and between effects can be written as

$$\begin{aligned}
y_{ij} = {} & (\beta_1 + \zeta_j) + \beta_2 x_{2j} + \beta_3 x_{3j} + \beta_4 (x_{4ij} - \overline{x}_{4\cdot j}) + \beta_5 (x_{5ij} - \overline{x}_{5\cdot j}) + \beta_6 (L_{ij} - \overline{L}_{\cdot j}) \\
& + \beta_8 E_j + \beta_9 \overline{x}_{4\cdot j} + \beta_{10} \overline{x}_{5\cdot j} + \beta_{11} \overline{L}_{\cdot j} + \epsilon_{ij}
\end{aligned} \tag{5.1}$$

where β_7 is missing because `yeart` was removed.

We first construct the subject means of the time-varying covariates `union`, `married`, and `exper` by using the commands

```
. egen mn_union = mean(union), by(nr)
. egen mn_married = mean(married), by(nr)
. egen mn_exper = mean(exper), by(nr)
```

We then construct the occasion-specific deviations from the subject means:

```
. generate dev_union = union - mn_union
. generate dev_married = married - mn_married
. generate dev_exper = exper - mn_exper
```

We fit model (5.1) by ML using `xtmixed` with the `vce(robust)` option, which has the advantage that the estimated standard errors are valid even if the level-1 errors are heteroskedastic or autocorrelated:

```
. xtmixed lwage black hisp dev_union dev_married dev_exper educt
> mn_union mn_married mn_exper || nr:, mle vce(robust)

Mixed-effects regression                    Number of obs      =      4360
Group variable: nr                          Number of groups   =       545

                                            Obs per group: min =         8
                                                           avg =       8.0
                                                           max =         8

                                            Wald chi2(9)       =    597.05
Log pseudolikelihood = -2206.1344           Prob > chi2        =    0.0000

                       (Std. Err. adjusted for 545 clusters in nr)
```

lwage	Coef.	Robust Std. Err.	z	P>|z|	[95% Conf. Interval]	
black	-.1414313	.0508596	-2.78	0.005	-.2411144	-.0417482
hisp	.0100387	.0384598	0.26	0.794	-.065341	.0854184
dev_union	.083791	.0231021	3.63	0.000	.0385117	.1290702
dev_married	.0610384	.0212003	2.88	0.004	.0194866	.1025902
dev_exper	.0598672	.0033705	17.76	0.000	.0532611	.0664734
educt	.0912614	.0111498	8.19	0.000	.0694082	.1131147
mn_union	.2587162	.0425155	6.09	0.000	.1753873	.3420451
mn_married	.1416358	.0400085	3.54	0.000	.0632206	.2200509
mn_exper	.0278124	.011318	2.46	0.014	.0056296	.0499952
_cons	1.378695	.0753624	18.29	0.000	1.230988	1.526403

Random-effects Parameters	Estimate	Robust Std. Err.	[95% Conf. Interval]	
nr: Identity				
sd(_cons)	.3224087	.0133976	.2971907	.3497665
sd(Residual)	.3533891	.0132959	.3282672	.3804336

The within effects of the time-varying covariates `union`, `married`, and `exper` are consistently estimated, as long as relevant time-varying covariates are not omitted from the model (and the functional form is correct). Importantly, this is true whether the time-constant covariates are exogenous or not.

Unfortunately, this approach produces inconsistent estimates for the effects of the time-constant covariates `black`, `hisp`, and `educt`, even if they are exogenous. The estimator for the random-intercept variance is also inconsistent.

We can test the joint hypothesis that all the between and within effects are equal using the test command (which in this case is robust against heteroskedastic or auto-correlated level-1 errors because we used the vce(robust) option for xtmixed):

```
. test (dev_union=mn_union) (dev_married=mn_married) (dev_exper=mn_exper)
 ( 1)  [lwage]dev_union - [lwage]mn_union = 0
 ( 2)  [lwage]dev_married - [lwage]mn_married = 0
 ( 3)  [lwage]dev_exper - [lwage]mn_exper = 0
           chi2(  3) =    21.22
         Prob > chi2 =    0.0001
```

We conclude that the between and within effects are significantly different from each other, which suggests that one or more of the time-varying covariates are endogenous. This test is numerically identical to a simultaneous test that the coefficients for all the cluster means are zero in a random-intercept model that includes the cluster means of the time-varying covariates as well as the noncentered covariates (the approach taken in section 3.7.4). The test is also asymptotically equivalent to the Hausman test discussed in section 3.7.6.

As a next step, it is useful to consider the evidence against exogeneity for each of the time-varying covariates by performing separate tests of equal between and within effects, for instance, using lincom:

```
. lincom dev_union-mn_union
 ( 1)  [lwage]dev_union - [lwage]mn_union = 0
```

lwage	Coef.	Std. Err.	z	P>|z|	[95% Conf. Interval]
(1)	-.1749252	.0474094	-3.69	0.000	-.267846 -.0820045

```
. lincom dev_married-mn_married
 ( 1)  [lwage]dev_married - [lwage]mn_married = 0
```

lwage	Coef.	Std. Err.	z	P>|z|	[95% Conf. Interval]
(1)	-.0805974	.0452476	-1.78	0.075	-.1692811 .0080864

```
. lincom dev_exper-mn_exper
 ( 1)  [lwage]dev_exper - [lwage]mn_exper = 0
```

lwage	Coef.	Std. Err.	z	P>|z|	[95% Conf. Interval]
(1)	.0320548	.0119636	2.68	0.007	.0086067 .055503

We see that there is most evidence against union being exogenous. This is not surprising, because one might expect the random intercept to be correlated with union membership. For instance, if the random intercept is interpreted as unmeasured ability, then high ability subjects might have higher mean earnings than expected given their observed covariates and be more likely to be union members. Interestingly, the research focus of Vella and Verbeek (1998), who made the wage-panel data available, was to develop an approach for handling the endogeneity of `union`.

5.3.2 Consistent estimation of effects of endogenous time-varying and endogenous time-constant covariates

A limitation of the approach with different within and between effects is that the coefficients of time-constant exogenous or endogenous covariates are not consistently estimated. Fortunately, Hausman and Taylor (1981) developed a method, implemented in Stata's `xthtaylor` command, that makes it possible to fit models with some endogenous time-constant covariates in addition to some endogenous time-varying covariates [still assuming that $E(\epsilon_{ij}|\mathbf{X}_j, \zeta_j) = 0$ so that all covariates are uncorrelated with ϵ_{ij}].

A requirement for using the Hausman–Taylor approach is that both time-varying and time-constant covariates can be classified as either endogenous or exogenous (correlated or not with the random intercept ζ_j). We hence have four kinds of covariates:

1. Exogenous time-varying covariates x_{ij}
2. Endogenous time-varying covariates x_{ij}^{end}
3. Exogenous time-constant covariates x_j
4. Endogenous time-constant covariates x_j^{end}

Furthermore, a necessary condition for identification is that there are at least as many exogenous time-varying covariates as there are endogenous time-constant covariates.

The basic idea of the Hausman–Taylor method is as follows. First, consistent estimates of the within effects for the time-varying covariates (and the variance of the level-1 residual) are obtained using a standard fixed-effects estimator (see sections 3.7.2 and 5.4). Residuals are then produced by predicting the response using uncentered time-varying covariates and the estimated coefficients from the first step. The subject-mean residuals are then regressed on the time-constant covariates, using the exogenous covariates as *instrumental variables*, to obtain consistent estimates of the coefficients for the time-constant covariates (if you are unfamiliar with instrumental-variables estimation you may want to consult display 5.1, which provides the basic ideas).

Consider a simple linear model for cross-sectional data (no clustering):

$$y_i \;=\; \beta_1 + \beta_2 x_i + \epsilon_{yi} \quad (1)$$

We want to estimate β_2, but x_i is endogenous because it is correlated with ϵ_{yi}, and the ordinary least squares (OLS) estimator $\widehat{\beta}_2^{\text{OLS}}$ is therefore inconsistent.

We can overcome this problem and obtain a consistent estimator of β_2 if we can find an *instrumental variable* (IV) z_i. The requirements for an instrumental variable are that it is 1) correlated with x_i and 2) uncorrelated with ϵ_{yi}. An example is the problem of estimating the effect of smoking x_i on health y_i. There may be omitted covariates, such as socioeconomic status, that affect both smoking and health, and hence x_i may be endogenous. A possible instrumental variable z_i in this case is the price of cigarettes (if this varies in the sample) because price affects smoking but not health. To understand the instrumental-variable estimator, consider the model shown in the path diagram below:

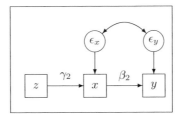

This model also involves the regression (treated as a linear projection)

$$x_i \;=\; \gamma_1 + \gamma_2 z_i + \epsilon_{xi} \qquad (2)$$

A double-headed arrow connects ϵ_{xi} and ϵ_{yi} because x_i is endogenous, whereas z_i and ϵ_{yi} are not connected because of the second property above of an instrumental variable. The reduced-form model for y_i is

$$y_i = \beta_1 + \beta_2 \underbrace{(\gamma_1 + \gamma_2 z_i + \epsilon_{xi})}_{x_i} + \epsilon_{yi} = (\beta_1 + \beta_2\gamma_1) + (\beta_2\gamma_2)z_i + \underbrace{\beta_2\epsilon_{xi} + \epsilon_{yi}}_{\epsilon_i^*} \quad (3)$$

The following OLS estimators are consistent because ϵ_{xi} and ϵ_i^* are uncorrelated with z_i:

$$(2)\ x \text{ on } z:\ \ \widehat{\gamma}_2^{\text{OLS}} \;=\; \frac{\text{Cov}(x,z)}{\text{Var}(z)} \qquad (3)\ y \text{ on } z:\ \ \widehat{\beta_2\gamma_2}^{\text{OLS}} \;=\; \frac{\text{Cov}(y,z)}{\text{Var}(z)}$$

From these estimators, we see that a consistent (but not unbiased) estimator for β_2 is

$$\widehat{\beta}_2^{\text{IV}} \;=\; \frac{\widehat{\beta_2\gamma_2}^{\text{OLS}}}{\widehat{\gamma}_2^{\text{OLS}}} = \frac{\text{Cov}(y,z)}{\text{Cov}(x,z)}$$

In practice, this estimator can be obtained using two-stage least squares (2SLS), where x_i is first regressed on z_i and then the prediction $\widehat{x}_i$ from this regression is used as a covariate in the model for y_i instead of the original x_i. This method can also be generalized to linear models with several covariates, possibly with multiple instruments; see Wooldridge (2010, sec. 5.1).

A problem with instrumental variables is that they are often weak in the sense that $\text{Cor}(z_i, x_i)$ is small, and in this case the instrumental-variable estimator is inefficient.

Display 5.1: Instrumental-variables estimation

We could stop here because we have consistent estimates of all regression parameters, but Hausman and Taylor proceed by using further instrumental-variable methods to obtain a more efficient estimator and perform valid statistical inference using that estimator. Specifically, a new set of residuals is formed by plugging in the estimated regression parameters from the previous stages, and simple moment estimators are used to estimate θ and ψ based on these residuals. Using these estimated variance parameters a so-called generalized least squares (GLS) transform is performed to make the responses uncorrelated across occasions, whereby standard instrumental-variables estimation becomes feasible. The Hausman–Taylor estimator is finally obtained by using three sets of instruments: the deviations from the cluster means of the time-varying covariates, the cluster means of exogenous time-varying covariates, and the exogenous time-constant covariates.

If the random-intercept model is correctly specified (including the designation of exogenous and endogenous covariates), the Hausman–Taylor method will produce consistent and asymptotically efficient estimators for all model parameters, including the regression coefficients of time-constant covariates and the random-intercept variance.

Subsequent work has improved on the finite-sample efficiency of the Hausman–Taylor estimator by including more instruments. Stata provides one of these approaches, due to Amemiya and MaCurdy (1986), which is invoked by the `amacurdy` option of the `xthtaylor` command.

As pointed out in the previous section, the most obvious candidate for an endogenous time-varying covariate is `union`, and we will stick to this here. Regarding endogenous time-constant covariates, we have chosen `educt` because the random intercept is likely to be correlated with education. For instance, if the random intercept is interpreted as unmeasured ability, then high ability subjects might have higher earnings than expected given their observed covariates and higher education, leading to a positive correlation between education and the random intercept. Conversely, it could be that high ability subjects are less educated (perhaps because they know that they do not need high education to succeed), producing a negative correlation. To illustrate the Hausman–Taylor approach, the covariates `married` and `exper` are treated as exogenous time-varying covariates, whereas `black` and `hisp` are treated as exogenous time-constant covariates.

The resulting random-intercept model with `union` x_{4ij}^{end} and `educt` E_j^{end} as endogenous covariates can be written as (we omit `yeart` because it is collinear with `exper` after mean-centering in the first step of the Hausman–Taylor estimator)

$$y_{ij} = (\beta_1 + \zeta_j) + \beta_2 x_{2j} + \beta_3 x_{3j} + \beta_4 x_{4ij}^{\text{end}} + \beta_5 x_{5ij} + \beta_6 L_{ij} + \beta_8 E_j^{\text{end}} + \epsilon_{ij}$$

where ϵ_{ij} is a level-1 residual with mean zero and variance θ.

We use the `xthtaylor` command to fit the model, specifying the endogenous covariates in the `endog()` option (Stata keeps track of whether covariates are time varying or time constant):

```
. quietly xtset nr

. xthtaylor lwage black hisp union married exper educt, endog(union educt)

Hausman-Taylor estimation                    Number of obs      =       4360
Group variable: nr                           Number of groups   =        545

                                             Obs per group: min =          8
                                                            avg =          8
                                                            max =          8

Random effects u_i ~ i.i.d.                  Wald chi2(6)       =     801.99
                                             Prob > chi2        =     0.0000
```

lwage	Coef.	Std. Err.	z	P>|z|	[95% Conf. Interval]	
TVexogenous						
married	.0715925	.0168862	4.24	0.000	.0384962	.1046888
exper	.0593827	.0025624	23.17	0.000	.0543605	.0644049
TVendogenous						
union	.0844763	.0194014	4.35	0.000	.0464503	.1225023
TIexogenous						
black	-.1255779	.049431	-2.54	0.011	-.2224608	-.028695
hisp	.0513715	.0459801	1.12	0.264	-.0387478	.1414908
TIendogenous						
educt	.1424144	.0161925	8.80	0.000	.1106777	.174151
_cons	1.249935	.0237772	52.57	0.000	1.203332	1.296537
sigma_u	.33790949					
sigma_e	.35338912					
rho	.47761912	(fraction of variance due to u_i)				

```
Note:  TV refers to time varying; TI refers to time invariant.
```

The prefixes `TV` and `TI` in the headings in the table of coefficients stand for "time varying" and "time invariant" (same as time-constant), respectively.

Although only `union` is treated as an endogenous time-varying covariate here, the estimated coefficients of time-varying covariates are quite similar to those produced by the approach based on different within and between effects in the previous section (where all time-varying covariates were treated as endogenous but no time-constant covariates were treated as endogenous). The estimate of the covariate designated as endogenous time constant, `educt`, is considerably higher than when it was treated as exogenous. This suggests that there is a negative correlation between education and the random intercept.

The estimates produced by `xthtaylor` are highly dependent on which covariates are designated as endogenous. Hence, subject-matter considerations regarding endogeneity should be combined with sensitivity analyses to try to ensure sensible estimates. The Hausman–Taylor estimator produces consistent and asymptotically efficient estimates of the coefficients of endogenous time-varying covariates if the model specification is correct. A Hausman test can therefore be used to compare the estimates for the time-varying covariates with their consistent (but possibly inefficient) counterparts from the fixed-effects approach (see exercises 5.7 and 5.8).

5.4 Fixed-intercept model

In the fixed-intercept model, often called the fixed-effects model in econometrics, subject-specific intercepts are treated as fixed, unknown parameters α_j. The model can be written as

$$y_{ij} = \alpha_j + \beta_2 x_{2j} + \beta_3 x_{3j} + \beta_4 x_{4ij} + \beta_5 x_{5ij} + \beta_6 L_{ij} + \beta_7 P_i + \beta_8 E_j + \epsilon_{ij}$$

It is assumed that given the covariates $\mathbf{X}_j$ of a subject, the level-1 residual ϵ_{ij} has mean zero, variance θ, and is uncorrelated across occasions and subjects. Because the subject-specific intercepts are treated as fixed, this model relaxes all assumptions made in the random-intercept model regarding the subject-specific intercepts.

As previously discussed in section 3.7.2, we can fit this model by OLS by including dummy variables for each subject in the model and omitting the overall constant (using the `noconstant` option of `regress`). This produces consistent estimates of the coefficients β_4, β_5, and β_6 of the time-varying covariates `union`, `married`, and `exper`, as long as relevant time-varying covariates (confounders) are not omitted from the model. These estimated within effects are numerically identical to the corresponding ML estimates for the random-intercept model (5.1) with different within and between effects.

However, the coefficients β_2, β_3, and β_8 for the time-constant covariates `black`, `hisp`, and `educt` cannot be estimated because all between-subject variability is explained by the fixed intercepts. To see this, think of the subject-specific effects as coefficients of dummy variables for subjects. The problem is that the time-constant covariates are perfectly collinear with these dummy variables. It also turns out that the coefficient β_7 of the time-varying covariate `yeart` cannot be estimated because `yeart` differs from `exper` (L_{ij}) only by a subject-specific constant t_j: $\text{yeart}_i = t_j + \text{exper}_{ij}$, where t_j is the time (in years since 1980) at which the subject entered the labor market. The linear combination `yeart`$-$`exper` is therefore collinear with dummy variables for subjects.

The α_j are not consistently estimated if the number of occasions remains fixed and the number of subjects increases. This is an incidental parameter problem that is due to the number of parameters α_j increasing as the number of subjects increases. Usually, the subject-specific effects are not of interest, and they can be eliminated by subject-mean centering the responses and covariates (which is what the `xtreg` command with the `fe` option does). From this perspective, α_j can also be viewed as random intercepts.

Because these intercepts are eliminated, it is not necessary to assume that they are uncorrelated with covariates or that they have a constant variance or a normal distribution. The random intercepts could also be correlated across subjects, for instance, because of clustering in states.

To see this, we first obtain the subject means of the model

$$y_{ij} = \alpha_j + \beta_2 x_{2j} + \beta_3 x_{3j} + \beta_4 x_{4ij} + \beta_5 x_{5ij} + \beta_6 L_{ij} + \beta_7 P_i + \beta_8 E_j + \epsilon_{ij}$$

which are

$$\overline{y}_{\cdot j} = \alpha_j + \beta_2 x_{2j} + \beta_3 x_{3j} + \beta_4 \overline{x}_{4\cdot j} + \beta_5 \overline{x}_{5\cdot j} + \beta_6 \overline{L}_{\cdot j} + \beta_7 \overline{P}_{\cdot} + \beta_8 E_j + \overline{\epsilon}_{\cdot j}$$

Subtracting these subject means from the model, we get

$$y_{ij} - \overline{y}_{\cdot j} = \beta_4 (x_{4ij} - \overline{x}_{4\cdot j}) + \beta_5 (x_{5ij} - \overline{x}_{5\cdot j}) + \beta_6 (L_{ij} - \overline{L}_{\cdot j}) + \underline{\beta_7 (P_i - \overline{P}_{\cdot})} + \epsilon_{ij} - \overline{\epsilon}_{\cdot j}$$

The subject-specific intercepts α_j are eliminated from this model as desired, but the coefficients for the time-constant covariates cannot be estimated because the terms involving these are also eliminated. As pointed out above, yeart has to be omitted because the mean-centered yeart $(P_i - \overline{P}_{\cdot})$ is identical to the mean-centered exper $(L_{ij} - \overline{L}_{\cdot j})$. The estimated coefficient of exper will then be an estimate of the sum $\beta_6 + \beta_7$.

Alternatively, we could eliminate the intercepts α_j by taking first-differences:

$$y_{ij} - y_{i-1,j} = (\beta_6 + \beta_7) + \beta_4 (x_{4ij} - x_{4,i-1,j}) + \beta_5 (x_{5ij} - x_{5,i-1,j}) + \epsilon_{ij} - \epsilon_{i-1,j} \quad (5.2)$$

The terms involving time-constant variables are again eliminated. Because $L_{ij} - L_{i-1,j}$ = 1 and $P_i - P_{i-1} = 1$, $\beta_6 + \beta_7$ becomes an intercept. The differencing approach is preferable if the residuals ϵ_{ij} in the fixed-effects model are correlated over time (see exercise 5.4, question 2.b.iii).

As discussed in section 3.7.2, the estimates based on the fixed-intercept approach represent the within-subject effects of the covariates. A great advantage of these estimates is that they are not susceptible to bias due to omitted subject-level covariates (level-2 endogeneity). Each subject truly serves as its own control when using this approach.

Keep in mind, however, that the fixed-effects approach is no panacea. It requires sufficient within-subject variability of the response and covariates to obtain reliable estimates (this is one of the reasons for investigating the within and between variability of variables using xtsum). In practice, the consistency of the within estimator is often purchased at the cost of large mean squared errors and low power. Griliches and Hausman (1986) also point out that measurement error bias is likely to be exacerbated using the conventional fixed-effects approach as compared with the random-effects approach. Finally, as is the case for the random-effects approach, the problem of level-1 endogeneity—where covariates are correlated with the level-1 residual ϵ_{ij}—is not addressed.

5.4.1 Using xtreg or regress with a differencing operator

The fixed-intercept model can be fit by OLS using `xtreg` with the `fe` option (where we have not included covariates whose effects cannot be estimated):

```
. quietly xtset nr
. xtreg lwage union married exper, fe
Fixed-effects (within) regression              Number of obs      =      4360
Group variable: nr                             Number of groups   =       545

R-sq:  within  = 0.1672                         Obs per group: min =         8
       between = 0.0001                                        avg =       8.0
       overall = 0.0513                                        max =         8

                                               F(3,3812)          =    255.03
corr(u_i, Xb)  = -0.1575                        Prob > F           =    0.0000

-------------------------------------------------------------------------------
       lwage |      Coef.   Std. Err.      t    P>|t|     [95% Conf. Interval]
-------------+-----------------------------------------------------------------
       union |   .083791    .019414     4.32   0.000     .045728    .1218539
     married |  .0610384   .0182929     3.34   0.001    .0251736    .0969032
       exper |  .0598672   .0025835    23.17   0.000     .054802    .0649325
       _cons |  1.211888   .0169244    71.61   0.000    1.178706     1.24507
-------------+-----------------------------------------------------------------
     sigma_u |  .40514496
     sigma_e |  .35352815
         rho |  .56772216   (fraction of variance due to u_i)
-------------------------------------------------------------------------------
F test that all u_i=0:     F(544, 3812) =    10.08             Prob > F = 0.0000
```

We store the estimates using

```
. estimates store fi
```

and report them under "Fixed intercept" in table 5.1. (`sigma_u` in the output is the sample standard deviation of the estimated intercepts $\widehat{\alpha}_j$.)

Table 5.1: Estimates for subject-specific models for wage-panel data

| | Subject-specific intercepts | | | | | | | | Subject-specific slopes | | | |
| | Random intercept | | Random int. & w/b† | | Hausman–Taylor | | Fixed intercept | | Random coefficient | | Fixed coefficient | |
	Est	(SE)	Est	(SE)	Est	(SE)	Est	(SE)	Est	(SE)	Est	(SE)
Fixed part												
β_1 [_cons]	1.32	(0.04)	1.37	(0.02)	1.25	(0.02)	1.21	(0.02)	1.31	(0.04)		
β_2 [black]	−0.13	(0.05)	−0.14	(0.05)	−0.13	(0.05)			−0.14	(0.05)		
β_3 [hisp]	0.02	(0.04)	0.01	(0.04)	0.05	(0.05)			0.01	(0.04)		
β_4 [union]	0.11	(0.02)	0.08	(0.02)	0.08	(0.02)	0.08	(0.02)	0.11	(0.02)		
β_5 [married]	0.08	(0.02)	0.06	(0.02)	0.07	(0.02)	0.06	(0.02)	0.08	(0.02)		
β_6 [exper]	0.03	(0.01)							0.04	(0.01)	0.04	(0.02)
β_7 [yeart]	0.03	(0.01)							0.02	(0.01)	0.04	(0.03)
$\beta_6 + \beta_7$			0.06	(0.00)	0.06	(0.00)	0.06	(0.00)				
β_8 [educt]	0.09	(0.01)	0.09	(0.01)	0.14	(0.02)			0.10	(0.01)	0.07	(0.01)
Random part												
$\sqrt{\psi_{11}}$	0.33		0.32		0.34				0.45			
$\sqrt{\psi_{22}}$									0.05			
ρ_{21}									−0.68			
$\sqrt{\theta}$ or res. SD	0.35		0.35		0.35		0.35		0.33		0.47‡	

† Separate within and between effects; coefficients of cluster means not shown
‡ Estimated standard deviation (SD) of first-differenced residual

We see that union membership, being married, and having more experience are all beneficial for wages, according to the fitted model. For instance, each extra year of experience is associated with an estimated increase in the expectation of log hourly wage of 0.06, controlling for the other covariates. In other words, mean hourly wages increase 6% [= 100%{exp(1.06) − 1}] for each extra year of experience. However, this actually represents the combined effect of experience and period, which cannot be disentangled here. Interestingly, the estimated coefficients of **exper** and **yeart** for the random-intercept model approximately add up to the present coefficient of **exper**. In contrast to the random-intercept model, cohort effects have been controlled for because they are subject specific. For a given subject, becoming a member of a union increases his or her mean hourly wages by about 8%, controlling for the other covariates. Importantly, this is a within effect and differs from the between effect that compares different subjects who are either union members or not. If a subject becomes married, this increases his or her mean hourly wages by about 6% according to the fitted model, controlling for the other covariates.

The above estimates are obtained by subject-mean centering. For estimation using first-differencing, it is very convenient to use Stata's time-series operators, where the required differences are simply obtained by using the prefix "D." (for first-differencing) for the variables included in the estimation command.

We can fit the first-differenced model by OLS by typing

```
regress D.lwage D.union D.married
```

or using the more compact syntax

```
regress D.(lwage union married)
```

(output not shown).

The Hausman test described in section 3.7.6 can be used to investigate endogeneity of the time-varying covariates by comparing the estimates from the fixed-intercept model (stored as **fi**) with those from the random-intercept model. We cannot use the estimates for the random-intercept model obtained in section 5.2 because that model included **yeart** as a covariate and was fit by ML and not by feasible generalized least squares (FGLS), as required by the **hausman** command. We therefore first fit the appropriate random-intercept model by FGLS using **xtreg** with the **re** option:

```
. quietly xtset nr
. quietly xtreg lwage black hisp union married exper educt, re
```

We then store the estimates using

```
. estimates store ri2
```

and perform a Hausman test:

```
. hausman fi ri2
                    ──── Coefficients ────
                     (b)          (B)          (b-B)      sqrt(diag(V_b-V_B))
                     fi           ri2          Difference       S.E.
      ─────────────────────────────────────────────────────────────────────
         union    .083791      .1100027       -.0262118       .0074711
       married    .0610384     .0757698       -.0147314       .0073449
         exper    .0598672     .0579462        .001921        .0006418
      ─────────────────────────────────────────────────────────────────────
                         b = consistent under Ho and Ha; obtained from xtreg
             B = inconsistent under Ha, efficient under Ho; obtained from xtreg
      Test:  Ho:  difference in coefficients not systematic
                       chi2(3) = (b-B)'[(V_b-V_B)^(-1)](b-B)
                               =        22.69
                     Prob>chi2 =        0.0000
```

As expected, the Hausman statistic is similar to the test statistic for the joint hypothesis of equal within and between effects, shown in section 5.3.1. An advantage of the latter test (based on robust standard errors) is that it can be used even if there are heteroskedastic or autocorrelated level-1 errors.

5.4.2 ❖ Using anova

Using the language of experimental design, subjects can be viewed as blocks or plots to which different treatments are applied. In a split-plot design, some treatments are applied to the entire plots (subjects)—these are the whole-plot, between-subject, or level-2 variables (black, hisp, and educt). The plots are split into subplots (subjects at different occasions) to which other treatments can be applied—these are the split-plot, within-subject, or level-1 variables (union, married, exper, and yeart).

One-way analysis of variance (ANOVA) was briefly discussed in section 1.4, where we showed how the total sum of squares is partitioned into the sum of squares attributed to a factor (a categorical covariate) and the sum of squared errors. An F statistic is then constructed by dividing the mean squares due to the factor by the mean squared error.

ANOVA with continuous covariates is often called ANCOVA (analysis of covariance). In a split-plot design, the important thing to remember is that a different mean squared error is used in the denominator of the F statistic for testing the whole-plot (between-subject) variables than for the split-plot (within-subject) variables. For the between-subject variables, the denominator is given by the (unique or partial) mean squares due to subjects. For the within-subject variables, the denominator is given by the mean squared error after allowing for fixed effects of subjects. Subjects are often viewed as random, and an estimator of the between-subject variance can be derived from the mean squares, but the estimators of the effects of within-subject variables are fixed-effects estimators.

In Stata's `anova` command, a categorical explanatory variable is entered directly instead of the corresponding dummy variables. We must therefore first construct the categorical variable `ethnic` (with values 0: white; 1: black; and 2: Hispanic) from the dummy variables `black` and `hisp`:

```
. generate ethnic = black*1 + hisp*2
```

For simplicity, we initially fit the model with only one time-constant variable, `ethnic`, omitting `educt` (after first increasing `matsize`):

```
. set matsize 800
. anova lwage ethnic / nr|ethnic union married c.exper, dropemptycells
```

| | | Number of obs = | 4360 | R-squared | =   0.6147 |
| | | Root MSE | = .353528 | Adj R-squared | =   0.5594 |

Source	Partial SS	df	MS	F	Prob > F
Model	760.097675	547	1.38957527	11.12	0.0000
ethnic	15.5566098	2	7.77830491	6.46	0.0017
nr|ethnic	652.885925	542	1.20458658		
union	2.32814565	1	2.32814565	18.63	0.0000
married	1.39152302	1	1.39152302	11.13	0.0009
exper	67.1116346	1	67.1116346	536.97	0.0000
Residual	476.431967	3812	.124982153		
Total	1236.52964	4359	.283672779		

(Without the `dropemptycells` option, the command would require a very large matrix size.)

The model includes the main effects `ethnic`, `nr|ethnic`, `union`, `married`, and `exper`, where `nr|ethnic` denotes that subjects are nested within ethnicities in the sense that each subject can have only one ethnicity. The term `ethnic` represents the main effect of ethnicity, whereas the term `nr|ethnic` represents the main effect of subject, where subject is nested in ethnicities. The prefix `c.` in `c.exper` denotes that `exper` should be treated as continuous by assuming the conditional expectation of the log wages to be linearly related to `exper`. The purpose of the forward slash, `/`, is to declare that the denominator for the F statistic(s) for the preceding term(s) is the mean square for subjects (nested in ethnicity) and not the mean squared error for the entire model, after subtracting the sums of squares due to all terms, including subjects nested within ethnicities. Using the latter "within-subject" mean squared error in the F test for `ethnic` would produce a p-value that is too small and ignores the longitudinal nature of the data. The within-subject mean squared error is used for the denominator in the F statistic for the time-varying variables `union`, `married`, and `exper`.

The F-test statistic for `ethnic` is $F(2, 542) = 6.46$, with $p = 0.002$. This can be interpreted as a test for the between-subject effect of ethnicity after removing the within-subject effects of the time-varying variables. The F tests for the within-subject variables are equivalent to the t-tests obtained using `xtreg` with the `fe` option (see page 259). Because each F test has one numerator degree of freedom, taking the square roots of the F statistics gives the corresponding t statistics:

```
. display sqrt(18.63)
4.3162484
. display sqrt(11.13)
3.3361655
. display sqrt(536.97)
23.172613
```

Including further time-constant variables is relatively straightforward, but remember that subject is now nested in the combinations of the values of these variables. For ethnicity and education, this means that every ethnicity may be combined with every possible number of years of education. The syntax for nesting within such a cross-classification is `nr|ethnic#educt`, and the syntax for the full model is

```
anova lwage ethnic c.educt / nr|ethnic#educt union married c.exper, dropemptycells
```

The ANOVA model assumes that the responses are uncorrelated given the explanatory variables (which include the factor subject). This assumption, together with constant variance, implies *compound symmetry* of the variance–covariance matrix of the responses given the explanatory variables (but not given the factor subject) when subject is treated as random. Compound symmetry means that the responses have the same conditional variances across occasions and the same conditional covariances between all pairs of occasions (not necessarily positive). This covariance structure is the same as for a random-intercept model except that the covariances can also be negative.

A less strict assumption that all pairwise differences between responses have the same variance, called *sphericity*, is sufficient for the F test for within-subject variables to be valid. When this assumption is violated, the `repeated()` option can be used in the `anova` command to correct the p-values for within-subject variables. Unfortunately, this option works only for categorical within-subject variables (but `exper` is continuous). By default, it also requires that there be only one observation per cell in the cross-tabulation of within-subject variables for each subject. So, omitting `exper` because it is continuous would not help because a subject can have a given combination of the values of `union` and `married`—for example, 0, 0—for several waves of data.

The version of repeated-measures ANOVA discussed here is sometimes referred to as *univariate* or as applicable to a split-plot design. There is also a multivariate version, called multivariate analysis of variance (MANOVA) and implemented in Stata's `manova` command, that specifies an unstructured covariance matrix for the repeated measures (see section 6.3.1). A great disadvantage of that approach is that it uses listwise deletion, dropping entire subjects if one or more of their responses are missing. Furthermore, the MANOVA approach requires that the within-subject variables take on identical values for all subjects; for instance, the variable `yeart` can be used but `union` cannot.

5.5 Random-coefficient model

We could in principle include random coefficients for any of the time-varying variables in the wage-panel data to allow the effect of these variables to vary between subjects (but remember the warnings in section 4.10). The data provide no information on subject-specific effects of time-constant variables x_j, and it therefore does not make sense to include random coefficients for these variables unless we want to model heteroskedasticity (see section 7.5.2)

It may well be that different subjects' wages increase at different rates with each extra year of experience. We can investigate this by including a random coefficient ζ_{2j} of labor-market experience L_{ij} in the model

$$y_{ij} = \beta_1 + \beta_2 x_{2j} + \beta_3 x_{3j} + \beta_4 x_{4ij} + \beta_5 x_{5ij} + \beta_6 L_{ij} + \beta_7 P_i + \beta_8 E_j + \zeta_{1j} + \zeta_{2j} L_{ij} + \epsilon_{ij}$$
$$= (\beta_1 + \zeta_{1j}) + \beta_2 x_{2j} + \beta_3 x_{3j} + \beta_4 x_{4ij} + \beta_5 x_{5ij} + (\beta_6 + \zeta_{2j}) L_{ij} + \beta_7 P_i + \beta_8 E_j + \epsilon_{ij}$$

Given the covariates $\mathbf{X}_j$ for a subject, the random intercept ζ_{1j} has mean zero and variance ψ_{11}, the random slope ζ_{2j} has mean zero and variance ψ_{22}, and the covariance between ζ_{1j} and ζ_{2j} is ψ_{21}. The random effects ζ_{1j} and ζ_{2j} are uncorrelated across subjects. Given the covariates and random effects, the level-1 residuals ϵ_{ij} have zero means, variance θ, and are mutually uncorrelated across both occasions and subjects.

We fit the random-coefficient model by ML using `xtmixed` with the `mle` option:

```
. xtmixed lwage black hisp union married exper yeart educt || nr: exper,
> covariance(unstructured) mle
```

```
Mixed-effects ML regression                    Number of obs      =      4360
Group variable: nr                             Number of groups   =       545

                                               Obs per group: min =         8
                                                              avg =       8.0
                                                              max =         8

                                               Wald chi2(7)       =    573.88
Log likelihood = -2130.4677                    Prob > chi2        =    0.0000
```

lwage	Coef.	Std. Err.	z	P>|z|	[95% Conf. Interval]	
black	-.139996	.0489058	-2.86	0.004	-.2358496	-.0441423
hisp	.009267	.0437623	0.21	0.832	-.0765055	.0950396
union	.1098184	.017896	6.14	0.000	.0747429	.144894
married	.0757788	.0173732	4.36	0.000	.041728	.1098296
exper	.0418495	.0119737	3.50	0.000	.0183815	.0653175
yeart	.0171964	.0118898	1.45	0.148	-.0061072	.0405001
educt	.097203	.0109324	8.89	0.000	.0757758	.1186302
_cons	1.307388	.0404852	32.29	0.000	1.228039	1.386738

Random-effects Parameters	Estimate	Std. Err.	[95% Conf.	Interval]
nr: Unstructured				
sd(exper)	.0539497	.0030854	.048229	.0603489
sd(_cons)	.4514402	.0215257	.411162	.4956641
corr(exper,_cons)	-.6801072	.0348441	-.7426584	-.6057911
sd(Residual)	.3266336	.0040591	.318774	.3346871

LR test vs. linear regression: chi2(3) = 1715.54   Prob > chi2 = 0.0000

Note: LR test is conservative and provided only for reference.

Then we store the estimates:

```
. estimates store rc
```

The estimates were also shown under "Random coefficient" in table 5.1. We note that the coefficient for exper varies quite considerably, with 95% of the effects being in the range from about -0.06 to 0.15 ($0.0418495 \pm 1.96 \times 0.0539497$). Negative effects of exper are therefore not uncommon.

We can perform a likelihood-ratio test comparing the random-intercept and random-coefficient models:

```
. lrtest ri rc
Likelihood-ratio test                           LR chi2(2)  =    167.78
(Assumption: ri nested in rc)                   Prob > chi2 =    0.0000
```

Note: The reported degrees of freedom assumes the null hypothesis is not on the boundary of the parameter space. If this is not true, then the reported test is conservative.

The correct p-value from this test should be obtained by using a 50:50 mixture of a $\chi^2(1)$ and a $\chi^2(2)$ as null distribution (see section 4.6), which makes no difference to the conclusion that the random-intercept model is rejected in favor of the random-coefficient model.

We could alternatively have included a random slope for yeart. However, if different subjects grow at different rates after 1980 (when yeart=0), it makes sense to assume that they have grown at different rates ever since they entered the labor market, which could be before 1980. The variance in wages in the year 1980 must then be larger for those who entered the labor market long before that time (and have grown at different rates for a long time) than for those who entered recently. However, a model with a random coefficient for yeart would assume a constant variance when yeart is zero.

5.6 Fixed-coefficient model

The fixed-effects version of the model with subject-specific intercepts and slopes is

$$
\begin{aligned}
y_{ij} &= \beta_1 + \beta_2 x_{2j} + \beta_3 x_{3j} + \beta_4 x_{4ij} + \beta_5 x_{5ij} + \beta_6 L_{ij} + \beta_7 P_i \\
&\quad + \beta_8 E_j + \alpha_{1j} + \alpha_{2j} L_{ij} + \epsilon_{ij} \\
&= (\beta_1 + \alpha_{1j}) + \beta_2 x_{2j} + \beta_3 x_{3j} + \beta_4 x_{4ij} + \beta_5 x_{5ij} + (\beta_6 + \alpha_{2j}) L_{ij} \\
&\quad + \beta_7 P_i + \beta_8 E_j + \epsilon_{ij}
\end{aligned}
$$

where α_{1j} and α_{2j} are fixed subject-specific intercept and slope parameters, respectively. Conditional on the covariates $\mathbf{X}_j$, ϵ_{ij} has mean zero, variance θ, and is uncorrelated across occasions and subjects. The estimated coefficients represent within-subject effects of the covariates and are not susceptible to bias due to omitted subject-level covariates.

If the covariate that has a subject-specific slope increases by constant amounts across occasions (as does L_{ij}) and if there are no missing data, first-differences can be used to turn the slope into an intercept. Specifically, we obtain

$$
\begin{aligned}
y_{ij} - y_{i-1,j} &= (\beta_6 + \beta_7 + \alpha_{2j}) + \beta_4 (x_{4ij} - x_{4,i-1,j}) + \beta_5 (x_{5ij} - x_{5,i-1,j}) \\
&\quad + \epsilon_{ij} - \epsilon_{i-1,j}
\end{aligned}
$$

The original intercepts α_{1j} disappear as do the terms involving time-constant covariates. Because $L_{ij} - L_{i-1,j} = 1$, the original subject-specific slopes α_{2j} now become subject-specific intercepts. The coefficient β_6 for L_{ij} becomes part of a fixed intercept, as does the coefficient β_7 because $P_i - P_{i-1} = 1$. The resulting fixed subject-specific intercept model can then be fit using the approach described in section 5.4 to produce consistent estimates of the coefficients β_4 and β_5 for the time-varying covariates union and married. As for the first-difference approach to fitting fixed-intercept models, this approach also handles serially correlated (nonexchangeable) residuals to some extent.

To implement the first-difference approach, we once again use Stata's time-series operators, where first-differences are obtained by using the prefix "D." for the variables. We can then use xtreg, fe to fit the fixed-coefficient model (we only include the covariates union and married, whose coefficients can be estimated by this approach here):

```
. quietly xtset nr yeart

. xtreg D.(lwage union married), fe
```

| Fixed-effects (within) regression | | | | Number of obs | | = | 3815 |
| Group variable: nr | | | | Number of groups | | = | 545 |

```
R-sq:  within  = 0.0020                      Obs per group: min =        7
       between = 0.0073                                     avg =      7.0
       overall = 0.0022                                     max =        7

                                             F(2,3268)           =     3.26
corr(u_i, Xb)  = 0.0075                       Prob > F           =   0.0387
```

D.lwage	Coef.	Std. Err.	t	P>|t|	[95% Conf.	Interval]
union						
D1.	.0402949	.0212656	1.89	0.058	-.0014003	.0819901
married						
D1.	.0427839	.0250491	1.71	0.088	-.0063296	.0918974
_cons	.0648842	.0077594	8.36	0.000	.0496704	.080098
sigma_u	.08632664					
sigma_e	.46977158					
rho	.03266576	(fraction of variance due to u_i)				

```
F test that all u_i=0:     F(544, 3268) =      0.24           Prob > F = 1.0000
```

The estimated coefficients of **union** and **married** (reported under "Fixed coefficient" in table 5.1) are considerably smaller than the corresponding estimates in the random-coefficient model. This is likely to be because of subject-level confounding in the latter model. The estimated intercept is quite close to the sum of $\widehat{\beta}_6$ and $\widehat{\beta}_7$ (the estimated coefficients for L_{ij} and P_i) from the random-coefficient model.

Alternatively, both the fixed intercept α_{1j} and the fixed slope α_{2j} can be eliminated by *double*-differencing, for instance, using $(y_{ij}-y_{i-1,j}) - (y_{i-1,j}-y_{i-2,j})$, and analogous double-differencing of the covariates. Estimation can then proceed by simply using OLS for the resulting regression model. Again, serially correlated residuals are handled to some extent by this approach.

Using the **regress** command, this approach can be implemented by using the prefix "D2." (for double-differencing) for the variables included in the model:

```
regress D2.(lwage union married)
```

Wooldridge (2010, sec. 11.7.2) discusses estimation of models with fixed subject-specific slopes for several covariates, also accommodating covariates that do not necessarily change by the same amount from occasion to occasion for every subject.

5.7   Lagged-response or dynamic models

5.7.1   Conventional lagged-response model

We now consider *lagged-response* models where responses at previous occasions are treated as covariates (usually in addition to other covariates). Lagged-response models are sometimes referred to as autoregressive-response models, Markov models, or conditional autoregressive models in statistics. In econometrics, the term *dynamic models* is invariably used. Note that when there are more than two occasions, lagged-response models are different from models that include the response at the first occasion (baseline) as a covariate for subsequent responses, as is often done, for instance, in clinical trials.

The most prevalent lagged-response model is the autoregressive lag-1 (AR(1)) model, where the current response y_{ij} is regressed on the previous response $y_{i-1,j}$:

$$
\begin{aligned}
y_{ij} \;=\; & \beta_1 + \gamma y_{i-1,j} + \beta_2 x_{2j} + \beta_3 x_{3j} + \beta_4 x_{4ij} \\
& + \beta_5 x_{5ij} + \beta_6 L_{ij} + \beta_7 P_i + \beta_8 E_j + \epsilon_{ij}
\end{aligned}
\tag{5.3}
$$

Here γ is the coefficient associated with the lagged response. It is assumed that given $\mathbf{z}_{ij}$, ϵ_{ij} has zero mean and variance σ^2, where $\mathbf{z}_{ij} = (\mathbf{x}'_{ij}, y_{i-1,j})'$ contains all covariates for subject j at occasion i. There is no conditional cross-sectional or longitudinal correlation between the residuals given the covariates; $\mathrm{Cov}(\epsilon_{ij}, \epsilon_{i'j'} | \mathbf{z}_{ij}, \mathbf{z}_{i'j'}) = 0$. A stationary model results if the process has been ongoing long before the first occasion in the dataset and $|\gamma| < 1$. A graphical representation of an AR(1) lagged-response model is shown in figure 5.1.

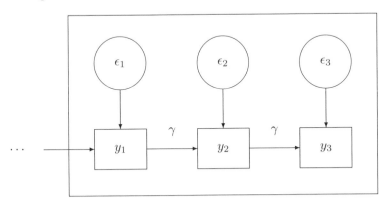

Figure 5.1: Path diagram of AR(1) lagged-response model

Lagged-response models should be used only if it really makes sense to control the effects of the other covariates for the previous response or if the effect of the lagged response is itself of scientific interest. For instance, in the context of the wage-panel data, it may make some sense to investigate the effect of previous wage on the current wage, because having experienced a high salary in the past could place subjects in a good bargaining position.

The previous response or "lag-1 response" can be obtained in Stata using

```
. by nr (yeart), sort: generate lag1 = lwage[_n-1]
(545 missing values generated)
```

On the right-hand side of the command, we see that the lag is produced by referring to the previous observation of lwage using the row counter _n minus 1. The by prefix command is used with the subject identifier nr to cause the counter _n to be reset to 1 every time nr increases. Otherwise, the lag-1 response for the second subject in the data would be the last response of the first subject. For this command to work, we must also sort by nr as specified by the sort option. yeart is also given in parentheses to sort the data by yeart within nr before defining the counter so that the counter increases in tandem with yeart. (If yeart were not in parentheses, the counter would reset to 1 each time yeart increases.)

To see the result of this command, we list the first nine observations (all panel waves or occasions for subject 13 and the first wave for subject 17):

```
. sort nr yeart
. list nr yeart lwage lag1 in 1/9, clean noobs
    nr    yeart        lwage        lag1
    13        0      1.19754           .
    13        1      1.85306     1.19754
    13        2     1.344462     1.85306
    13        3     1.433213    1.344462
    13        4     1.568125    1.433213
    13        5     1.699891    1.568125
    13        6    -.7202626    1.699891
    13        7     1.669188   -.7202626
    17        0     1.675962           .
```

Obviously, there is no lag-1 response for the first wave of data when yeart is 0, and lag1 is hence missing at the first occasion.

We can now fit the lagged-response model by OLS using the `regress` command:

```
. regress lwage lag1 black hisp union married exper yeart educt

     Source |       SS       df       MS              Number of obs =    3815
------------+------------------------------           F(  8,  3806) =  379.23
      Model |  455.410767      8  56.9263459           Prob > F      =  0.0000
   Residual |  571.325295   3806  .150111743           R-squared     =  0.4436
------------+------------------------------           Adj R-squared =  0.4424
      Total |  1026.73606   3814  .269201904           Root MSE      =  .38744

------------------------------------------------------------------------------
      lwage |      Coef.   Std. Err.      t    P>|t|     [95% Conf. Interval]
------------+-----------------------------------------------------------------
       lag1 |   .5572532   .0128673    43.31   0.000     .5320257    .5824807
      black |  -.0721592   .0204291    -3.53   0.000    -.1122123   -.0321062
       hisp |   .0092496   .0179206     0.52   0.606    -.0258854    .0443846
      union |   .0755261   .0149509     5.05   0.000     .0462135    .1048388
    married |   .0482021   .0132629     3.63   0.000      .022199    .0742051
      exper |   .0028574   .0047477     0.60   0.547    -.0064509    .0121657
      yeart |   .0175801   .0056328     3.12   0.002     .0065365    .0286237
      educt |   .0389675   .0046551     8.37   0.000     .0298409    .0480942
      _cons |   .6683362   .0253468    26.37   0.000     .6186417    .7180308
------------------------------------------------------------------------------
```

A more elegant (but perhaps less transparent) way of including the lagged response as a covariate, without explicitly constructing it as above, is to use Stata's very powerful time-series operators within the estimation command.

If the data have been `xtset`, we can refer to the lag-1 of a variable simply by using the prefix `L.` before the relevant variable name (and `L2.` for lag-2 etc.), so the results presented above could alternatively have been produced by the commands

```
xtset nr
regress lwage L.lwage black hisp union married exper yeart educt
```

The estimates for the lagged-response model are placed under "OLS" in table 5.2. We see that $\hat{\gamma} = 0.56$ and that all the other estimated regression coefficients are now closer to zero than for the models considered in table 5.1. Such a change in estimated effects is typical in lagged-response models because the estimates now have a different interpretation, as the estimated effects of the covariates on the response *after controlling for the previous response*. Reexpressing the model as

$$y_{i,j} - \gamma y_{i-1,j} \;=\; \beta_1 + \beta_2 x_{2j} + \beta_3 x_{3j} + \beta_4 x_{4ij} + \beta_5 x_{5ij} + \beta_6 L_{ij} + \beta_7 P_i + \beta_8 E_j + \epsilon_{ij}$$

it is clear that this is similar to a change-score approach because the left-hand side becomes $y_{i,j} - y_{i-1,j}$ if $\gamma = 1$.

Table 5.2: Estimates for AR(1) lagged-response models for wage-panel data

| | Exogenous lag | | Endogenous lag | | | |
| | OLS | | Anderson–Hsiao IV | | Arellano–Bond GMM | |
	Est	(SE)	Est	(SE)	Est	(SE)
Fixed part						
β_1 [_cons]	0.67	(0.03)				
β_2 [black]	-0.07	(0.02)				
β_3 [hisp]	0.01	(0.02)				
β_4 [union]	0.08	(0.01)	0.02	(0.02)	0.02	(0.03)
β_5 [married]	0.05	(0.01)	0.04	(0.03)	0.04	(0.02)
β_6 [exper]	0.00	(0.00)				
β_7 [yeart]	0.02	(0.01)				
$\beta_6 + \beta_7$			0.05	(0.01)	0.04	(0.00)
β_8 [educt]	0.04	(0.00)				
γ	0.56	(0.01)	0.12	(0.04)	0.13	(0.04)
Random part						
$\sqrt{\theta}$	0.39		0.44			

Model (5.3) makes the strong assumption that all within-subject dependence is due to the lagged response. If the true model also includes a subject-specific intercept, the estimators of the regression coefficients are likely to be inconsistent, as discussed in the next section. The lagged-response model is only sensible if the occasions are approximately equally spaced in time. Otherwise, it would be strange to assume that the lagged response has the same effect on the current response regardless of the time interval between them. Also remember that the sample size is reduced when using a lagged-response approach because lags are missing for the first occasion (here 545 observations are lost). Furthermore, the problem of missing data becomes exacerbated because not just the missing response itself is discarded but also the subsequent response because its lagged response is missing.

A natural extension of the AR(1) model considered above is the general AR(k) model that includes k lagged responses as covariates. For instance, in the AR(2) model, the current response y_{ij} is regressed on both the previous response $y_{i-1,j}$ and the response preceding the previous response, $y_{i-2,j}$. Another extension of the lagged-response model is the *antedependence model*, where the coefficient associated with a lagged response is occasion specific γ_i (accommodating unequal spacing in time). The antedependence model could have several lags, each with occasion-specific coefficients.

5.7.2 ❖ **Lagged-response model with subject-specific intercepts**

We now specify a model for the wage-panel data that includes both a random intercept ζ_j and a lagged response $y_{i-1,j}$ as a covariate (in addition to the original covariates):

$$
\begin{aligned}
y_{ij} &= (\beta_1 + \zeta_j) + \gamma y_{i-1,j} + \beta_2 x_{2j} + \beta_3 x_{3j} + \beta_4 x_{4ij} \\
&\quad + \beta_5 x_{5ij} + \beta_6 L_{ij} + \beta_7 P_i + \beta_8 E_j + \epsilon_{ij}
\end{aligned} \tag{5.4}
$$

where γ is still the coefficient of the lagged response. It is assumed that given $\mathbf{z}_{ij}$ and ζ_j, the ϵ_{ij} have zero mean, variance σ^2, and are uncorrelated across subjects and occasions.

A useful feature of such models is that they can be used to distinguish between two competing explanations of within-subject dependence: *unobserved heterogeneity* (represented by the random intercepts) or *state dependence* (represented by the lagged responses). For instance, the within-subject dependence of salaries over time, over and above that explained by observed covariates, may be due to some subjects being especially gifted and thus more valued or subjects with high previous salaries having a better bargaining position for future salaries.

The model assumptions discussed in section 3.3.2 imply that covariates at other occasions than i do not affect the response y_{ij} given the random intercept ζ_j (see section 3.3.3). Because this contradicts the lagged-response model where $y_{i-1,j}$ is treated as a covariate, we can instead assume that covariates other than the lagged response are "strictly exogenous given the random intercept" as before but that the lagged response is "sequentially exogenous given the random intercept":

$$
E(\epsilon_{ij}|\zeta_j, \mathbf{X}_j, y_{i-1,j}, \ldots, y_{1j}) = 0
$$

where $\mathbf{X}_j$ now represents all observed covariates, apart from the lagged response, for all occasions i for subject j. Together with model (5.4), this implies that

$$
\begin{aligned}
E(y_{ij}|\zeta_j, \mathbf{X}_j, y_{i-1,j}, \ldots, y_{1j}) &= \beta_1 + \gamma y_{i-1,j} + \beta_2 x_{2j} + \beta_3 x_{3j} + \beta_4 x_{4ij} \\
&\quad + \beta_5 x_{5ij} + \beta_6 L_{ij} + \beta_7 P_i + \beta_8 E_j + \zeta_j \\
&= E(y_{ij}|\zeta_j, \mathbf{x}_{ij}, y_{i-1,j})
\end{aligned}
$$

Hence, after conditioning on the random intercept ζ_j, past values of the covariates, apart from the lagged response, do not affect the mean response at an occasion.

It would be tempting to fit the model using any of the commands for random-intercept models, such as `xtreg` and `xtmixed`, by simply including the lagged responses as covariates. However, a major problem with that approach is that it would produce inconsistent estimates of the regression coefficients because lagged responses such as $y_{i-1,j}$, which are included as covariates, are correlated with the random intercept ζ_j. This is because all responses are affected by the random intercept. When fitting the model using a standard random-intercept model that includes a lagged response, as demonstrated above, we are falsely assuming that the first or initial response (the 1980 response in the wage-panel data) is not affected by the random intercept. Inconsistent parameter estimates are produced by this *initial-conditions problem*.

A standard fixed-intercept approach does not address the initial-conditions problem and will also give inconsistent estimators. To see this, it is instructive to consider estimating the parameters of model (5.4) by using a first-differences approach (assuming that there are at least three occasions in the data). It follows from that model that the model for the response $y_{i-1,j}$ at occasion $i-1$ becomes

$$y_{i-1,j} = (\beta_1 + \zeta_j) + \gamma y_{i-2,j} + \beta_2 x_{2j} + \beta_3 x_{3j} + \beta_4 x_{4,i-1,j}$$
$$+ \beta_5 x_{5,i-1,j} + \beta_6 L_{i-1,j} + \beta_7 P_{i-1} + \beta_8 E_j + \epsilon_{i-1,j}$$

Taking the difference between the models for y_{ij} and $y_{i-1,j}$, we obtain

$$y_{ij} - y_{i-1,j} = \gamma(y_{i-1,j} - y_{i-2,j}) + \beta_4(x_{4ij} - x_{4,i-1,j}) + \beta_5(x_{5ij} - x_{5,i-1,j})$$
$$+ \beta_6 + \beta_7 + \epsilon_{ij} - \epsilon_{i-1,j} \tag{5.5}$$

which sweeps out the ζ_j as desired (as well as the terms corresponding to the time-constant covariates). The intercept in this model becomes the sum of the coefficients β_6 of exper and β_7 of yeart. The first term on the right-hand side is called a lagged first-difference; the first-difference of the response is $y_{ij} - y_{i-1,j}$, and taking the lag of this difference gives the lagged difference $y_{i-1,j} - y_{i-2,j}$.

Unfortunately, this attempt to address one endogeneity problem creates another endogeneity problem. Specifically, the lagged difference of the responses $y_{i-1,j} - y_{i-2,j}$, which is a covariate here, becomes correlated with the residual $\epsilon_{ij} - \epsilon_{i-1,j}$, because $y_{i-1,j}$ (which is part of the lagged difference) is obviously correlated with its own error term $\epsilon_{i-1,j}$. Hence, proceeding by fitting the model by OLS would lead to inconsistent estimates, in contrast to the case without lagged responses, (5.2).

Anderson and Hsiao (1981, 1982) pointed out that either the second lag of the response $y_{i-2,j}$ or the second lag of the *first-difference* of the responses $y_{i-2,j} - y_{i-3,j}$ can be used as an *instrumental variable* (IV) for the endogenous covariate $y_{i-1,j} - y_{i-2,j}$, because both these variables are correlated with $y_{i-1,j} - y_{i-2,j}$ but uncorrelated with the error term $\epsilon_{ij} - \epsilon_{i-1,j}$ (you may want to consult display 5.1 on page 254 for the basic ideas of instrumental-variables estimation). The instrumental-variables estimator suggested by Anderson and Hsiao is consistent for the coefficients of the time-varying covariates and the coefficient of the lagged response (and for a simple function of the level-1 residual variance, as we will see below), but it does not provide estimators for the coefficients of the time-constant covariates or the random-intercept variance. However, Hsiao (2003, sec. 4.3.3.c) demonstrates how consistent estimators can also be obtained for the level-2 residual variance ψ and the coefficients for time-constant covariates.

The estimator used by Stata's dedicated panel-data command xtivreg, fd for instrumental-variables estimation uses the second lag of the difference as an instrument, but because the performance of this estimator can be problematic, the use of the second lag of the response as instrument is recommended instead. For this purpose, we can use Stata's ivregress command, combined with Stata's powerful time-series operators for specifying the required lags of responses and differences of lagged responses (after using the xtset command to define the variables representing subjects and occasions).

In Stata's time-series operators, the D. prefix before a variable name produces a first-difference (it can also be used for a list of variables, as in D.(union married) above), the LD. prefix produces a lagged difference, and the L2. prefix produces a second lag (see table 5.3 for an overview).

Table 5.3: Prefix for different lags and lagged differences in Stata's time-series operators

#	Lag L#.	Lagged Difference L#D.
0	y_{ij} (lag-0)	$y_{ij} - y_{i-1,j}$ (first-difference)
1	$y_{i-1,j}$ (lag-1)	$y_{i-1,j} - y_{i-2,j}$ (lagged difference)
2	$y_{i-2,j}$ (lag-2)	$y_{i-2,j} - y_{i-3,j}$ (lag-2 difference)
3	$y_{i-3,j}$ (lag-3)	$y_{i-3,j} - y_{i-4,j}$ (lag-3 difference)

We can fit model (5.5) using the Anderson–Hsiao approach, with the second lag of the responses $y_{i-2,j}$ (L2.lwage) as instrumental variable for the lagged difference (LD.lwage) by including the expression (LD.lwage = L2.lwage) in the ivregress command:

```
. quietly xtset nr yeart
. ivregress 2sls D.lwage D.(union married) (LD.lwage = L2.lwage)
Instrumental variables (2SLS) regression      Number of obs =     3270
                                               Wald chi2(3)  =    14.38
                                               Prob > chi2   =   0.0024
                                               R-squared     =        .
                                               Root MSE      =    .4429
```

D.lwage	Coef.	Std. Err.	z	P>|z|	[95% Conf. Interval]	
lwage LD.	.1150229	.0382794	3.00	0.003	.0399966	.1900492
union D1.	.0246217	.0217933	1.13	0.259	-.0180925	.0673359
married D1.	.0444829	.0251453	1.77	0.077	-.004801	.0937668
_cons	.0486734	.0082742	5.88	0.000	.0324564	.0648904

```
Instrumented:  LD.lwage
Instruments:   D.union D.married L2.lwage
```

As shown at the bottom of the output, the instrumental-variables estimator has used
L2.wage as instrumental variable for the lagged difference of log wage LD.lwage. The
variables D.union and D.married are used as instruments for themselves because union
and married are assumed to be exogenous variables. Note that $2 \times 545 = 1090$ observations are lost when using this estimator.

The estimates were shown under "Anderson–Hsiao IV" in table 5.2. A striking
feature is that the estimated coefficient of the lagged response ($\widehat{\gamma} = 0.12$) is considerably
reduced compared with the corresponding estimate previously obtained for the lagged-
response model without random effects ($\widehat{\gamma} = 0.56$). This confirms that the subject-
specific intercept is positively correlated with the lagged response as expected. The
Root MSE of 0.44 reported in the output represents the estimated standard deviation
of the differenced level-1 errors $\epsilon_{ij} - \epsilon_{i-1,j}$. Because ϵ_{ij} and $\epsilon_{i-1,j}$ are assumed to
be uncorrelated, the variance of this difference is equal to the sum of the variances
of each term 2θ. It follows that the estimated variance of the level-1 error becomes
$\widehat{\theta} = 0.44^2/2 = 0.19$. The reduced magnitude of the estimated γ and θ is due to
the current model accommodating dependence among the responses in two different
ways: by incorporating a random intercept that is shared among the responses and by
conditioning on previous responses.

It has been suggested to use more lags as instruments than in the Anderson–Hsiao
approach to increase efficiency. Arellano and Bond (1991), among others, apply an
extension of instrumental-variables estimation called generalized method of moments
(GMM) for this purpose, and this approach is implemented in Stata's xtabond, twostep
command. Fitting model (5.5) by the Arellano–Bond approach is accomplished by using
the lags(1) option (for a lag-1 response) and the noconstant option (for only using
instruments for the differenced model). We also use the vce(robust) option because
the estimated standard errors from GMM otherwise tend to be underestimated:

```
. xtabond lwage union married exper, lags(1) twostep noconstant vce(robust)
Arellano-Bond dynamic panel-data estimation   Number of obs      =      3270
Group variable: nr                            Number of groups   =       545
Time variable: yeart
                                              Obs per group:   min =         6
                                                               avg =         6
                                                               max =         6
Number of instruments =     24                Wald chi2(4)       =    352.02
                                              Prob > chi2        =    0.0000
Two-step results
                                   (Std. Err. adjusted for clustering on nr)
```

lwage	Coef.	WC-Robust Std. Err.	z	P>|z|	[95% Conf. Interval]	
lwage L1.	.1304084	.0373708	3.49	0.000	.057163	.2036538
union	.015568	.0252765	0.62	0.538	-.033973	.0651089
married	.0448827	.0224378	2.00	0.045	.0009055	.0888599
exper	.0447144	.0041293	10.83	0.000	.0366212	.0528077

```
Instruments for differenced equation
        GMM-type: L(2/.).lwage
        Standard: D.union D.married D.exper
```

The estimates were shown under "Arellano–Bond GMM" in table 5.2. The coefficient for the lagged response is now estimated as $\widehat{\gamma} = 0.13$. The estimated coefficient for exper, 0.0447144, corresponds to the intercept in the Anderson–Hsiao approach, estimated as 0.048. This is because exper increases by the constant 1 between adjacent occasions. Both estimates are estimates of the sum of the effects of exper and yeart, $\beta_6 + \beta_7$. We see from the output that 24 instruments are used.[1]

The Anderson–Hsiao and the Arellano–Bond approaches both produce estimates ($\widehat{\gamma} = 0.12$ and $\widehat{\gamma} = 0.13$, respectively) that are considerably lower than the $\widehat{\gamma} = 0.56$ reported for the naïve lagged-response model without a subject-specific intercept.

Even more instruments than those used by Arellano and Bond are used in Stata's xtdpdsys command to increase efficiency. However, a potential problem, particularly when many instruments are used, is *weak instruments* that have low correlations with the instrumented lagged differences of the responses. In this case, GMM estimators may suffer from quite severe finite-sample bias.

---

1. In 1987, the six instruments $\text{lwage}_{1985,j}$ to $\text{lwage}_{1980,j}$ are used for $\text{lwage}_{1986,j} - \text{lwage}_{1985,j}$. In 1986, the five instruments $\text{lwage}_{1984,j}$ to $\text{lwage}_{1980,j}$ are used for $\text{lwage}_{1985,j} - \text{lwage}_{1984,j}$. In 1985, the four instruments $\text{lwage}_{1983,j}$ to $\text{lwage}_{1980,j}$ are used for $\text{lwage}_{1984,j} - \text{lwage}_{1983,j}$. In 1984, the three instruments $\text{lwage}_{1982,j}$ to $\text{lwage}_{1980,j}$ are used for $\text{lwage}_{1983,j} - \text{lwage}_{1982,j}$. In 1983, the two instruments $\text{lwage}_{1981,j}$ to $\text{lwage}_{1980,j}$ are used for $\text{lwage}_{1982,j} - \text{lwage}_{1981,j}$. And in 1982, the instrument $\text{lwage}_{1980,j}$ is used for $\text{lwage}_{1981,j}$. In addition, the exogenous variables D.union, D.married, and D.exper are used as instruments for themselves.

All the estimators discussed so far for lagged-response models have assumed that the residuals ϵ_{ij}, and hence the differenced residuals $\epsilon_{ij} - \epsilon_{i-1,j}$, are uncorrelated. If this assumption is violated, these estimators become inconsistent. We can test the assumption of no correlation between differenced residuals up to lag-2 (the default) using `xtabond`'s postestimation command `estat abond`. Stata's `xtdpd` command implements a GMM estimator that accommodates correlated residuals at the cost of more convoluted command statements.

Endogeneity also becomes an issue if the response at the first occasion (baseline) is included as a covariate at all subsequent occasions instead of lagged responses. However, if the baseline is viewed as a *proxy* for omitted covariates (as when using a pretest when studying educational achievement), the target of inference would not be the causal effect of baseline and this kind of endogeneity should be ignored.

5.8 Missing data and dropout

A ubiquitous problem in longitudinal and panel modeling is missing data. When data for a subject are missing from some time onward, the situation is called *dropout* or *attrition*. Often subjects do not exhibit monotone missingness patterns like this but rather *intermittent missingness* where, for instance, a response is given at an occasion, missing at the next occasion, but given again at a future occasion.

All the methods discussed so far use data for those subjects j and occasions i where neither the response y_{ij} nor the covariates $\mathbf{x}_{ij}$ are missing. This is in contrast to more old-fashioned approaches to longitudinal data, such as MANOVA, where subjects with any missing responses or covariates are discarded altogether, an approach often referred to as "listwise deletion" or "complete-case analysis". Using all available data does not waste information and is less susceptible to bias.

Using ML estimation, as we did for random-intercept and random-coefficient models, has the advantage that consistency (estimates approaching parameter values in large samples) is retained for correctly specified models, as long as the missing data are *missing at random* (MAR). This means that the probability of being missing may only depend on the covariates $\mathbf{x}_{ij}$ or responses at previous occasions (or future occasions, although this seems strange), but not on the responses we would have observed had they not been missing.

In practice, dependence of missingness on previous responses is only allowed if missingness is monotone, meaning that subjects drop out (never return) when missing an occasion. For intermittent missing data, if missingness at occasion 2, say, can depend on the response at occasion 1, this becomes a problem for subjects for whom the response at occasion 1 is missing (because for them missingness at occasion 2 depends on a response that is not observed). In the case of monotone missing data or dropout, this problem does not occur because the response at occasion 2 is missing with certainty if the response at occasion 1 is missing.

In a random-intercept model, an example of *not missing at random* (NMAR) is when missingness depends on the random intercept. In contrast, such missingness is MAR in a fixed-intercept model because missingness can be viewed as depending on subjects, and dummy variables for subjects are included as covariates, making missingness covariate-dependent. (This still holds if the fixed-effects estimator is implemented by mean-centering or first-differencing, because the estimates are identical.)

5.8.1 ❖ Maximum likelihood estimation under MAR: A simulation

We now use a simulation to investigate how well ML estimation works when data are MAR.

We first simulate complete data from a random-intercept model,

$$y_{ij} = 2 + \zeta_j + \epsilon_{ij}, \qquad \zeta_j \sim N(0,1), \; \epsilon_{ij} \sim N(0,1) \tag{5.6}$$

for $J = 100{,}000$ subjects and $n_j = 2$ occasions. The random intercepts ζ_j are independent across subjects, the level-1 residuals ϵ_{ij} are independent across subjects and occasions, and ζ_j and ϵ_{ij} are independent.

It is convenient to simulate the data in wide form, generating a row of data for each subject with variables `y1` and `y2` for the two occasions:

```
. clear
. set obs 100000
. set seed 123123123
. generate zeta = rnormal(0,1)
. generate y1 = 2 + zeta + rnormal(0,1)
. generate y2 = 2 + zeta + rnormal(0,1)
```

Here the `rnormal(0,1)` function draws a pseudorandom variable from a standard normal distribution (as required for ζ_j, ϵ_{1j}, and ϵ_{2j}). We have set the seed of the pseudorandom-number generator to an arbitrary number so that you can get the same results as we do, but you can also try other seeds if you like.

Among those subjects who have a value of `y1` greater than 2, we now randomly sample about 90% of subjects and replace their `y2` with a missing value:

```
. replace y2 = . if y1>2 & runiform()<0.9
(45104 real changes made, 45104 to missing)
```

where `runiform()` generates pseudorandom numbers from a uniform distribution. Here we used the fact that the probability that a (pseudo)random number with a uniform distribution (on the interval from 0 to 1) is less than 0.9 is 0.9. This selection mechanism is of course rather extreme but nicely illustrates the ideas. The responses are MAR because the probability of dropout or attrition depends only on the previous response.

The sample means at the two occasions are

```
. tabstat y1 y2
    stats |        y1         y2
----------+----------------------
     mean | 1.997182   1.532937
```

We see that the sample mean at the second occasion is 1.53, much lower than the population mean of 2. This is because subjects with larger than average values at occasion 1 ($y_{1j} > 2$) are likely to have larger than average values at occasion 2 as well (because of an intraclass correlation of 0.5), but about 90% of these values are missing at occasion 2.

To see if such a difference in estimated means is also found using ML estimation, we fit a slightly more general model than the true model (5.6), which allows the means at the two occasions to be different:

$$y_{ij} \;=\; \beta_1 x_{1i} + \beta_2 x_{2i} + \zeta_j + \epsilon_{ij} \tag{5.7}$$

where x_{1i} is a dummy variable for occasion $i = 1$ and x_{2i} is a dummy variable for occasion $i = 2$. According to the true model (5.6) used to generate the complete data, the population means at both occasions are $\beta_1 = \beta_2 = 2$.

We must first reshape the data and generate the dummy variables:

```
. generate id = _n
. reshape long y, i(id) j(occasion)
(note: j = 1 2)
Data                               wide   ->   long
----------------------------------------------------------------
Number of obs.                   100000   ->   200000
Number of variables                   4   ->        4
j variable (2 values)                    ->   occasion
xij variables:
                                  y1 y2   ->   y
----------------------------------------------------------------

. quietly tabulate occasion, generate(occ)
```

Now we can use the **xtreg** command to fit the random-intercept model (5.7) using the **noconstant** option because the model does not include an intercept:

```
. quietly xtset id

. xtreg y occ1 occ2, noconstant mle
Random-effects ML regression                    Number of obs      =    154896
Group variable: id                              Number of groups   =    100000

Random effects u_i ~ Gaussian                   Obs per group: min =         1
                                                               avg =       1.5
                                                               max =         2

                                                Wald chi2(2)       = 223725.11
Log likelihood  = -265985.81                    Prob > chi2        =    0.0000
```

y	Coef.	Std. Err.	z	P>|z|	[95% Conf. Interval]	
occ1	1.997182	.004485	445.30	0.000	1.988392	2.005973
occ2	1.998877	.0066523	300.48	0.000	1.985839	2.011916
/sigma_u	1.003776	.0049132			.9941921	1.013452
/sigma_e	1.001969	.0033497			.9954253	1.008556
rho	.5009008	.0036862			.4936763	.508125

```
Likelihood-ratio test of sigma_u=0: chibar2(01)= 9845.55 Prob>=chibar2 = 0.000
```

All parameter estimates, including $\widehat{\beta}_2$, are almost identical to the true parameter values. The standard error for $\widehat{\beta}_2$ is larger than for $\widehat{\beta}_1$ because of the missing values at the second occasion. The reason these estimates are so good is that the random-intercept model takes the intraclass correlation into account.

In comparison, consider now using pooled OLS (see *Introduction to models for longitudinal and panel data (part III)*) to fit the analogous regression model

$$y_{ij} = \beta_1 x_{1i} + \beta_2 x_{2i} + \xi_{ij}$$

This is accomplished by using the following **regress** command, where we have used the **vce(cluster id)** option to obtain robust standard errors taking the dependence into account:

```
. regress y occ1 occ2, noconstant vce(cluster id)
Linear regression                               Number of obs  =    154896
                                                F(  2, 99999)  =         .
                                                Prob > F       =    0.0000
                                                R-squared      =    0.6378
                                                Root MSE       =     1.391

                       (Std. Err. adjusted for 100000 clusters in id)
```

y	Coef.	Robust Std. Err.	t	P>|t|	[95% Conf. Interval]	
occ1	1.997182	.0044854	445.26	0.000	1.988391	2.005973
occ2	1.532937	.005718	268.09	0.000	1.52173	1.544144

In contrast to ML, the pooled OLS estimator of β_2 is based on assuming zero intraclass correlation and is just the sample mean at the second occasion with $\widehat{\beta}_2^{\text{OLS}} = 1.54$.

We see that the pooled OLS estimator can perform very badly when missingness at an occasion depends on responses at other occasions. Falsely assuming that units are uncorrelated is unproblematic for the pooled OLS estimator if missingness only depends on covariates and not on responses.

In general, it is important to get the covariance structure right (in addition to the mean structure) when there are missing data. In this case, ML produces consistent estimators if responses are MAR. See also exercise 6.6.

5.9   Summary and further reading

In this chapter, we focused on models with random or fixed subject-specific effects. The random-effects models can be extended to allow level-1 residuals to be correlated or heteroskedastic (see section 6.4). Chapter 7 is dedicated to random-effects models that include random slopes of time.

A weakness of conventional random-effects approaches is the possibility of subject-level confounding. For time-varying covariates, this problem can be eliminated by extending random-effects models to include subject means of time-varying covariates or alternatively by using fixed-effects approaches. However, neither approach gives consistent estimators of coefficients of time-constant covariates, not even for exogenous time-constant covariates. The Hausman–Taylor approach can be used to obtain consistent estimators of all model parameters if we can correctly classify the covariates as being exogenous or endogenous.

Dynamic or lagged-response models should in general be used only if dependence on previous responses is of substantive interest. Such models can generally not be fit consistently by simply including the lagged response as a covariate because the (total) residual is correlated with the lagged response whenever the correct model includes subject-specific effects (whenever there are time-constant omitted variables). We discussed methods addressing this problem.

Good, but somewhat demanding, books on the topics covered in this chapter include Cameron and Trivedi (2005, chap. 21–22), Wooldridge (2010, chap. 10–11), and Frees (2004), with the first two written from an explicit econometric perspective. Hsiao (2003) and Baltagi (2008) are technical books on the econometrics of panel data. Random-effects models are also treated in the books by the biostatisticians Fitzmaurice, Laird, and Ware (2011) and Verbeke and Molenberghs (2000).

The exercises cover almost all topics discussed in this chapter. Specifically, exercises 5.1, 5.2, 5.5, and 5.6 are about random- and fixed-intercept models, and exercises 5.4 and 5.5 involve lagged-response models. Exercise 5.3 introduces the difference-in-difference approach, not covered in this chapter, that is often used for estimating treatment effects in quasi-experiments or natural experiments. See also exercises 3.2 and 3.3 in chapter 3 for further examples with random intercepts; see also exercises 6.2 and 6.3 in chapter 6, as well as the exercises of chapter 7, for further examples with random intercepts and slopes.

5.10 Exercises

5.1 Tax-preparer data

Frees (2004) analyzed panel or repeated-measurement data on tax returns filed by 258 taxpayers for the years 1982, 1983, 1984, 1986, and 1987. (The data come from the Statistics of Income panel of individual returns.)

The dataset `taxprep.dta` has the following variables:

- `subject`: subject identifier
- `time`: identifier for the panel wave or occasion
- `lntax`: natural logarithm of tax liability in 1983 dollars
- `prep`: dummy variable for using a tax preparer for the tax return
- `ms`: dummy variable for being married
- `hh`: dummy variable for being the head of the household
- `depend`: number of dependents claimed by taxpayer
- `age`: dummy variable for being at least 65 years old
- `lntpi`: natural logarithm of the sum of all positive income line items on return in 1983 dollars
- `mr`: marginal tax rate computed on total income minus exemptions and standard deductions
- `emp`: dummy variable for schedule C or F being present on the return, a proxy for self-employment income

  1. Fit a random-intercept model for `lntax` with all the subsequent variables given above as covariates and with a random intercept for subjects.
  2. Obtain between and within estimates for all the covariates using `xtreg` with the `fe` and `be` options. Compare the estimates for the effect of tax preparer.
  3. Perform a Hausman specification test.
  4. Obtain histograms for the level-1 and level-2 residuals. Do the normality assumptions appear plausible?

5.2 Antisocial-behavior data

Allison (2005) considered a sample of 581 children who were interviewed in 1990, 1992, and 1994 as part of the U.S. National Longitudinal Survey of Youth. The children were between 8 and 10 years old in 1990.

The dataset `antisocial.dta` includes the following variables:

- `id`: child identifier (j)
- `occ`: year of interview (90, 92, 94)
- `anti`: a measure of the child's antisocial behavior (y_{ij}) (higher values mean more antisocial behavior)
- `pov`: dummy variable for child being from a poor family (x_{2ij}) (1: poor; 0: otherwise); this covariate varies both between and within children

- momage: mother's age at birth of child in years (x_{3j})
- female: dummy variable for child being female (x_{4j}) (1: female; 0: male)
- childage: child's age in years in 1990 (x_{5j})
- hispanic: dummy variable for child being Hispanic (x_{6j}) (1: Hispanic; 0: otherwise)
- black: dummy variable for child being black (x_{7j}) (1: black; 0: otherwise)
- momwork: dummy variable for mother being employed in 1990 (x_{8j}) (1: employed; 0: not employed)
- married: dummy variable for mother being married in 1990 (x_{9j}) (1: married; 0: otherwise)

1. Fit the random-intercept model

$$y_{ij} = \beta_1 + \beta_2 x_{2ij} + \beta_3 x_{3j} + \beta_4 x_{4j} + \beta_5 x_{5j} + \beta_6 x_{6j} + \beta_7 x_{7j} + \beta_8 x_{8j} + \beta_9 x_{9j}$$
$$+ \zeta_j + \epsilon_{ij}$$

   (with the usual assumptions) and interpret the estimated regression coefficients that are significant at the 5% level.
2. Test the null hypothesis that the random-intercept variance is zero.
3. State the expression for the residual intraclass correlation in terms of the model parameters and give the estimate.
4. Replace x_{2ij} in the random-intercept model with the covariates

$$x_{10,j} = \frac{1}{n_j} \sum_{i=1}^{n_j} x_{2ij}$$

   where n_j is the number of units in cluster j, and

$$x_{11,ij} = x_{2ij} - x_{10,j}$$

   Fit the resulting model and interpret the estimates $\widehat{\beta}_{10}$ and $\widehat{\beta}_{11}$ corresponding to $x_{10,j}$ and $x_{11,ij}$, respectively.
5. Test the null hypothesis $\beta_{10} = \beta_{11}$ against the alternative $\beta_{10} \neq \beta_{11}$. Explain why this test can be interpreted as an endogeneity test.

See also exercise 6.1.

5.3 Unemployment-claims data I [Solutions]

Papke (1994) analyzed panel data on Indiana's enterprise zone program, which provided tax credits for areas with high unemployment and high poverty levels. One of the purposes was to investigate whether inclusion in an enterprise zone would reduce the number of unemployment claims. Here we consider data from 1983 and 1984 on 22 unemployment claims offices (serving a zone and the surrounding city), 6 of which were included as enterprise zones in 1984.

The dataset `papke_did.dta`, extracted from a dataset supplied by Wooldridge (2010), contains the following variables from the 1983 and 1984 waves of the panel study:

- `city`: unemployment claims office identifier (j)
- `year`: year (i)
- `luclms`: logarithm of number of unemployment claims (y_{ij})
- `ez`: dummy variable for unemployment claims office being in an enterprise zone (x_{ij})

1. Use a "posttest-only design with nonequivalent groups", which is based on comparing those receiving the intervention with those not receiving the intervention at the second occasion only.

   a. Use an appropriate t test to test the hypothesis of no intervention effect on the log-transformed number of unemployment claims in 1984.

   b. Consider the model

   $$\ln(y_{2j}) = \beta_1 + \beta_2 x_{2j} + \epsilon_{2j}$$

   where the usual assumptions are made. Estimate the intervention effect and test the null hypothesis that there is no intervention effect.

2. Use a "one-group pretest–posttest design", which is based on comparing the second occasion (posttest) with the first occasion (pretest) for the intervention group only. To do this, first construct a new variable for intervention group, taking the value 1 if an unemployment claims office is ever in an enterprise zone and 0 for the control group (consider using `egen`).

   a. Use an appropriate t test to test the hypothesis of no intervention effect on the log-transformed number of unemployment claims. (It may be useful to reshape the data to wide form for the t test and then reshape them to long form again for the next questions.)

   b. For the intervention group, consider the model

   $$\ln(y_{ij}) = \beta_1 + \alpha_j + \beta_2 x_{ij} + \epsilon_{ij}$$

   where α_j is an office-specific parameter (fixed effect). Estimate the intervention effect and test the null hypothesis that there is no intervention effect.

3. Discuss the pros and cons of the "posttest-only design with nonequivalent groups" and the "one-group pretest–posttest design".

4. Use an "untreated control group design with dependent pretest and posttest samples", which is based on data from both occasions and both intervention groups.

   a. Find the difference between the following two differences:
      i. the difference in the sample means of `luclms` for the intervention group between 1984 and 1983

ii. the difference in the sample means of `luclms` for the control group between 1984 and 1983

The resulting estimator is called the *difference-in-difference estimator* and is commonly used for the analysis of intervention effects in quasi-experiments and natural experiments.

b. Consider the model

$$\ln(y_{ij}) = \beta_1 + \alpha_j + \tau z_i + \beta_2 x_{ij} + \epsilon_{ij}$$

where α_j is an office-specific parameter (fixed effect) and τ is the coefficient of a dummy variable z_i for 1984. Estimate the intervention effect and test the null hypothesis that there is no intervention effect. Note that the estimate $\widehat{\beta}_2$ is identical to the difference-in-difference estimate. The advantage of using a model is that statistical inference regarding the intervention effect is straightforward, as is extension to many occasions, several intervention groups, and inclusion of extra covariates.

c. What are the advantages of using the "untreated control group design with dependent pretest and posttest samples" compared with the "posttest-only design with nonequivalent groups" and the "one-group pretest–posttest design"?

The full dataset is used in exercises 5.4 and 7.8.

5.4 Unemployment-claims data II [Solutions]

The full dataset used by Papke (1994) has annual panel waves from 1980 to 1988 on 22 unemployment claims offices, 6 of which were included as enterprise zones in 1984 and 4 in 1985. A subset of this dataset was used in exercise 5.3.

The dataset `ezunem.dta` supplied by Wooldridge (2010) contains the following variables:

- `city`: unemployment claims office identifier (j)
- `year`: year (i)
- `uclms`: number of unemployment claims (y_{ij})
- `ez`: dummy variable for office being in an enterprise zone (x_{2ij})
- `t`: time 1, ..., 9 (x_{3i})

1. Use the `xtset` command to specify the variables representing the clusters and units for this application. This enables you to use Stata's time-series operators, which should be used within the estimation commands in this exercise. Interpret the output.

2. Consider the fixed-intercept model

$$\ln(y_{ij}) = \tau_i + \beta_2 x_{2ij} + \alpha_j + \epsilon_{ij}$$

where τ_i and α_j are year-specific and office-specific parameters, respectively. (Use dummy variables for years to include τ_i in the model.) This gives the difference-in-difference estimator for more than two panel waves (see exercise 5.3).

   a. Fit the model using `xtreg` with the `fe` option.

   b. Fit the first-difference version of the model using OLS.

      i. Do the estimates of the intervention effect differ much?

      ii. Papke (1994) actually assumed a linear trend of year instead of year-specific intercepts as specified above. Write down the first-difference version of Papke's model.

      iii. ❖ A random walk is the special case of an AR(1) process where $\alpha = 1$. Show that the first-difference approach accommodates a random walk for the residuals ϵ_{ij}.

3. Fit the lagged-response model

$$\ln(y_{ij}) = \tau_i + \beta_2 x_{2ij} + \gamma \ln(y_{i-1,j}) + \epsilon_{ij}$$

where γ is the regression coefficient for the lagged response $\ln(y_{i-1,j})$. Compare the estimated intervention effect with that for the fixed-intercept model. Interpret β_2 in the two models.

4. Consider a lagged-response model with an office-specific intercept b_j:

$$\ln(y_{ij}) = \tau_i + \beta_2 x_{2ij} + \gamma \ln(y_{i-1,j}) + b_j + \epsilon_{ij}$$

   a. Treat b_j as a random intercept, and fit a random-intercept model by ML using `xtmixed`. Are there any problems associated with this random-intercept model?

   b. Fit the model using the Anderson–Hsiao approach with the second lag of the response as an instrumental variable. Compare the estimated intervention effect with that from step 4a.

   c. Papke (1994) used the Anderson–Hsiao approach with the second lag of the first-difference of the response as an instrumental variable. Does the choice of instruments matter in this case?

Growth-curve models are applied to this dataset in exercise 7.8.

5.5 Hours-worked data

In this exercise, we use panel data on 532 men with annual panel waves from 1979 to 1988 from Ziliak (1997). The dataset can be downloaded from the website of *Journal of Business and Economic Statistics* and accompanies the book by Cameron and Trivedi (2005).

The file `hours.dta` contains the following variables:

- `id`: identification number for man (j)
- `year`: years 1979–1988 of panel wave (i)
- `lnhr`: natural logarithm of annual hours worked (y_{ij})
- `lnwg`: natural logarithm of hourly wage in U.S. dollars (x_{2ij})

- kids: number of children (x_{3ij})
- ageh: age of man (x_{4ij})
- agesq: age of man squared (x_{5ij})
- disab: dummy variable for having a disability (x_{6ij})

The model we will consider is of the form

$$ y_{ij} \ = \ \beta_1 + \beta_2 x_{2ij} + \beta_3 x_{3ij} + \beta_4 x_{4ij} + \beta_5 x_{5ij} + \beta_2 x_{6ij} + c_j + \epsilon_{ij} $$

where c_j is a man-specific intercept and ϵ_{ij} is a level-1 residual with mean zero. The main interest concerns β_2, which is called the "intertemporal substitution wage elasticity of labor supply" by economists. (See display 6.2 for information on elasticities.)

1. First have a look at the data:
   a. Use xtdescribe to describe the longitudinal structure of the dataset. Are the data balanced?
   b. Use xtsum to investigate the within, between, and total variability of the response and the covariates. Are there any covariates that only have within variability, and are there any covariates that only have between variability?

2. Treating c_j as a random intercept, use FGLS to fit a random-intercept model. Store the estimates. What is the estimated residual intraclass correlation of log hours worked?

3. Now treat c_j as a fixed intercept.

   a. Fit the model by using xtreg with the fe option. Store the estimates.
   b. Estimate the regression coefficients by using OLS for the transformed model where all variables are first-differences.

   Do the estimates of the regression coefficients differ appreciably?

4. Perform a Hausman test to compare the fixed-intercept model with the random-intercept model.

   a. What is the null hypothesis of the Hausman test? What is the alternative hypothesis?
   b. What is the conclusion at the 5% significance level?

5. Use FGLS to fit a random-intercept model where the cluster means of all covariates and the deviations from these cluster means are included as covariates, making sure to obtain robust standard errors. Test the null hypothesis that the coefficients of the cluster means equal the coefficients of the deviations from the cluster means at the 5% significance level.

   a. How do you interpret this test?
   b. Is your conclusion the same as for the Hausman test?
   c. Are there any advantages of this test compared with the Hausman test?

6. Now also include the lag-1 of the response, $y_{i-1,j}$, as a covariate in the model.

    a. Fit a lagged-response model with a subject-specific intercept using the Anderson–Hsiao method and obtain robust standard errors.

    b. Does the estimate of β_2 change much compared with the estimate for the fixed-intercept model?

    c. Is the coefficient of the lagged response significant at the 5% level?

5.6 Cognitive-style data

Here we consider classic repeated-measures data described and made available by Broota (1989). The data have also been analyzed by Crowder and Hand (1990), Everitt (1995), and others. Subjects selected at random from a large group of potential subjects were identified as having either field-independent or field-dependent cognitive styles. They read two types of words (color and form names) under three cue conditions: normal, congruent, and incongruent. The order in which the six reading tasks were carried out was randomized. The response variable is the time in milliseconds taken to read the stimulus words.

The data `cogstyle.dta` are in wide form and the variables are

- `subj`: subject identifier
- `dependent`: dummy variable for being field dependent
- `rfn`: form word, normal cue
- `rfc`: form word, congruent cue
- `rfi`: form word, incongruent cue
- `rcn`: color word, normal cue
- `rcc`: color word, congruent cue
- `rci`: color word, incongruent cue

1. Reshape the data to long form so that there are six rows per subject with variables `word` and `cue` identifying the word type and cue condition, respectively. (Use `reshape` twice, first to obtain three rows of data for form and color words and then to stack the word types into a single variable. In each `reshape` command, use the `string` option to handle the fact that word type and cue condition are indicated by letters instead of numbers in the variable names.)

2. Create numeric variables `words` and `cues` for words and cues, with cues labeled 1 for normal, 2 for congruent, and 3 for incongruent. Take the logarithm of the response variable after subtracting 134 and call the new variable `lnr`.

3. Produce box plots using the command

```
graph box lnr, over(cues) over(words) asyvars by(dependent) legend(row(1))
```

4. Fit a random-intercept model with three main effects, three two-way interactions, and one three-way interaction among the three covariates `dependent`, `words`, and `cues`. Use `xtmixed` with factor variables.

5. Test whether the three-way interaction is significant at the 5% level, and fit the model without the three-way interaction.

6. Test all pairwise interactions at the 5% level, and fit the model with non-significant two-way interactions removed.

7. Also remove any main effects that are not significant for variables not involved in any two-way interactions. For this final model, obtain predicted means for all combinations of the values of the covariates included in the model using the `margins` command. Run the command `marginsplot, xdim(cues)` to produce a graph of these means with confidence intervals.

5.7 Returns-to-schooling data

In this exercise, we consider data from the Panel Study of Income Dynamics used by Cornwell and Rupert (1988). The subjects are 595 heads of household who were between 18 and 65 years old in 1976 and report a positive wage in some private, nonfarm employment in 1976–1982, the years included in the dataset. The main research question concerns the causal effect of years of schooling on wage.

The datafile `returns.dta` (based on the data supplied by Baltagi [2008]) contains the following variables:

- `nr`: person identifier
- `year`: year of survey
- `lwage`: log hourly wage in U.S. dollars
- `exp`: years of full-time work experience
- `wks`: weeks worked
- `occ`: dummy variable for having blue-collar occupation
- `ind`: dummy variable for working in manufacturing industry
- `south`: dummy variable for residing in the south of the U.S.A.
- `smsa`: dummy variable for residing in standard metropolitan statistical area
- `ms`: dummy variable for being married
- `union`: dummy variable for being a member of a union (that is, wage being set in collective bargaining agreement)
- `fem`: dummy variable for being female
- `ed`: years of schooling
- `blk`: dummy variable for being black

1. The variables from `exp` to `blk` above are used as covariates. Also generate a new covariate, `exp2`, which is equal to `exp` squared. Which of the covariates are time varying and which are time constant?

2. Fit a fixed-intercept model that includes all covariates and store the estimates. Does this analysis address the research question?

3. Fit a random-intercept model that includes all covariates using FGLS and store the estimates. Which kind of information is used to estimate the effect of years of schooling in this analysis?

4. Perform a Hausman test to investigate if the random intercepts are correlated with the covariates. Which of the models in steps 2 and 3 are preferable according to the test?

5. Following Cornwell and Rupert (1988), suppose that the covariates can be partitioned in the following way: wks, south, smsa, ms, exp, exp2, occ, ind, and union are exogenous covariates, and fem, blk, and ed are endogenous covariates. Fit a random-intercept model that allows for endogenous time-varying and endogenous time-constant covariates using the Hausman–Taylor estimator and store the estimates. Has the estimated effect of years of schooling changed appreciably compared with step 3 (you need not perform a formal test to answer this question)?

6. Perform a Hausman test to assess whether the partitioning into exogenous and endogenous covariates in step 5 appears to be appropriate. What do you conclude?

7. Fit the same model as in step 5 but now with the amacurdy option. Do you gain much efficiency compared with the Hausman–Taylor estimator?

5.8 Hausman test for Hausman–Taylor estimator

Consider the data in wagepan.dta from the U.S. National Longitudinal Survey of Youth 1979 that we introduced on page 229 and have used throughout the current chapter. Perform a Hausman test to assess whether the partitioning of covariates into time-varying exogenous covariates, time-varying exogenous covariates, time-constant exogenous covariates, and time-constant endogenous covariates used in section 5.3.2 seems reasonable.

6 Marginal models

6.1 Introduction

Instead of specifying multilevel linear models for longitudinal data—from which we can *derive* the marginal or population-averaged expectations, variances, and covariances of the responses (averaged over the random effects but conditional on the observed covariates)—we can *directly specify* a model for the marginal expectations and the marginal covariance matrix.

In both multilevel and marginal linear models, the population-averaged relationship between the response variable and the covariates is the fixed part of the model. The regression coefficients therefore have the same meaning, and the estimates tend to be similar if the fixed part of the model is the same. (As we will see in volume 2, this is not the case for most other generalized linear mixed models.)

However, in contrast to marginal models, multilevel models also provide subject-specific relationships. For instance, for a model with a random-intercept ζ_{1j} and a random slope ζ_{2j} of x_{2ij}, the subject-specific relationships are $E(y_{ij}|\mathbf{x}_{ij}, \zeta_{1j}, \zeta_{2j}) = E(y_{ij}|\mathbf{x}_{ij}) + \zeta_{1j} + \zeta_{2j}x_{2ij}$. In the *multilevel approach*, the focus is on modeling such subject-specific relationships and how they vary around the population average, as summarized by the covariance matrix of the random effects. Although it is possible to derive a marginal covariance matrix, that is, the covariance matrix of the total residuals $\xi_{ij} = \zeta_{1j} + \zeta_{2j}x_{2ij} + \epsilon_{ij}$, that matrix is generally not of interest in multilevel modeling. In contrast, in the *marginal approach*, we are only interested in the marginal residual covariance matrix and the marginal relationship $E(y_{ij}|\mathbf{x}_{ij})$ between the response variable and the covariates.

6.2 Mean structure

The *mean structure* is the fixed part of the model because this part represents the mean of the response variable given the covariates, as a function of the regression coefficients.

In this chapter, we will consider marginal models for the wage-panel data described in *Introduction to models for longitudinal and panel data (part III)*. The marginal model can be written as

$$y_{ij} = \beta_1 + \beta_2 x_{2j} + \beta_3 x_{3j} + \beta_4 x_{4ij} + \beta_5 x_{5ij} + \beta_6 L_{ij} + \beta_7 P_i + \beta_8 E_j + \xi_{ij}$$

where the (total) residuals ξ_{ij} have zero means given the covariates.

The mean structure of the marginal model for each of the $n_j = 8$ occasions i for subject j is then

$$E(y_{ij}|\mathbf{x}_{ij}) = \beta_1 + \beta_2 x_{2j} + \beta_3 x_{3j} + \beta_4 x_{4ij} + \beta_5 x_{5ij} + \beta_6 L_{ij} + \beta_7 P_i + \beta_8 E_j$$

We will keep this mean structure for the wage-panel data throughout the chapter.

6.3   Covariance structures

In this section, we will make different assumptions regarding the covariance matrix of the $n_j = 8$ residuals for a subject j in the wage-panel data. To write these structures in a compact form using matrices, we think of the responses and total residuals for subject j as being assembled in column vectors:

$$\mathbf{y}_j = \begin{bmatrix} y_{1j} \\ y_{2j} \\ \vdots \\ y_{8j} \end{bmatrix} \qquad \boldsymbol{\xi}_j = \begin{bmatrix} \xi_{1j} \\ \xi_{2j} \\ \vdots \\ \xi_{8j} \end{bmatrix}$$

The covariate matrix $\mathbf{X}_j$ has the covariates $(x_{2j}, x_{3j}, x_{4ij}, x_{5ij}, L_{ij}, P_i, E_j)$ for subject j at occasion i placed in row i.

Different structures will be specified for the residual covariance matrix $\mathbf{V}_j$ (conditional on the covariates):

$$\mathbf{V}_j \equiv \begin{bmatrix} \mathrm{Var}(\xi_{1j}|\mathbf{X}_j) \\ \mathrm{Cov}(\xi_{2j},\xi_{1j}|\mathbf{X}_j) & \mathrm{Var}(\xi_{2j}|\mathbf{X}_j) \\ \mathrm{Cov}(\xi_{3j},\xi_{1j}|\mathbf{X}_j) & \mathrm{Cov}(\xi_{3j},\xi_{2j}|\mathbf{X}_j) & \mathrm{Var}(\xi_{3j}|\mathbf{X}_j) \\ \vdots & \vdots & \vdots & \ddots \\ \mathrm{Cov}(\xi_{8j},\xi_{1j}|\mathbf{X}_j) & \mathrm{Cov}(\xi_{8j},\xi_{2j}|\mathbf{X}_j) & \mathrm{Cov}(\xi_{8j},\xi_{3j}|\mathbf{X}_j) & \cdots & \mathrm{Var}(\xi_{8j}|\mathbf{X}_j) \end{bmatrix}$$

Sometimes we use the expression "within-subject" residual covariance matrix to emphasize that it is the covariance matrix for residuals of the same subject and not across subjects. The covariances are longitudinal, not cross-sectional. This matrix is also the covariance matrix of the responses given the covariates:

$$\mathrm{Cov}(\mathbf{y}_j|\mathbf{X}_j) \;=\; \mathrm{Cov}(\boldsymbol{\xi}_j|\mathbf{X}_j) \;=\; \mathbf{V}_j$$

Viewed as a function of model parameters (and sometimes of covariates), this is the *covariance structure* of the marginal model.

We will generally not present estimated residual covariance matrices directly, but rather will show the standard deviations (the square roots of the diagonal elements of the covariance matrix) and the correlation matrix (each covariance divided by the product of the relevant standard deviations). We have already seen an example of this on page 243.

Possible assumptions about the residual within-subject covariance matrix would be that all variances are equal and all covariances are equal. In clustered data, such a structure is the only appropriate choice if units within clusters are exchangeable. For instance, for students nested in schools, in the absence of student-level covariates, the students are exchangeable and there is no reason to believe that the responses for one pair of students are more or less correlated than those for another pair of students within the same school. The exchangeability assumption would generally be inappropriate for longitudinal data because the occasions are ordered in time.

The ordering and also the timing and spacing of occasions is relevant when considering the covariance structure. Usually, we would expect pairs of responses to be more correlated if the time interval between them is small than if the interval is large. When describing covariance structures, we will therefore refer to the time t_{ij} associated with occasion i for subject j, where t_{ij} is the (time-varying) time scale of interest, such as age, period, or time since some event of interest.

The residual covariance matrix is typically assumed to be the same for all subjects; $\mathbf{V}_j = \mathbf{V}$. Generally, such models are appropriate only if the data are quite balanced, in the sense that occasion i means approximately the same thing across subjects, which we denote as $t_{ij} = t_i$. An example of imbalance would be if the timing of occasions varies considerably between subjects so that it does not make sense to assume that all subjects have the same covariance between a given pair of occasions (such as occasions 1 and 2). For some subjects, the occasions may be distant in time and for others they may be close. Some covariance structures require *constant spacing* of occasions for each pair of occasions across subjects, which is denoted as $t_{ij} - t_{i-1,j} = \Delta$. In this case, the timing of the first occasion t_{1j} can differ between subjects. Other models allow the residual covariance matrix to differ between subjects. In these models (for example, random-coefficient models), the variances and covariances are typically functions of the subject-specific timing of occasions or other covariates.

Some popular covariance structures for marginal modeling are shown in table 6.1. In addition to showing the different structures, the table shows the corresponding number of parameters as a function of the number of occasions n, whether balance is required ($t_{ij} = t_i$) and whether equal spacing is required ($t_{ij} - t_{i-1,j} = \Delta$) for the structure to make sense.

Table 6.1: Common marginal covariance structures for longitudinal data ($n = 3$). The number of parameters and requirements for timing of occasions are also given. (In Stata, missing data are allowed for all structures.) Whenever the variance is constant, it is denoted σ^2 and factored out.

Unstructured $\{n(n+1)/2 \text{ parameters}\}$ $(t_{ij} = t_i)$

$$
\begin{bmatrix}
V_{11} & & \\
V_{21} & V_{22} & \\
V_{31} & V_{32} & V_{33}
\end{bmatrix}
$$

Random-intercept $(\psi_{11} \geq 0)$ **or**
Compound symmetric or Exchangeable (2 parameters) (any t_{ij})

$$
\begin{bmatrix}
\psi_{11} + \theta & & \\
\psi_{11} & \psi_{11} + \theta & \\
\psi_{11} & \psi_{11} & \psi_{11} + \theta
\end{bmatrix}
\quad \text{or} \quad
\sigma^2
\begin{bmatrix}
1 & & \\
\rho & 1 & \\
\rho & \rho & 1
\end{bmatrix}
$$

Random-coefficient (intercept and slope, t=0,1,2) (4 parameters) (any t_{ij})

$$
\begin{bmatrix}
\psi_{11} + \theta & & \\
\psi_{11} + \psi_{21} & \psi_{11} + 2\psi_{21} + \psi_{22} + \theta & \\
\psi_{11} + 2\psi_{21} & \psi_{11} + 3\psi_{21} + 2\psi_{22} & \psi_{11} + 4\psi_{21} + 4\psi_{22} + \theta
\end{bmatrix}
$$

Autoregressive residual AR(1) (2 parameters) $(t_{ij} - t_{i-1,j} = \Delta)$

$$
\sigma^2
\begin{bmatrix}
1 & & \\
\alpha & 1 & \\
\alpha^2 & \alpha & 1
\end{bmatrix}
$$

Exponential(1) (2 parameters) (any t_{ij})

$$
\sigma^2
\begin{bmatrix}
1 & & \\
\exp(-\phi|t_{2j} - t_{1j}|) & 1 & \\
\exp(-\phi|t_{3j} - t_{1j}|) & \exp(-\phi|t_{3j} - t_{2j}|) & 1
\end{bmatrix}
$$

Moving average MA(1) (2 parameters) $(t_{ij} - t_{i-1,j} = \Delta)$

$$
\sigma^2
\begin{bmatrix}
1 & & \\
\phi/(1 + \phi^2) & 1 & \\
0 & \phi/(1 + \phi^2) & 1
\end{bmatrix}
$$

Banded(1) $(2n - 1 \text{ parameters})$ $(t_{ij} = t_i)$

$$
\begin{bmatrix}
V_{11} & & \\
V_{21} & V_{22} & \\
0 & V_{32} & V_{33}
\end{bmatrix}
$$

Table 6.1: Common marginal covariance structures for longitudinal data ($n = 3$). The number of parameters and requirements for timing of occasions are also given. (In Stata, missing data are allowed for all structures.) Whenever the variance is constant, it is denoted σ^2 and factored out. (cont.)

Toeplitz(1)   (2 parameters)   ($t_{ij} - t_{i-1,j} = \Delta$)

$$\sigma^2 \begin{bmatrix} 1 & & \\ \rho_1 & 1 & \\ 0 & \rho_1 & 1 \end{bmatrix}$$

Independence   (n parameters)   ($t_{ij} = t_i$)

$$\begin{bmatrix} V_{11} & & \\ 0 & V_{22} & \\ 0 & 0 & V_{33} \end{bmatrix}$$

We will describe each of these structures in detail in the following subsections. When reading these subsections, it may be useful to refer to figure 6.1, which shows how the models are related to each other assuming balance and constant spacing. An arrow pointing from model A to model B means that model B is nested in (a special case of) model A. Therefore, any model that can be reached by following arrows from model A is nested in model A. For example, the identity structure (zero correlations and equal variances) is nested in the exchangeable structure, which is nested in the unstructured model. In fact, all models are nested in the unstructured model, and the identity structure is nested in all models.

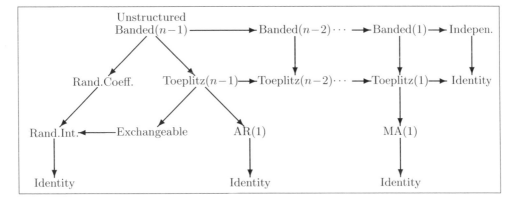

Figure 6.1: Relationships between covariance structures assuming balance and constant spacing; arrows point from a more general model to a model nested within it

In this chapter, we assume that $E(\xi_{ij}|\mathbf{X}_j) = 0$. This kind of exogeneity assumption is invariably (usually implicitly) made in the literature on marginal modeling, possibly because of the popularity of this approach in clinical trials where treatments are exogenous per design since subjects are randomized to treatments. Alternatively, the coefficients are not interpreted causally, but the mean structure is viewed merely as a linear projection (see section 1.13).

When the model is fit by maximum likelihood (ML), as we will do throughout most of this chapter, the likelihood is based on assuming multivariate normality of the total residuals. When this assumption is violated, point estimates of regression coefficients remain consistent (if the mean structure is correctly specified) as do model-based standard errors (if the covariance structure is also correctly specified). The residual covariance matrix is also consistently estimated if its structure, and the mean structure, are correctly specified. It is not widely known that as long as the distribution of the residuals is symmetric (for example, normal), ML and feasible generalized least squares (FGLS) estimators of regression coefficients are not only consistent but also small-sample unbiased, even if the covariance structure is incorrectly specified.

Model-based standard errors are valid only if the covariance structure is correctly specified. Furthermore, even for the point estimates of the regression coefficients, consistency relies on correct specification of the covariance structure if responses are missing, and missingness depends on observed responses at other occasions (see section 5.8 and exercise 6.6). Therefore, it is important to specify a good (close to correct) covariance structure if there are missing data. A good choice of covariance structure also improves efficiency (precision of estimation) for regression coefficients. Finally, understanding the nature of the longitudinal dependence may be of interest in its own right.

6.3.1   Unstructured covariance matrix

The most obvious approach is not to impose any structure on the residual covariance matrix and to estimate the variance at each occasion and the covariances between each pair of occasions freely. This is the most general example of a covariance structure that is constant across subjects. If there are n occasions, such an *unstructured* covariance matrix has $n(n + 1)/2$ parameters—namely, n variances and $n(n - 1)/2$ unique covariances [keeping in mind that $\text{Cov}(\xi_{ij}, \xi_{i'j}) = \text{Cov}(\xi_{i'j}, \xi_{ij})$, that is, covariance matrices are symmetric].

An unstructured covariance matrix should be used only if $t_{ij} = t_i$ (or in practice if the equality holds approximately and the time variable is rounded to obtain the same integer values across subjects), but gaps due to missing data are permitted. However, there must be sufficient numbers of subjects who have nonmissing responses at each occasion to make it feasible to estimate a separate variance parameter for each occasion and separate covariance parameters for each pair of occasions.

We will fit the model to the wage-panel data described in *Introduction to models for longitudinal and panel data (part III)*. We first read in the data and generate the necessary variables:

```
. use http://www.stata-press.com/data/mlmus3/wagepan
. generate educt = educ - 12
. generate yeart = year - 1980
```

Previously, we have used Stata's `xtmixed` command for multilevel modeling and now use `xtmixed` for marginal modeling. As for multilevel models, the double pipe, ||, is used to separate the fixed and random parts of the model. The cluster identifier is given, followed by a colon, to specify the clusters (here subjects) within which the residuals should be correlated. By default, `xtmixed` includes a random intercept, so we use the `noconstant` option in the random part to omit the random intercept. What would ordinarily be the level-1 residual ϵ_{ij} then becomes the total residual ξ_{ij}. The `residuals()` option allows covariance structures to be specified for ϵ_{ij} and hence for ξ_{ij}. In contrast, covariance structures for random effects (such as ζ_{1j} and ζ_{2j}) are specified by using the `covariance()` option.

We now use `xtmixed` with the `noconstant` and `mle` options for ML estimation of marginal models. The `residuals(unstructured, t())` option is used to specify an unstructured covariance matrix where `t()` is used to specify the time variable t_i, which should be integer-valued here. For brevity, we display the random part of the model only and omit cluster information by using the `nofetable` and `nogroup` options (the model takes a long time to estimate).

```
. xtmixed lwage black hisp union married exper yeart educt || nr:,
> noconstant residuals(unstructured, t(yeart)) nofetable nogroup mle
```

Mixed-effects ML regression   Number of obs   =   4360
                                               Wald chi2(7)   =   569.40
Log likelihood = -1977.7961   Prob > chi2   =   0.0000

Random-effects Parameters	Estimate	Std. Err.	[95% Conf.	Interval]
nr: (empty)				
Residual: Unstructured				
sd(e0)	.5310578	.0163	.5000524	.5639856
sd(e1)	.5032696	.0153426	.4740793	.5342571
sd(e2)	.4615847	.0140538	.4348456	.489968
sd(e3)	.4481479	.0136054	.4222597	.4756232
sd(e4)	.4990527	.0151449	.4702349	.5296367
sd(e5)	.4878995	.0148655	.4596165	.5179228
sd(e6)	.4859753	.0147864	.4578417	.5158377
sd(e7)	.4332586	.013296	.4079671	.4601181
corr(e0,e1)	.3930551	.036518	.3192001	.4621645
corr(e0,e2)	.3699349	.0372008	.2948549	.4404751
corr(e0,e3)	.3413282	.0381116	.2645994	.4137627
corr(e0,e4)	.2489353	.0403531	.168344	.3262208
corr(e0,e5)	.2813188	.0396332	.2019437	.3570224
corr(e0,e6)	.2216649	.04092	.1401321	.3002119
corr(e0,e7)	.2272449	.0407984	.1459159	.3055229
corr(e1,e2)	.5634612	.0293651	.5031855	.618281
corr(e1,e3)	.5320892	.0308314	.4689796	.5897928
corr(e1,e4)	.4559925	.0340303	.3868009	.5200785
corr(e1,e5)	.4125282	.0355852	.340453	.4797838
corr(e1,e6)	.3359342	.0381591	.259151	.4084978
corr(e1,e7)	.4156282	.0355712	.3435559	.4828333
corr(e2,e3)	.6392672	.0254606	.5866517	.6865085
corr(e2,e4)	.5816768	.0283895	.523317	.6346054
corr(e2,e5)	.5309919	.030854	.4678444	.5887455
corr(e2,e6)	.466369	.0335636	.3980685	.5295275
corr(e2,e7)	.4295472	.0350594	.3584261	.495711
corr(e3,e4)	.6351057	.025611	.5822074	.6826498
corr(e3,e5)	.5756103	.0286975	.5166483	.6291385
corr(e3,e6)	.4851861	.0329002	.4181174	.5469937
corr(e3,e7)	.5045518	.0320256	.439159	.5646266
corr(e4,e5)	.6217451	.0263582	.5673651	.6707257
corr(e4,e6)	.512188	.0317281	.4473563	.5716651
corr(e4,e7)	.5413647	.0303945	.4790993	.5982091
corr(e5,e6)	.5838479	.0283346	.5255848	.6366597
corr(e5,e7)	.629591	.0259946	.5759177	.6778598
corr(e6,e7)	.6453735	.0251813	.5933006	.6920676

LR test vs. linear regression: chi2(35) =   2020.88   Prob > chi2 = 0.0000

Note: The reported degrees of freedom assumes the null hypothesis is not on
      the boundary of the parameter space. If this is not true, then the
      reported test is conservative.

```
. estimates store un
```

The `variance` option could have been used to display estimated variances and co-variances instead of standard deviations and correlations. We can use the program `xtmixed_corr` by Bobby Gutierrez (2011)[1] to display the model-implied standard deviations and correlation matrix for one of the subjects (here the same for all subjects). The program can be downloaded using

```
ssc install xtmixed_corr, replace
```

and run using

```
. xtmixed_corr
Standard deviations and correlations for nr = 13:
Standard Deviations:
     yeart |     0       1       2       3       4       5       6       7

        sd |  0.531   0.503   0.462   0.448   0.499   0.488   0.486   0.433

Correlations:
     yeart |     0       1       2       3       4       5       6       7

         0 |  1.000
         1 |  0.393   1.000
         2 |  0.370   0.563   1.000
         3 |  0.341   0.532   0.639   1.000
         4 |  0.249   0.456   0.582   0.635   1.000
         5 |  0.281   0.413   0.531   0.576   0.622   1.000
         6 |  0.222   0.336   0.466   0.485   0.512   0.584   1.000
         7 |  0.227   0.416   0.430   0.505   0.541   0.630   0.645   1.000
```

Because the number and timing of occasions can vary between subjects in some datasets, `xtmixed_corr` displays the results for the first subject in the data (here subject 13) by default. The `at()` option can be used to specify a different subject (see also section 6.4.3).

The estimated residual standard deviations and correlation matrix are also shown in the top left panel of figure 6.2.

---

1. We thank Bobby for writing this program as a result of seeing an earlier draft of our chapter where we presented Mata code to produce this output.

Unstructured

```
|| nr:, noconstant
   residuals(unstructured, t(yeart))
```

[.53   .50   .46   .45   .50   .49   .49   .43]

```
⎡  1                                        ⎤
⎢ .39    1                                  ⎥
⎢ .37   .56    1                            ⎥
⎢ .34   .53   .64    1                      ⎥
⎢ .25   .46   .58   .64    1                ⎥
⎢ .28   .41   .53   .58   .62    1          ⎥
⎢ .22   .34   .47   .49   .51   .58    1    ⎥
⎣ .23   .42   .43   .50   .54   .63   .65  1⎦
```

Random-intercept

```
|| nr:
```

[.48   .48   .48   .48   .48   .48   .48   .48]

```
⎡  1                                        ⎤
⎢ .46    1                                  ⎥
⎢ .46   .46    1                            ⎥
⎢ .46   .46   .46    1                      ⎥
⎢ .46   .46   .46   .46    1                ⎥
⎢ .46   .46   .46   .46   .46    1          ⎥
⎢ .46   .46   .46   .46   .46   .46    1    ⎥
⎣ .46   .46   .46   .46   .46   .46   .46  1⎦
```

Random-coefficient

```
|| nr:  yeart, cov(unstructured)
```

[.49   .48   .47   .46   .47   .48   .49   .51]

```
⎡  1                                        ⎤
⎢ .55    1                                  ⎥
⎢ .52   .52    1                            ⎥
⎢ .48   .49   .50    1                      ⎥
⎢ .43   .46   .48   .50    1                ⎥
⎢ .38   .42   .46   .49   .52    1          ⎥
⎢ .33   .38   .43   .47   .51   .54    1    ⎥
⎣ .28   .34   .40   .45   .50   .54   .58  1⎦
```

AR(1)

```
|| nr:, noconstant
   residuals(ar 1, t(yeart))
```

[.48   .48   .48   .48   .48   .48   .48   .48]

```
⎡  1                                        ⎤
⎢ .58    1                                  ⎥
⎢ .33   .58    1                            ⎥
⎢ .19   .33   .58    1                      ⎥
⎢ .11   .19   .33   .58    1                ⎥
⎢ .06   .11   .19   .33   .58    1          ⎥
⎢ .04   .06   .11   .19   .33   .58    1    ⎥
⎣ .02   .04   .06   .11   .19   .33   .58  1⎦
```

MA(1)

```
|| nr:, noconstant
   residuals(ma 1, t(yeart))
```

[.46   .46   .46   .46   .46   .46   .46   .46]

```
⎡  1                                        ⎤
⎢ .36    1                                  ⎥
⎢  0    .36    1                            ⎥
⎢  0     0    .36    1                      ⎥
⎢  0     0     0    .36    1                ⎥
⎢  0     0     0     0    .36    1          ⎥
⎢  0     0     0     0     0    .36    1    ⎥
⎣  0     0     0     0     0     0    .36  1⎦
```

Banded(1)

```
|| nr:, noconstant
   residuals(banded 1, t(yeart))
```

[.53   .49   .43   .42   .47   .48   .47   .43]

```
⎡  1                                        ⎤
⎢ .31    1                                  ⎥
⎢  0    .31    1                            ⎥
⎢  0     0    .44    1                      ⎥
⎢  0     0     0    .26    1                ⎥
⎢  0     0     0     0    .50    1          ⎥
⎢  0     0     0     0     0    .13    1    ⎥
⎣  0     0     0     0     0     0    .59  1⎦
```

Figure 6.2: Estimated residual standard deviations and correlation matrices from xtmixed

	Toeplitz(2)						
|| nr:, noconstant							
residuals(toeplitz 2, t(yeart))							

$$\begin{bmatrix} .46 & .46 & .46 & .46 & .46 & .46 & .46 & .46 \end{bmatrix}$$

$$\begin{bmatrix} 1 & & & & & & & \\ .44 & 1 & & & & & & \\ .26 & .44 & 1 & & & & & \\ 0 & .26 & .44 & 1 & & & & \\ 0 & 0 & .26 & .44 & 1 & & & \\ 0 & 0 & 0 & .26 & .44 & 1 & & \\ 0 & 0 & 0 & 0 & .26 & .44 & 1 & \\ 0 & 0 & 0 & 0 & 0 & .26 & .44 & 1 \end{bmatrix}$$

	Random-intercept & AR(1)						
|| nr:							
residuals(ar 1, t(yeart))							

$$\begin{bmatrix} .48 & .48 & .48 & .48 & .48 & .48 & .48 & .48 \end{bmatrix}$$

$$\begin{bmatrix} 1 & & & & & & & \\ .57 & 1 & & & & & & \\ .45 & .57 & 1 & & & & & \\ .41 & .45 & .57 & 1 & & & & \\ .40 & .41 & .45 & .57 & 1 & & & \\ .40 & .40 & .41 & .45 & .57 & 1 & & \\ .40 & .40 & .40 & .41 & .45 & .57 & 1 & \\ .40 & .40 & .40 & .40 & .41 & .45 & .57 & 1 \end{bmatrix}$$

Figure 6.2: Estimated residual standard deviations and correlation matrices from `xtmixed` (cont.)

Models with an unstructured covariance matrix can also be used when there are several different response variables, such as weight and height, instead of longitudinal data. In this case, it seems necessary to allow for different residual variances and covariances for different (pairs) of variables because the variables are measured in different units. In addition, the effects of covariates would typically be allowed to differ between the response variables (see exercise 6.5 for an example). In such situations, the models are referred to as multivariate regression models or seemingly unrelated regressions.

Specifying an unstructured covariance matrix may appear to be ideal because it seems to make no assumptions about the covariance structure. However, the covariance structure is not completely unrestricted because it is assumed to be constant across subjects. Also, unless there are relatively few occasions, the number of parameters becomes large (for example, 36 (co)variance parameters for the wage-panel data for $n = 8$ and 120 parameters for $n = 15$), which may lead to unreliable estimation. Therefore, structured covariance matrices are often used that depend on a smaller number of parameters.

The regression parameters will be more precisely estimated if a structured covariance matrix is specified, assuming that the structure is correct. Another reason for specifying structured covariance matrices is that unstructured covariance matrices cannot be used if different timings are associated with occasions for different subjects. For these two reasons, a large variety of covariance structures have been proposed.

6.3.2 Random-intercept or compound symmetric/exchangeable structure

In a random-intercept model, the total residual ξ_{ij} is partitioned as

$$\xi_{ij} = \zeta_{1j} + \epsilon_{ij}$$

where ζ_{1j} and ϵ_{ij} have zero means, variances ψ_{11} and θ, and are uncorrelated with each other. The ζ_{1j} are uncorrelated across subjects, and the ϵ_{ij} are uncorrelated across both subjects and occasions.

In section 3.3.1, we discussed the marginal residual covariance matrix implied by this model; see table 6.2. The marginal variances are $\psi_{11} + \theta$ at all occasions, and the covariances are ψ_{11} for all pairs of occasions. The corresponding correlation is often referred to as an intraclass correlation. This structure does not require balance or constant spacing of occasions.

Table 6.2: Conditional variances and covariances of total residuals for random-intercept model

Conditional or subject-specific			Marginal or population-averaged		
$\mathrm{Var}(\xi_{ij}\|\zeta_{1j})$	$=$	θ	$\mathrm{Var}(\xi_{ij})$	$=$	$\psi_{11} + \theta$
$\mathrm{Cov}(\xi_{ij}, \xi_{i'j}\|\zeta_{1j})$	$=$	0	$\mathrm{Cov}(\xi_{ij}, \xi_{i'j})$	$=$	ψ_{11}

When a random-intercept structure is used in the marginal modeling approach, it could be argued that ψ_{11} and θ need not both be positive as long as their sum $\psi_{11} + \theta$ is nonnegative. Only this sum is interpreted as a variance, and ψ_{11} is interpreted as a covariance, so ψ_{11} could be negative. This more general covariance structure, with possibly negative covariances, is called the *compound symmetric* or *exchangeable* structure. All variances are equal and all covariances (and thus correlations) are also equal.

The random-intercept model fit to the wage-panel data using `xtmixed` in section 5.2 can be viewed as a marginal model with a random-intercept structure. The residual intraclass correlation was estimated as 0.46, and the total standard deviation was estimated as $\sqrt{0.327^2 + 0.354^2} = 0.48$. The model therefore implies that all pairwise correlations are equal to 0.46 and all standard deviations are equal to 0.48 (as shown in the top-right panel of figure 6.2).

A model with the compound symmetric or exchangeable covariance structure can be fit in `xtmixed` with the `residuals(exchangeable)` option:

```
. quietly xtmixed lwage black hisp union married exper yeart educt || nr:,
> noconstant residuals(exchangeable) mle
```

where the `noconstant` option after the double pipe ensures that no random intercept is included. No `t()` option is needed here because the covariance structure does not depend on the order or timing of occasions. Because the estimated covariance is positive, all estimates are identical to those reported for the random-intercept model (see section 5.2 and see table 5.1 on page 260), and we therefore used the `quietly` prefix command to omit the output. We store the estimates as `ri` for later use:

```
. estimates store ri
```

6.3.3   Random-coefficient structure

Consider now a random-coefficient model with a random slope for a time-varying covariate, such as the time t_{ij} associated with occasion i for subject j. In this case, the total residual becomes

$$\xi_{ij} \;=\; \zeta_{1j} + \zeta_{2j}t_{ij} + \epsilon_{ij}$$

where the random effects ζ_{1j} and ζ_{2j} and the level-1 error ϵ_{ij} have zero means and variances ψ_{11}, ψ_{22}, and θ, respectively. The random effects ζ_{1j} and ζ_{2j} have covariance ψ_{21} within subjects and are uncorrelated across subjects. The level-1 errors are uncorrelated with the random effects and uncorrelated across subjects and occasions.

An overview of the relationship between conditional and marginal (with respect to ζ_{1j} and ζ_{2j}) variances and covariances of y_{ij} for this model is given in table 6.3. As discussed in section 4.4.2, the variances and covariances (as well as correlations) now depend on the variable having a random coefficient, here t_{ij}. Balance and equal spacing are not required. We see that the marginal covariance matrix differs between subjects if there is no balance, $t_{ij} \neq t_i$. (The random-coefficient structure is also shown in matrix form in display 6.1.)

Table 6.3: Conditional and marginal variances and covariances of total residuals for random-coefficient model

Conditional or subject-specific		Marginal or population-averaged	
$\mathrm{Var}(\xi_{ij}\|t_{ij},\zeta_{1j},\zeta_{2j})$	$=\theta$	$\mathrm{Var}(\xi_{ij}\|t_{ij})$	$=\psi_{11}+2\psi_{21}t_{ij}+\psi_{22}t_{ij}^{2}+\theta$
$\mathrm{Cov}(\xi_{ij},\xi_{i'j}\|t_{ij},t_{i'j},\zeta_{1j},\zeta_{2j})=0$		$\mathrm{Cov}(\xi_{ij},\xi_{i'j}\|t_{ij},t_{i'j})$	$=\psi_{11}+\psi_{21}(t_{ij}+t_{i'j})+\psi_{22}t_{ij}t_{i'j}$

We first write a general two-level linear mixed model for the n_j-dimensional vector of responses $\mathbf{y}_j$ for subject j as

$$\mathbf{y}_j \;=\; \mathbf{X}_j\boldsymbol{\beta} + \mathbf{Z}_j\boldsymbol{\zeta}_j + \boldsymbol{\epsilon}_j$$

where $\mathbf{X}_j$ is the $n_j \times p$ covariate matrix for the fixed part and $\mathbf{Z}_j$ is the $n_j \times q$ covariate matrix for the random part, consisting of rows $(1, z_{1ij}, \ldots, z_{q-1,ij})$, where z_{1ij}, etc., are covariates that have random slopes. Analogous to the p-dimensional fixed coefficient vector $\boldsymbol{\beta}$, there is a q-dimensional random coefficient vector $\boldsymbol{\zeta}_j = (\zeta_{1j}, \ldots, \zeta_{qj})'$, and $\boldsymbol{\epsilon}_j$ is an n_j-dimensional vector of level-1 residuals. Under the usual assumption that $\boldsymbol{\zeta}_j$ is uncorrelated with $\boldsymbol{\epsilon}_j$, we obtain

$$\mathrm{Cov}(\mathbf{y}_j|\mathbf{X}_j,\mathbf{Z}_j) \;=\; \mathbf{Z}_j\boldsymbol{\Psi}\mathbf{Z}_j' + \mathbf{R}_j$$

where $\boldsymbol{\Psi} = \mathrm{Cov}(\boldsymbol{\zeta}_j)$ and $\mathbf{R}_j = \mathrm{Cov}(\boldsymbol{\epsilon}_j)$, usually specified as $\mathbf{R}_j = \theta\mathbf{I}_{n_j}$ where $\mathbf{I}_{n_j}$ is the $n_j \times n_j$ identity matrix.

Display 6.1: Covariance structure induced by random-coefficient model

To get a better idea of the correlation pattern induced by random-coefficient models, consider examples of balanced longitudinal data. Figure 6.3 gives two examples with five time points and $t_{ij} = t_i$ given by 0, 1, 2, 3, and 4. The two examples differ from each other only in terms of the correlation between intercept and slope, equal to 0.2 on the left and -0.8 on the right.

As can be seen from the randomly drawn subject-specific regression lines for 10 subjects, the variances increase as a function of time on the left, whereas they change much less, first decreasing and then increasing, on the right. On the left, the correlations between adjacent time points (or lag-1 correlations) are 0.63 between occasions 1 and 2, 0.76 between occasions 2 and 3, then 0.84, and finally 0.90. Similarly, the lag-2 and lag-3 correlations increase over time. We see that the correlation between occasions 1 and 2 is greater than that between occasions 1 and 3. In general, at a given occasion, the correlations with other occasions decrease as the time lag increases. The correlation matrix on the right is very different, illustrating that the random-coefficient model can produce a wide range of different correlation patterns.

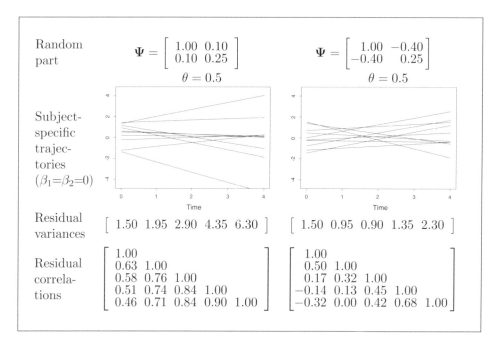

Figure 6.3: Illustration of marginal variances and correlations induced by random-coefficient models ($t = 0, 1, 2, 3, 4$)

When the random-coefficient structure is used in a marginal modeling approach, it can be argued that it is acceptable to relax the usual restriction that the covariance matrix of the random effects is positive semidefinite (for example, with nonnegative variances $0 \leq \psi_{11}$ and $0 \leq \psi_{22}$ and valid correlation $-1 \leq \rho_{21} \leq 1$) and that the level-1 residual variance is nonnegative as long as the marginal covariance matrix is positive semidefinite. In this case, the parameters are no longer interpreted as variances and covariances but merely as parameters structuring the residual covariance matrix. Relaxing the restrictions (not currently possible in Stata) makes the random-coefficient structure even more flexible but no longer interpretable as being induced by random intercepts and slopes.

A random-coefficient model with a random intercept and a random coefficient of yeart can be fit using xtmixed with the random part || nr: yeart and with the covariance(unstructured) option:

```
. xtmixed lwage black hisp union married exper yeart educt || nr: yeart,
> covariance(unstructured) nofetable nogroup mle
Mixed-effects ML regression              Number of obs      =      4360
                                         Wald chi2(7)       =    542.92
Log likelihood = -2120.9657              Prob > chi2        =    0.0000
```

Random-effects Parameters	Estimate	Std. Err.	[95% Conf. Interval]	
nr: Unstructured				
sd(yeart)	.0562206	.0031127	.0504392	.0626646
sd(_cons)	.3720286	.0151849	.3434261	.4030132
corr(yeart,_cons)	-.4626367	.050111	-.5550306	-.3589666
sd(Residual)	.3255862	.0040282	.3177859	.3335779

```
LR test vs. linear regression:       chi2(3) =   1734.54    Prob > chi2 = 0.0000
Note: LR test is conservative and provided only for reference.
. estimates store rc
```

The marginal standard deviations and correlation matrix can be obtained using

```
. xtmixed_corr
Standard deviations and correlations for nr = 13:
Standard Deviations:
```

obs	1	2	3	4	5	6	7	8
sd	0.494	0.478	0.467	0.463	0.466	0.476	0.492	0.514

Correlations:

obs	1	2	3	4	5	6	7	8
1	1.000							
2	0.545	1.000						
3	0.515	0.518	1.000					
4	0.477	0.493	0.503	1.000				
5	0.432	0.461	0.485	0.502	1.000			
6	0.382	0.423	0.460	0.491	0.516	1.000		
7	0.330	0.381	0.430	0.475	0.512	0.541	1.000	
8	0.278	0.339	0.398	0.454	0.503	0.544	0.575	1.000

and were shown in row 2, column 1 of figure 6.2 on page 302.

More complex covariance structures can be produced by adding random coefficients of the square, the cube, and possibly higher powers of yeart.

6.3.4 Autoregressive and exponential structures

Autoregressive residual covariance structures can be derived from a model where the residuals are regressed on themselves ("auto" in Greek means "self") in the sense that the residual at an occasion is regressed on residuals at previous or lagged occasions. This makes the correlations between the residuals fall off as the number of occasions between them increases.

A popular special case is the first-order or autoregressive lag-1 structure, AR(1). This structure is produced by a model where the residual at occasion i is regressed on the residual at the previous occasion $i-1$:

$$\xi_{ij} \;=\; \alpha\xi_{i-1,j} + v_{ij}, \qquad \mathrm{Cov}(\xi_{i-1,j}, v_{ij}) = 0, \quad E(v_{ij}) = 0, \quad \mathrm{Var}(v_{ij}) = \sigma_v^2$$

The disturbances v_{ij} are independent across occasions and subjects. An AR(1) process is displayed in figure 6.4.

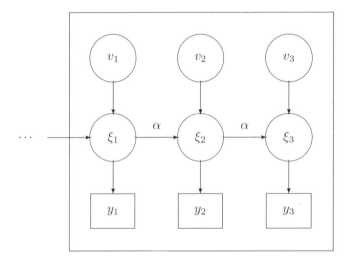

Figure 6.4: Path diagram of AR(1) process

The residual at a given occasion is determined partly by the residual at the previous occasion and partly by independent random noise. As a result of such inertia, where the present depends to some extent on the immediate past, AR(1) residuals change less from one occasion to the next than independent "white noise" residuals that have the same variance. This phenomenon is clearly seen in the simulated AR(1) and white noise time series shown in figure 6.5. In the figure, the parameters for the AR(1) process were set to $\alpha = 0.8$ and $\sigma_v^2 = 0.04$, whereas the white noise was independently distributed. Both processes were simulated from normal distributions with the same marginal variance.

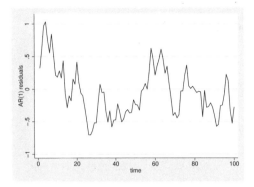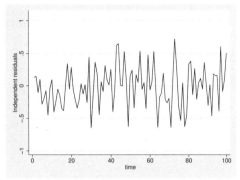

Figure 6.5: Simulated AR(1) process (left panel) and white noise (right panel) where both processes have the same mean and variance

The AR(1) model makes sense only if the time intervals between successive occasions are constant across time and across subjects, $t_{ij} - t_{i-1,j} = \Delta$ (although Stata can handle gaps, assumed to be unobserved realizations from the process). Higher-order autoregressive residual structures are produced by including further lags in the model.

A weakly or second-order *stationary* process for ξ_{ij} has expectations that are identical at all occasions, variances that are identical at all occasions, and covariances that are identical between all residuals at occasions a given time interval apart. An AR(1) process is stationary (and therefore also weakly stationary) if the process has been ongoing long before the first occasion in the dataset and $|\alpha| < 1$. For a stationary AR(1) process, the variances of the total residual at all occasions i are

$$\sigma^2 \equiv \text{Var}(\xi_{ij}) = \frac{\sigma_v^2}{1 - \alpha^2}$$

and the covariances and correlations between the residuals at occasions i and i' are

$$\text{Cov}(\xi_{ij}, \xi_{i'j}) = \sigma^2 \alpha^{|i-i'|} \qquad \text{and} \qquad \text{Cor}(\xi_{ij}, \xi_{i'j}) = \alpha^{|i-i'|}$$

respectively. We see that the correlations decrease as the time interval between the occasions increases.

For the wage-panel data, an AR(1) residual correlation structure can be fit using `xtmixed` with the `residuals(ar 1, t())` option. Here the time variable `yeart` is given in the `t()` option to determine the ordering and spacing of the occasions (the time variable must be integer valued). It is necessary to specify the cluster identifier `nr:` to identify the groups of observations that are to follow the specified structure. The `noconstant` option is used to omit the random intercept, which is otherwise included by default in `xtmixed`:

```
. xtmixed lwage black hisp union married exper yeart educt || nr:, noconstant
> residuals(ar 1, t(yeart)) nofetable nogroup mle
```

Mixed-effects ML regression   Number of obs   =   4360

                                    Wald chi2(7)   =   468.26

Log likelihood = -2237.6188   Prob > chi2   =   0.0000

Random-effects Parameters	Estimate	Std. Err.	[95% Conf. Interval]	
nr: (empty)				
Residual: AR(1)				
rho	.575936	.0126414	.5506306	.6001838
sd(e)	.4821976	.0069132	.4688365	.4959394

LR test vs. linear regression: chi2(1) =   1501.23   Prob > chi2 = 0.0000

Note: The reported degrees of freedom assumes the null hypothesis is not on
      the boundary of the parameter space. If this is not true, then the
      reported test is conservative.

```
. estimates store ar1
```

The standard deviation σ of the (total) residuals (called `sd(e)` in the output) is estimated as 0.48, and the autoregressive parameter α (called `rho` in the output) is estimated as 0.58. We can obtain the standard deviations and correlation matrix by using

```
. xtmixed_corr
```

Standard deviations and correlations for nr = 13:

Standard Deviations:

yeart	0	1	2	3	4	5	6	7
sd	0.482	0.482	0.482	0.482	0.482	0.482	0.482	0.482

Correlations:

yeart	0	1	2	3	4	5	6	7
0	1.000							
1	0.576	1.000						
2	0.332	0.576	1.000					
3	0.191	0.332	0.576	1.000				
4	0.110	0.191	0.332	0.576	1.000			
5	0.063	0.110	0.191	0.332	0.576	1.000		
6	0.036	0.063	0.110	0.191	0.332	0.576	1.000	
7	0.021	0.036	0.063	0.110	0.191	0.332	0.576	1.000

These standard deviations and correlations were shown in row 2 and column 2 of figure 6.2 on page 302.

When the occasions are irregularly spaced or the times t_{ij} associated with occasions i for subject j are nonintegers, a correlation structure of the form

$$\mathrm{Cor}(\xi_{ij}, \xi_{i'j}|t_{ij}, t_{i'j}) = \rho^{|t_{ij}-t_{i'j}|}$$

is sometimes specified. The covariance structure is no longer constant across subjects if the data are not balanced ($t_{ij} \neq t_i$). It is called the *exponential structure* in Stata and is fit by using the `residuals(exponential, t())` option in `xtmixed`. Because the wage-panel data are balanced, identical results are produced as for the AR(1) structure and therefore not shown here.

6.3.5 Moving-average residual structure

A first-order *moving-average*, MA(1), residual covariance structure can be derived from a model where the total residual depends on the current and previous value of a random variable u_{ij} that is uncorrelated between occasions.

$$\xi_{ij} = \phi u_{i-1,j} + u_{ij}, \qquad \text{Cov}(u_{i-1,j}, u_{ij}) = 0, \quad E(u_{ij}) = 0, \quad \text{Var}(u_{ij}) = \sigma_u^2$$

The process makes sense only if the time intervals are constant, $t_{ij} - t_{i-1,j} = \Delta$. The MA(1) process is displayed in figure 6.6.

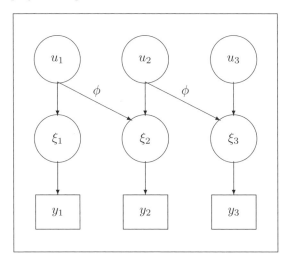

Figure 6.6: Path diagram of MA(1) process

Moving-average processes can be interpreted as arising from random shocks u_{ij}—such as strikes or government decisions or personal experiences—that affect residuals for a fixed number of occasions before disappearing. For the MA(1) process, the shock affects the current and next occasion. Higher-order moving-average structures allow shocks to affect more future occasions. An AR(1) process can be generated by letting the order k of the MA(k) process tend to infinity.

For the MA(1) process, the variance of the (total) residual at all occasions i becomes

$$\sigma^2 \equiv \text{Var}(\xi_{ij}) = \sigma_u^2 (1 + \phi^2)$$

The covariances and correlations between residuals at occasions i and i' become

$$\text{Cov}(\xi_{ij}, \xi_{i'j}) = \begin{cases} \sigma_u^2 \phi & \text{if } |i - i'| = 1 \\ 0 & \text{otherwise} \end{cases}$$

and

$$\operatorname{Cor}(\xi_{ij}, \xi_{i'j}) = \begin{cases} \phi/(1+\phi^2) & \text{if } |i-i'| = 1 \\ 0 & \text{otherwise} \end{cases}$$

respectively. We see that the process is weakly or second-order stationary.

For the wage-panel data, an MA(1) process can be estimated using `xtmixed` with the `residuals(ma 1, t())` option:

```
. xtmixed lwage black hisp union married exper yeart educt || nr:,
> noconstant residuals(ma 1, t(yeart)) nofetable nogroup mle
Mixed-effects ML regression                    Number of obs    =      4360
                                               Wald chi2(7)     =    707.90
Log likelihood =   -2522.49                    Prob > chi2      =    0.0000
```

Random-effects Parameters	Estimate	Std. Err.	[95% Conf. Interval]
nr: (empty)			
Residual: MA(1)			
theta1	.4219844	.013202	.3957695 .4475122
sd(e)	.4626877	.0052703	.4524726 .4731335

```
LR test vs. linear regression:       chi2(1) =    931.49   Prob > chi2 = 0.0000
Note: The reported degrees of freedom assumes the null hypothesis is not on
      the boundary of the parameter space.  If this is not true, then the
      reported test is conservative.
. estimates store ma1
```

The standard deviation of the (total) residual σ (called `sd(e)` in the output) is estimated as 0.46, and the weight parameter ϕ (called `theta1` in the output) is estimated as 0.42, corresponding to a correlation of 0.36 [$= 0.4219844/(1 + 0.4219844^2)$]. The standard deviations and correlation matrix can be obtained using

```
. xtmixed_corr
Standard deviations and correlations for nr = 13:
Standard Deviations:
```

yeart	0	1	2	3	4	5	6	7
sd	0.463	0.463	0.463	0.463	0.463	0.463	0.463	0.463

Correlations:

yeart	0	1	2	3	4	5	6	7
0	1.000							
1	0.358	1.000						
2	0.000	0.358	1.000					
3	0.000	0.000	0.358	1.000				
4	0.000	0.000	0.000	0.358	1.000			
5	0.000	0.000	0.000	0.000	0.358	1.000		
6	0.000	0.000	0.000	0.000	0.000	0.358	1.000	
7	0.000	0.000	0.000	0.000	0.000	0.000	0.358	1.000

and were shown in row 3 and column 1 of figure 6.2 on page 302.

6.3.6 Banded and Toeplitz structures

The *banded* covariance structure is a special case of the unstructured covariance structure with covariances between pairs of occasions further than some lag apart set to zero. For instance, the banded(1) structure allows all variances to be freely estimated as well as all covariances on the first off-diagonal (lag 1), whereas all other covariances are set to zero. A banded(2) structure also allows the second off-diagonal (lag 2) to be free, and so forth. See table 6.1 for an example of a banded(1) structure for three occasions. This structure only makes sense for balanced data with $t_{ij} = t_i$.

We fit a model with a banded(1) structure for the wage-panel data using xtmixed with the residuals(banded 1, t()) option:

```
. xtmixed lwage black hisp union married exper yeart educt || nr:, noconstant
> residuals(banded 1, t(yeart)) nofetable nogroup mle
Mixed-effects ML regression                    Number of obs     =      4360
                                               Wald chi2(7)      =    674.54
Log likelihood = -2470.0904                    Prob > chi2       =    0.0000
```

Random-effects Parameters	Estimate	Std. Err.	[95% Conf. Interval]	
nr: (empty)				
Residual: Banded(1)				
sd(e0)	.5298451	.01615	.4991185	.5624631
sd(e1)	.488302	.0141587	.4613252	.5168563
sd(e2)	.4336913	.0121559	.4105088	.4581831
sd(e3)	.4249331	.0121275	.4018161	.44938
sd(e4)	.4690608	.0133869	.4435432	.4960464
sd(e5)	.4758399	.0142071	.4487937	.5045161
sd(e6)	.4655073	.0141701	.4385467	.4941255
sd(e7)	.4323662	.0131472	.407351	.4589176
corr(e0,e1)	.3072858	.0401419	.2266593	.3837294
corr(e1,e2)	.3065635	.0437078	.2186103	.3895758
corr(e2,e3)	.4440175	.0453184	.3510051	.5283533
corr(e3,e4)	.262966	.045426	.1719201	.3495574
corr(e4,e5)	.499973	.0402668	.416992	.5746881
corr(e5,e6)	.1341619	.0375518	.0599544	.2068946
corr(e6,e7)	.594792	.0319827	.5284894	.6538825

```
LR test vs. linear regression:      chi2(14) =   1036.29    Prob > chi2 = 0.0000
Note: The reported degrees of freedom assumes the null hypothesis is not on
      the boundary of the parameter space.  If this is not true, then the
      reported test is conservative.

. estimates store ba1
```

The estimated residual standard deviations and correlations are

```
. xtmixed_corr
Standard deviations and correlations for nr = 13:
Standard Deviations:
         yeart |     0       1       2       3       4       5       6       7
---------------+----------------------------------------------------------------
            sd | 0.530   0.488   0.434   0.425   0.469   0.476   0.466   0.432

Correlations:
         yeart |     0       1       2       3       4       5       6       7
---------------+----------------------------------------------------------------
             0 | 1.000
             1 | 0.307   1.000
             2 | 0.000   0.307   1.000
             3 | 0.000   0.000   0.444   1.000
             4 | 0.000   0.000   0.000   0.263   1.000
             5 | 0.000   0.000   0.000   0.000   0.500   1.000
             6 | 0.000   0.000   0.000   0.000   0.000   0.134   1.000
             7 | 0.000   0.000   0.000   0.000   0.000   0.000   0.595   1.000
```

We see that only the correlations on the first off-diagonal of the correlation matrix are estimated (see also row 3 and column 2 of figure 6.2 on page 302).

Comparing these correlations with the first off-diagonal of the estimated unstructured correlation matrix, it is at first surprising that the correlations for the banded structure are all lower than in the unstructured case, because one might expect them to be identical. However, setting, for instance, the correlation between residuals at occasions 1 and 3 to zero is incompatible with a high correlation between occasions 1 and 2 and a high correlation between occasions 2 and 3. An extreme case would be where $\text{Cor}(\xi_{1j}, \xi_{2j}) = 1$ and $\text{Cor}(\xi_{2j}, \xi_{3j}) = 1$, from which it follows that $\text{Cor}(\xi_{1j}, \xi_{3j})$ is one and not zero. By setting all but lag-1 correlations to zero, we are therefore imposing a constraint on the lag-1 correlations. Formally, such constraints on the correlation matrix follow from the fact that such matrices are positive semidefinite.

The *Toeplitz* structure is similar to the banded structure in the sense that covariances outside the bands are set to zero. However, this structure constrains the parameters on the main diagonal to be equal and the parameters on each of the off-diagonals to be equal. For instance, the Toeplitz(1) structure constrains all variance parameters to be equal and all covariance parameters in the first off-diagonal to be equal, whereas all other covariances are set to zero; see table 6.1 for the case of three occasions. This structure appears to be equivalent to an MA(1) structure, but MA(1) forces the correlation to be at most 0.5, whereas in three dimensions, the Toeplitz(1) structure allows the correlation to be as large as $\sqrt{0.5} = 0.71$. The Toeplitz(2) structure extends Toeplitz(1) by specifying a common covariance parameter in the second off-diagonal instead of zeros, still setting all covariances on the third and further off-diagonals to zero. This structure makes sense only if the spacing between occasions is constant, $t_{ij} - t_{i-1,j} = \Delta$.

Because the estimated Toeplitz(1) covariance structure in the present application turns out to be identical to that previously reported for the MA(1) structure (because the correlation is estimated as 0.36, which is less than 0.5), we instead fit a Toeplitz(2) structure by using xtmixed with the residuals(toeplitz 2, t()) option:

```
. xtmixed lwage black hisp union married exper yeart educt || nr:, noconstant
> residuals(toeplitz 2, t(yeart)) nofetable nogroup mle
Mixed-effects ML regression                      Number of obs     =       4360
                                                 Wald chi2(7)      =     607.48
Log likelihood =  -2325.486                      Prob > chi2       =     0.0000
```

Random-effects Parameters	Estimate	Std. Err.	[95% Conf. Interval]	
nr: (empty)				
Residual: Toeplitz(2)				
rho1	.4359403	.012634	.4108515	.4603694
rho2	.258459	.0116724	.2354403	.281188
sd(e)	.4583503	.0055387	.4476222	.4693355

```
LR test vs. linear regression:        chi2(2) =   1325.50    Prob > chi2 = 0.0000
Note: The reported degrees of freedom assumes the null hypothesis is not on
      the boundary of the parameter space.  If this is not true, then the
      reported test is conservative.
. estimates store to2
```

Here rho_1 and rho_2 represent the common correlation on the first and second off-diagonal of the correlation matrix, respectively, and sd(e) represents the standard deviation parameter that is common for all occasions. The implied standard deviations and correlations are

```
. xtmixed_corr
Standard deviations and correlations for nr = 13:
Standard Deviations:
```

yeart	0	1	2	3	4	5	6	7
sd	0.458	0.458	0.458	0.458	0.458	0.458	0.458	0.458

Correlations:

yeart	0	1	2	3	4	5	6	7
0	1.000							
1	0.436	1.000						
2	0.258	0.436	1.000					
3	0.000	0.258	0.436	1.000				
4	0.000	0.000	0.258	0.436	1.000			
5	0.000	0.000	0.000	0.258	0.436	1.000		
6	0.000	0.000	0.000	0.000	0.258	0.436	1.000	
7	0.000	0.000	0.000	0.000	0.000	0.258	0.436	1.000

(see row 4 and column 1 of figure 6.2). Again the correlations are lower than in the unstructured case to satisfy positive semidefiniteness.

6.4   Hybrid and complex marginal models

6.4.1   Random effects and correlated level-1 residuals

In many of the covariance structures discussed so far, the correlations are zero for large lags or approach zero in the case of autoregressive processes. However, if there are unobserved subject-specific characteristics or individual differences that affect the response variable, we would expect nonzero correlations no matter how large the lag. More realistic covariance structures can often be produced by combining a random-intercept model with a covariance structure for the level-1 residuals.

An appealing and quite commonly used hybrid specification is a random-intercept model in which the level-1 residuals have a first-order autoregressive correlation structure. This produces a correlation matrix with serial correlations that do not approach zero as the time lag increases (when occasions are not equally spaced, an exponential structure can be used).

For the wage-panel data, we can fit a random-intercept model with AR(1) residuals using `xtmixed` by no longer including the `noconstant` option after the double pipe (as done repeatedly in this chapter) and by using the `residuals(ar 1, t())` option:

```
. xtmixed lwage black hisp union married exper yeart educt || nr:,
> residuals(ar 1, t(yeart)) nofetable nogroup mle
Mixed-effects ML regression                Number of obs    =        4360
                                           Wald chi2(7)     =      656.72
Log likelihood = -2123.5033                Prob > chi2      =      0.0000
```

Random-effects Parameters	Estimate	Std. Err.	[95% Conf. Interval]	
nr: Identity				
sd(_cons)	.3045101	.0124286	.2810994	.3298705
Residual: AR(1)				
rho	.2774511	.0213005	.2352016	.3186541
sd(e)	.3722109	.0054527	.3616757	.3830529

```
LR test vs. linear regression:      chi2(2) =  1729.46    Prob > chi2 = 0.0000
```
Note: LR test is conservative and provided only for reference.

```
. estimates store ri_ar1
```

The estimated marginal residual standard deviations and correlation matrix are obtained using

```
. xtmixed_corr
Standard deviations and correlations for nr = 13:
Standard Deviations:
```

yeart	0	1	2	3	4	5	6	7
sd	0.481	0.481	0.481	0.481	0.481	0.481	0.481	0.481

```
Correlations:
```

yeart	0	1	2	3	4	5	6	7
0	1.000							
1	0.567	1.000						
2	0.447	0.567	1.000					
3	0.414	0.447	0.567	1.000				
4	0.404	0.414	0.447	0.567	1.000			
5	0.402	0.404	0.414	0.447	0.567	1.000		
6	0.401	0.402	0.404	0.414	0.447	0.567	1.000	
7	0.401	0.401	0.402	0.404	0.414	0.447	0.567	1.000

and shown in bottom right panel of figure 6.2 on page 302 (obtained using the matrix expressions given in display 6.1, where $\mathbf{R}$ now has an AR(1) structure). As the lag increases, the estimated correlations approach $\widehat{\psi}_{11}/(\widehat{\psi}_{11}+\widehat{\theta}) = 0.40$ (see also exercise 6.8).

There are of course many different covariance structures for the level-1 residuals that could be specified for models with random intercepts and possibly random slopes of time. This offers great flexibility at the cost of potential identification problems if the implied covariance structure for the responses, given the observed covariates, becomes complex. For instance, with n occasions, a random-intercept model with a Toeplitz($n-1$) structure for the level-1 residuals is not identified. To see this, consider the covariance matrix of the total residual, which is the Toeplitz($n-1$) covariance matrix with ψ_{11} added to each element. We can add a constant to ψ_{11} and subtract the same constant from Toeplitz($n-1$) without changing the covariance matrix of the total residual, and therefore ψ_{11} is a redundant parameter. However, for Toeplitz($n-2$), ψ_{11} becomes the covariance $\text{Cov}(\xi_{1j}, \xi_{nj})$ between the first and last occasions and is therefore identified. Following this logic, Toeplitz and banded matrices of order $n-2$ or lower can be combined with a random intercept.

6.4.2 Heteroskedastic level-1 residuals over occasions

We now combine a random-coefficient structure with heteroskedastic level-1 residuals, letting the standard deviations of these residuals be occasion specific by using the `residuals(independent, by())` option:

```
. xtmixed lwage black hisp union married exper yeart educt || nr: yeart,
> covariance(unstructured) residuals(independent, by(yeart)) nofetable nogroup
> mle
```

Mixed-effects ML regression   Number of obs   =   4360
                                                Wald chi2(7)   =   548.74
Log likelihood = -2036.413   Prob > chi2   =   0.0000

Random-effects Parameters	Estimate	Std. Err.	[95% Conf. Interval]	
nr: Unstructured				
sd(yeart)	.053767	.0030945	.0480315	.0601873
sd(_cons)	.3804544	.0161599	.350064	.413483
corr(yeart,_cons)	-.472357	.0509171	-.5659588	-.3667326
Residual: Independent, by yeart				
0: sd(e)	.4568489	.0164572	.4257057	.4902705
1: sd(e)	.3558621	.0131617	.3309784	.3826165
2: sd(e)	.2880905	.010939	.2674289	.3103484
3: sd(e)	.2794221	.0100623	.2603802	.2998565
4: sd(e)	.3334089	.0112416	.312088	.3561864
5: sd(e)	.3089093	.0107731	.2884999	.3307625
6: sd(e)	.321404	.0116002	.2994535	.3449634
7: sd(e)	.2348136	.0125032	.2115432	.2606439

LR test vs. linear regression: chi2(10) =   1903.64   Prob > chi2 = 0.0000
Note: LR test is conservative and provided only for reference.

```
. estimates store rc_het
```

The estimated occasion-specific standard deviations of the level-1 errors range from 0.23 to 0.46. The model-implied marginal residual standard deviations and correlations are

```
. xtmixed_corr
```

Standard deviations and correlations for nr = 13:

Standard Deviations:

obs	1	2	3	4	5	6	7	8
sd	0.595	0.505	0.448	0.437	0.474	0.465	0.486	0.454

Correlations:

obs	1	2	3	4	5	6	7	8
1	1.000							
2	0.450	1.000						
3	0.471	0.537	1.000					
4	0.446	0.520	0.581	1.000				
5	0.376	0.451	0.517	0.540	1.000			
6	0.349	0.432	0.509	0.546	0.525	1.000		
7	0.300	0.385	0.469	0.517	0.510	0.554	1.000	
8	0.286	0.382	0.483	0.548	0.554	0.616	0.637	1.000

6.4.3   Heteroskedastic level-1 residuals over groups

We can also let the level-1 residuals have different variances for different groups of subjects, such as ethnic groups. Based on the dummy variables black and hisp, we first construct the categorical variable `ethnic`, taking the values 0 for white, 1 for black, and 2 for Hispanic:

```
. generate ethnic = black*1 + hisp*2
```

A random-intercept model with level-1 residuals that have different variances for ethnic groups can be fit by using `xtmixed` with the `residuals(independent, by())` option:

```
. xtmixed lwage black hisp union married exper yeart educt || nr:,
> residuals(independent, by(ethnic)) nofetable nogroup mle
Mixed-effects ML regression                  Number of obs    =       4360
                                             Wald chi2(7)     =     896.36
Log likelihood = -2213.0207                  Prob > chi2      =     0.0000
```

Random-effects Parameters	Estimate	Std. Err.	[95% Conf. Interval]	
nr: Identity				
sd(_cons)	.3270843	.0114203	.3054497	.3502513
Residual: Independent, by ethnic				
0: sd(e)	.3548746	.0047629	.3456612	.3643336
1: sd(e)	.3634011	.0122687	.3401332	.3882608
2: sd(e)	.3394479	.0098302	.3207177	.359272

```
LR test vs. linear regression:       chi2(3) =  1550.43    Prob > chi2 = 0.0000
Note: LR test is conservative and provided only for reference.
. estimates store ri_het2
```

We see that the estimated standard deviations (and hence variances) of the level-1 residuals are very similar for the different ethnic groups. To display the model-implied marginal residual standard deviations and correlations for the three groups, we first find one subject per group, for example, the subject with the smallest identifier in the variable `nr`,

```
. table ethnic, contents(min nr)
```

ethnic	min(nr)
0	13
1	383
2	1142

and then display the results for these three subjects by using the `at()` option in `xtmixed_corr` (within each group all subjects have the same standard deviations and the same correlations). For white, we obtain

```
. xtmixed_corr, at(nr=13)
```

Standard deviations and correlations for nr = 13:

Standard Deviations:

obs	1	2	3	4	5	6	7	8
sd	0.483	0.483	0.483	0.483	0.483	0.483	0.483	0.483

Correlations:

obs	1	2	3	4	5	6	7	8
1	1.000							
2	0.459	1.000						
3	0.459	0.459	1.000					
4	0.459	0.459	0.459	1.000				
5	0.459	0.459	0.459	0.459	1.000			
6	0.459	0.459	0.459	0.459	0.459	1.000		
7	0.459	0.459	0.459	0.459	0.459	0.459	1.000	
8	0.459	0.459	0.459	0.459	0.459	0.459	0.459	1.000

For black, we get

```
. xtmixed_corr, at(nr=383)
```

Standard deviations and correlations for nr = 383:

Standard Deviations:

obs	1	2	3	4	5	6	7	8
sd	0.489	0.489	0.489	0.489	0.489	0.489	0.489	0.489

Correlations:

obs	1	2	3	4	5	6	7	8
1	1.000							
2	0.448	1.000						
3	0.448	0.448	1.000					
4	0.448	0.448	0.448	1.000				
5	0.448	0.448	0.448	0.448	1.000			
6	0.448	0.448	0.448	0.448	0.448	1.000		
7	0.448	0.448	0.448	0.448	0.448	0.448	1.000	
8	0.448	0.448	0.448	0.448	0.448	0.448	0.448	1.000

For Hispanic, we have

```
. xtmixed_corr, at(nr=1142)
Standard deviations and correlations for nr = 1142:
Standard Deviations:
        obs |    1      2      3      4      5      6      7      8
         sd | 0.471  0.471  0.471  0.471  0.471  0.471  0.471  0.471
Correlations:
        obs |    1      2      3      4      5      6      7      8
          1 | 1.000
          2 | 0.481  1.000
          3 | 0.481  0.481  1.000
          4 | 0.481  0.481  0.481  1.000
          5 | 0.481  0.481  0.481  0.481  1.000
          6 | 0.481  0.481  0.481  0.481  0.481  1.000
          7 | 0.481  0.481  0.481  0.481  0.481  0.481  1.000
          8 | 0.481  0.481  0.481  0.481  0.481  0.481  0.481  1.000
```

6.4.4 Different covariance matrices over groups

Stata can also fit marginal models where the covariance matrix is different for groups of subjects. We illustrate this by fitting an AR(1) structure stratified by ethnicity, using xtmixed with the residuals(ar(1), t() by()) option:

```
. xtmixed lwage black hisp union married exper yeart educt || nr:, noconstant
> residuals(ar 1, t(yeart) by(ethnic)) nofetable nogroup mle
Mixed-effects ML regression              Number of obs     =      4360
                                         Wald chi2(7)      =    472.11
Log likelihood = -2233.2874              Prob > chi2       =    0.0000

---------------------------------------------------------------------------
  Random-effects Parameters  |  Estimate   Std. Err.    [95% Conf. Interval]
-----------------------------+---------------------------------------------
nr:              (empty)     |
-----------------------------+---------------------------------------------
Residual: AR(1),             |
    by ethnic                |
          0:   rho           |  .5894316   .0144551    .5603749    .6170392
               sd(e)         |  .485829    .0082504    .4699246    .5022716
                             |
          1:   rho           |  .5600764   .0374796    .4822492    .6291319
               sd(e)         |  .5025142   .020865     .4632393    .5451191
                             |
          2:   rho           |  .5114246   .0352765    .4390287    .5772225
               sd(e)         |  .4470734   .015431     .4178296    .478364
---------------------------------------------------------------------------
LR test vs. linear regression:      chi2(5) =  1509.90   Prob > chi2 = 0.0000
Note: The reported degrees of freedom assumes the null hypothesis is not on
      the boundary of the parameter space.  If this is not true, then the
      reported test is conservative.
```

The estimated parameters of the AR(1) structure are somewhat smaller for Hispanics (group 2) than for whites and blacks. We could again display the group-specific marginal residual standard deviations and correlations using

```
xtmixed_corr, at(nr=13)
xtmixed_corr, at(nr=383)
xtmixed_corr, at(nr=1142)
```

Random effects whose covariance matrix differs between groups can also be accommodated, but we defer discussion of this to section 7.5.2 in the chapter on growth curves.

6.5   Comparing the fit of marginal models

We have fit a large number of marginal models with the same mean structure (fixed part of the model) and a wide range of covariance structures.

For balanced data, an informal comparison of the fit of these different models consists of eyeballing the estimated standard deviations and correlations for each model to see how much they differ from the corresponding estimates for the unstructured case. This way, it can be judged how reasonable the restrictions are. For instance, in figure 6.2, it is immediately apparent that the restrictions imposed by the MA(1) structure are unreasonable. Correlations for lag greater than 1 are set to zero though the corresponding estimates for the unstructured case are as large as 0.63 and never smaller than 0.22. A correlation as high as 0.63 is unlikely to occur by chance.

All covariance structures that are constant across subjects (the only restriction for the unstructured case) can be obtained by imposing restrictions on the unstructured model and are hence nested in the unstructured model. (For nesting relationships between other pairs of models, refer to figure 6.1.) Therefore, we could conduct a likelihood-ratio test to compare a structured model, for example, MA(1), to the unstructured model:

```
. lrtest un ma1

Likelihood-ratio test                              LR chi2(34) =    1089.39
(Assumption: ma1 nested in un)                     Prob > chi2 =     0.0000
Note: The reported degrees of freedom assumes the null hypothesis is not on
      the boundary of the parameter space.  If this is not true, then the
      reported test is conservative.
```

As expected, the MA(1) structure is rejected at any reasonable significance level. The note at the end of the output is a useful reminder of the problem of testing a null hypothesis that restricts parameters to lie on the boundary of the parameter space. We have discussed this problem before in the context of testing the null hypothesis that a variance parameter is zero in a random-intercept or random-coefficient model. For the null hypothesis just tested here, there does not appear to be a boundary issue, but sometimes this is not transparent.

Performing likelihood-ratio tests to compare each of the structures with the unstructured model, we reject each of the structures at the 5% level. However, it should be kept in mind that any structure that is not exactly correct will be rejected as the sample size and hence the power of the tests increases. It could be argued that models should be simplifications of reality (parsimonious, with few parameters) as long as prominent features are not smoothed away (good fit or large log likelihood). Therefore, other criteria for model selection have been proposed that are not based on statistical significance.

For a given dataset, the fit can usually be improved by increasing the number of parameters that are estimated. There is therefore an unavoidable trade-off between desired model fit and undesired model complexity. Information criteria, such as AIC and BIC (described below), have been developed in an attempt to provide a rational trade-off between fit and complexity. The idea is to choose the model with the smallest "badness", where badness is defined as twice the log likelihood, a measure of misfit, plus a penalty for model complexity (a function of the number of model parameters).

An additional reason for using information criteria in favor of likelihood-ratio tests is that they can be used to choose between models that are not nested.

We can obtain information criteria for the models we have fit (ignoring models that allow covariance matrices to differ between ethnic groups) by using `estimates stats` as follows:

```
. estimates stats un ri rc ar1 ma1 ba1 to2 rc_het
```

Model	Obs	ll(null)	ll(model)	df	AIC	BIC
un	4360	.	-1977.796	44	4043.592	4324.322
ri	4360	.	-2214.357	10	4448.714	4512.517
rc	4360	.	-2120.966	12	4265.931	4342.494
ar1	4360	.	-2237.619	10	4495.238	4559.04
ma1	4360	.	-2522.49	10	5064.98	5128.782
ba1	4360	.	-2470.09	23	4986.181	5132.926
to2	4360	.	-2325.486	11	4672.972	4743.155
rc_het	4360	.	-2036.413	19	4110.826	4232.05

Note: N=Obs used in calculating BIC; see [R] BIC note

In addition to the log likelihoods, `ll(model)`, and the number of estimated parameters, `df`, the output shows the Akaike information criterion (AIC) and the Bayesian information criterion (BIC) for each of the models.

The AIC for a particular model is defined as

$$\text{AIC} \ = \ -2\log\text{likelihood} + 2\,k$$

where "log likelihood" is the maximum log likelihood for the model and k is the number of parameters estimated. BIC is defined as

$$\text{BIC} \ = \ -2\log\text{likelihood} + \ln(N)\,k$$

where N is the number of observations. For both the AIC and the BIC, the model that has the smallest value of the criterion is preferred, but the best model according to the AIC may differ from the best model according to the BIC.

For the wage-panel data, the unstructured model (with 44 parameters) is best according to the AIC, whereas the random-coefficient model with heteroskedastic level-1 residuals (with 19 parameters) is best according to the BIC. In general, the BIC tends to select more parsimonious models than does the AIC.

When two models have the same number of parameters, both criteria will lead to the same choice between the models. Note that we have fit some models [for example, the banded(1) model] that are clearly bad in the sense that they have a lower log likelihood than a model with fewer parameters (for example, the random-intercept model). In such cases, any sensible criterion would prefer the simpler and better-fitting model.

Whenever log likelihoods or information criteria are being compared, the log likelihood must be based on the same data and therefore the same number of observations. As shown in the output, the number of observations analyzed for each model was 4,360. (When comparing models with different sets of covariates, different missingness patterns for different covariates will often result in different estimation samples for different models.)

A theoretical justification for the AIC can be given in terms of cross-validation, and the BIC can be motivated as an approximation to Bayes factors used by Bayesians for model selection. Unfortunately, the AIC and BIC given above were derived for models with independent observations, not for longitudinal or clustered data, and there does not appear to be any theoretical justification for their use in longitudinal (or multilevel) models. For instance, the BIC depends on the sample size N, but it is not clear what the sample size should be for clustered data. Should it be the number of clusters J, the total number of observations $\sum_{j=1}^{J} n_j$ (used by Stata), or something in between? Nevertheless, AIC and BIC are quite commonly used for model selection in longitudinal and multilevel modeling. In this setting, the criteria are best viewed as informal indices of lack of fit, with the smallest value suggesting the preferred model.

In this chapter on marginal modeling, we have concentrated on specifying the correct covariance structure. However, research questions usually concern the relationships between response and explanatory variables, so regression coefficients are the main focus. These coefficients can be estimated consistently even if the covariance structure is misspecified, as long as the mean structure (fixed part of the model) is correctly specified. (This is true here because the models are linear and there are no missing data; see section 5.8 and exercise 6.6.) However, in finite samples, the estimates will of course differ.

We can obtain a table of estimated regression coefficients and their standard errors for the models compared here by using the following `estimates table` command:

```
. estimates table un ri rc ar1 ma1 ba1 to2 ri_ar1 rc_het,
> b(%4.3f) se(%4.3f) keep(lwage:)
```

Variable	un	ri	rc	ar1	ma1	ba1	to2	ri_ar1	rc_het
black	-0.134	-0.134	-0.130	-0.133	-0.137	-0.141	-0.131	-0.133	-0.141
	0.047	0.048	0.048	0.039	0.029	0.029	0.032	0.048	0.048
hisp	0.017	0.017	0.017	0.020	0.018	0.019	0.014	0.019	0.020
	0.042	0.043	0.043	0.035	0.025	0.025	0.028	0.042	0.043
union	0.094	0.111	0.111	0.099	0.136	0.127	0.131	0.097	0.103
	0.017	0.018	0.018	0.018	0.018	0.017	0.018	0.018	0.017
married	0.077	0.075	0.076	0.081	0.100	0.093	0.092	0.074	0.075
	0.017	0.017	0.017	0.019	0.017	0.017	0.018	0.018	0.017
exper	0.036	0.033	0.036	0.035	0.033	0.031	0.033	0.034	0.031
	0.011	0.011	0.011	0.009	0.007	0.007	0.007	0.011	0.011
yeart	0.022	0.026	0.023	0.026	0.026	0.027	0.027	0.026	0.026
	0.011	0.011	0.012	0.010	0.008	0.008	0.008	0.011	0.012
educt	0.096	0.095	0.095	0.095	0.094	0.094	0.094	0.095	0.095
	0.010	0.011	0.011	0.009	0.006	0.006	0.007	0.011	0.011
_cons	1.313	1.317	1.308	1.300	1.301	1.314	1.298	1.313	1.334
	0.037	0.037	0.038	0.033	0.025	0.026	0.028	0.038	0.039

```
legend: b/se
```

The overall impression is that the estimates of the regression coefficients (first line for each covariate) are quite similar for the different covariance structures.

However, the model-based standard errors rely on correct specification of the covariance structure. We see that the estimated standard errors (second line for each covariate) differ somewhat across covariance structures and appear to be larger for the unstructured, random-intercept, and random-coefficient structures than for the other structures. We could of course use robust standard errors by specifying the vce(robust) option in the xtmixed commands. However, these standard errors can perform poorly in small samples (when there are few subjects).

Because inferences for regression coefficients depend on the specified covariance structure, it is usually recommended to select the covariance structure before selecting the mean structure. However, the *residual* covariance structure depends on the residuals and hence on the specified mean structure. Therefore it is often recommended to include all potentially relevant covariates and interactions when selecting the covariance structure and then keep the chosen covariance structure when refining the mean structure.

6.6   Generalized estimating equations (GEE)

Recall from section 3.10.1 that for a known residual covariance matrix, maximum likelihood (ML) estimation of the regression coefficients can be accomplished by generalized least squares (GLS). Feasible generalized least squares (FGLS) substitutes estimates for the covariance matrix based on residuals from pooled ordinary least squares (OLS). In iteratively reweighted generalized least squares (IGLS), estimation proceeds by iterating between estimating the covariance matrix based on residuals from FGLS and using FGLS based on this covariance matrix to estimate new regression coefficients and hence new residuals, until convergence is achieved.

For linear marginal models with continuous responses, the xtgee command for generalized estimating equations (GEE) works essentially in the same way as IGLS, iterating between estimation of the residual covariance matrix and the regression coefficients. The only difference between GEE and IGLS is the estimator for the residual correlation matrix, which is called the "working correlation matrix" in GEE. The residual variance is assumed to be constant across occasions.

Stata's xtgee command offers all the correlation structures corresponding to the covariance structures discussed in this chapter except the random-coefficient and moving-average structures. Note that banded correlation structures are called "nonstationary" and Toeplitz correlation structures are called "stationary" in xtgee. Specifying the "independent" structure simply produces the pooled OLS estimator.

The correlation structure is what is specified in GEE, not the covariance structure. Even in the unstructured case, the variances are all constrained to be equal in GEE, which may be very restrictive. For instance, when modeling children's weights as a function of age, it seems obvious that their weights are less variable at age 0 than they are at age 5.

We now briefly demonstrate how a marginal model with an AR(1) working correlation structure can be fit using xtgee with the corr(ar 1) option:

```
. quietly xtset nr yeart

. xtgee lwage black hisp union married exper yeart educt, corr(ar 1) vce(robust)

GEE population-averaged model              Number of obs      =      4360
Group and time vars:            nr yeart   Number of groups   =       545
Link:                           identity   Obs per group: min =         8
Family:                         Gaussian                  avg =       8.0
Correlation:                       AR(1)                   max =         8
                                           Wald chi2(7)       =    547.37
Scale parameter:                .2322738   Prob > chi2        =    0.0000

                                       (Std. Err. adjusted for clustering on nr)
```

lwage	Coef.	Robust Std. Err.	z	P>|z|	[95% Conf. Interval]	
black	-.1329348	.0487983	-2.72	0.006	-.2285776	-.0372919
hisp	.0195104	.0398453	0.49	0.624	-.058585	.0976058
union	.0995969	.0195432	5.10	0.000	.061293	.1379008
married	.0813371	.0190996	4.26	0.000	.0439026	.1187715
exper	.0351937	.0107019	3.29	0.001	.0142183	.0561691
yeart	.025618	.0113301	2.26	0.024	.0034115	.0478246
educt	.0952535	.0107629	8.85	0.000	.0741586	.1163485
_cons	1.300357	.0381552	34.08	0.000	1.225574	1.37514

As is invariably done in practice, we have also used the vce(robust) option to produce robust standard errors based on the sandwich estimator. Robust standard errors for the estimated regression coefficients do not rely on correct specification of the correlation structure, although the mean structure must be correctly specified. The estimates of the regression coefficients are nearly identical to the ML estimates with the same covariance structure, but the standard errors are a little different. With the vce(robust) option in xtmixed,

```
xtmixed lwage black hisp union married exper yeart educt || nr:, noconstant
    residuals(ar 1, t(yeart)) vce(robust)
```

the standard errors are practically identical to those produced by xtgee with the vce(robust) option above.

The estimated correlation matrix of the residuals or working correlation matrix can be displayed using

```
. estat wcorrelation, format(%4.3f)

Estimated within-nr correlation matrix R:

         c1     c2     c3     c4     c5     c6     c7     c8

   r1 │ 1.000
   r2 │ 0.575  1.000
   r3 │ 0.330  0.575  1.000
   r4 │ 0.190  0.330  0.575  1.000
   r5 │ 0.109  0.190  0.330  0.575  1.000
   r6 │ 0.063  0.109  0.190  0.330  0.575  1.000
   r7 │ 0.036  0.063  0.109  0.190  0.330  0.575  1.000
   r8 │ 0.021  0.036  0.063  0.109  0.190  0.330  0.575  1.000
```

We see that the estimated correlation matrix is identical, to two decimal places, to the one using ML estimation shown in figure 6.2 (partly because of the balanced nature of the data). The estimated residual standard deviation is the square root of the scale parameter, here estimated as 0.48 ($= \sqrt{0.2322738}$).

Identical results can also be obtained using xtreg with the pa (for "population averaged") and vce(robust) (for "robust standard errors") options:

```
xtset nr yeart
xtreg lwage black hisp union married exper yeart educt, pa corr(ar 1) vce(robust)
```

followed by

```
matrix list e(R)
```

GEE was actually developed for noncontinuous responses, such as dichotomous responses and counts (see volume 2, sections 10.14.2 and 13.11.3), and is rarely used for continuous responses.

6.7 Marginal modeling with few units and many occasions

So far we have considered longitudinal data with short panels, where there are far more units (or subjects) than occasions. In data with long panels, there are nearly as many or more occasions than there are units. Political scientists tend to call such data "time-series–cross-sectional data". Usually, the units are states or countries, and cross-sectional correlations are expected between responses for different (possibly neighboring) units at a given occasion, which should be taken into account when estimating standard errors.

6.7.1   Is a highly organized labor market beneficial for economic growth?

The political scientist Garrett (1998) argues that a highly organized labor market is beneficial for economic growth when left-wing parties are powerful, whereas a less organized labor market is beneficial when right-wing parties are powerful.

The data analyzed here are from 14 OECD countries that were observed annually in the 25-year period from 1966 to 1990. The dataset `garrett.dta` contains the following variables:

- `country`: country identifier (j)
- `year`: year (i)
- `gdp`: annual real (adjusted for inflation) growth in gross domestic product (GDP) in percent (y_{ij})
- `oildep`: price of oil in U.S. dollars weighted by dependence on imported oil (as a proportion of all energy requirements) (x_{2ij})
- `demand`: overall real GDP growth in the OECD, weighted by national exposure to trade (x_{3ij})
- `leftpow`: left-wing power, an index that weights party groupings by their shares of legislative seats and cabinet portfolios; more left-wing power gives a higher index (x_{4ij})
- `organ`: degree of organization of the labor market; an index that is increasing in union density and major confederation share but decreasing in public sector share and the number of confederation-affiliated unions (x_{5ij})

To address the research question, we include `leftpow`, `organ`, and their interaction in the model. In addition, we control for `gdp`, `oildep`, and `demand`. The model then is

$$
\begin{aligned}
y_{ij} &= \beta_1 + \beta_2 x_{2ij} + \beta_3 x_{3ij} + \beta_4 x_{4ij} + \beta_5 x_{5ij} + \beta_6 x_{4ij} x_{5ij} + \xi_{ij} \\
&= \beta_1 + \beta_2 x_{2ij} + \beta_3 x_{3ij} + \beta_4 x_{4ij} + \underbrace{\left(\beta_5 + \beta_6 x_{4ij}\right)}_{\text{effect of } \texttt{organ}} x_{5ij} + \xi_{ij}
\end{aligned} \tag{6.1}
$$

where the total residual ξ_{ij} has zero mean given the covariates.

If Garrett (1998) is right, we would expect a negative coefficient β_5 of `organ` (a detrimental effect of organization when the left-wing power index `leftpow` is 0) and a positive interaction β_6 (an increase in the beneficial effect of `organ` as `leftpow` increases) such that for high values of `leftpow`, the slope of `organ`, $\beta_5 + \beta_6 x_{4ij}$, is positive.

The data are clearly long panel because there are as many as 25 occasions but only 14 countries.

6.7.2 Marginal modeling for long panels

Until now in this book, we have assumed that the number of units (or more generally clusters) is large and that there are relatively few occasions. It was therefore possible to use many occasion-specific parameters to model the longitudinal covariance structure. In this short panel case, large-sample properties (asymptotics), such as the consistency of estimators, refer to the number of units going to infinity for a fixed number of occasions.

Now we will consider models where the roles of units and occasions are reversed. Residuals for different units at an occasion are now allowed to be correlated, possibly with an unstructured covariance matrix. In addition to such cross-sectional correlations, some of the models can also handle longitudinal correlations by including unit-specific autocorrelation parameters (and unit-specific residual variances). There are now a large number of unit-specific parameters and few occasion-specific parameters. Estimation therefore requires a large number of occasions. Because the number of parameters increases when the number of units increases, large-sample results are now based on the number of occasions going to infinity for a fixed number of units.

To accommodate both cross-sectional and longitudinal dependence of the residuals, Parks (1967) proposed using FGLS to fit a model with unstructured cross-sectional covariances and an AR(1) structure longitudinally with unit-specific autocorrelation parameters. If there are many more occasions than units, we can use Stata's `xtgls` command to fit these models using FGLS or IGLS, the latter with the `igls` option.

The number of variance and covariance parameters becomes large when flexible covariance structures are specified both cross-sectionally and longitudinally. Unfortunately, Beck and Katz (1995) have shown that estimated standard errors from FGLS can exhibit downward finite-sample bias in this case. Instead of fitting the Parks model by FGLS, Beck and Katz advocate estimating regression parameters using pooled OLS, assuming both cross-sectional and longitudinal independence. "Panel-corrected standard errors" are then obtained, which are robust to misspecification of the cross-sectional correlation but are not robust to misspecification of the longitudinal correlation structure (independence). To accommodate longitudinal dependence in the random part of the model, Beck and Katz also consider a model with a common autocorrelation parameter for all units; they suggest estimating the regression parameters using pooled OLS after applying a Prais–Winsten transformation to make the observations uncorrelated across occasions.

6.7.3 Fitting marginal models for long panels in Stata

These methods are implemented in Stata's `xtpcse` command. To investigate the research question, we follow Garrett (1998) and fit time-series–cross-sectional models because we have a long panel with 25 occasions and 14 units.

First, we read in the data and construct the interaction term `left_org` between `leftpow` and `organ`:

```
. use http://www.stata-press.com/data/mlmus3/garrett
. generate left_org = leftpow*organ
```

We then `xtset` the data using

```
. xtset country year
       panel variable:  country (strongly balanced)
        time variable:  year, 1966 to 1990
                delta:  1 year
```

We note that the data are strongly balanced with observations for each year for every country. We are then ready to use the `xtpcse` command to fit model (6.1) by pooled OLS and obtain panel-corrected standard errors:

```
. xtpcse gdp oild demand leftpow organ left_org

Linear regression, correlated panels corrected standard errors (PCSEs)

Group variable:    country                 Number of obs      =       350
Time variable:     year                    Number of groups   =        14
Panels:            correlated (balanced)   Obs per group: min =        25
Autocorrelation:   no autocorrelation                     avg =        25
                                                          max =        25
Estimated covariances      =       105     R-squared          =    0.1410
Estimated autocorrelations =         0     Wald chi2(5)       =     42.01
Estimated coefficients     =         6     Prob > chi2        =    0.0000
```

		Panel-corrected				
gdp	Coef.	Std. Err.	z	P>|z|	[95% Conf.	Interval]
oildep	-15.2321	5.228694	-2.91	0.004	-25.48015	-4.98405
demand	.0049977	.0015394	3.25	0.001	.0019804	.0080149
leftpow	-1.483548	.2755847	-5.38	0.000	-2.023684	-.9434123
organ	-1.139716	.2234088	-5.10	0.000	-1.577589	-.7018423
left_org	.4547182	.0839526	5.42	0.000	.2901741	.6192624
_cons	5.919865	.583395	10.15	0.000	4.776432	7.063298

According to the fitted model, when left-wing power is zero, a unit increase in organization of the labor market produces a decrease of 1.14 in the mean growth of GDP (in percent), controlling for the other covariates. A unit increase in left-wing power decreases the mean by 1.48 when labor-market organization is zero, controlling for the other covariates. As the power of left-wing parties increases, the negative impact of organization of the labor market on GDP is reduced because the estimated interaction parameter associated with `left_org` is positive (equal to 0.45). When left-wing power is greater than 2.5 ($= 1.139716/0.4547183$), the effect of increasing organization of the labor market on growth becomes positive. Using a test based on panel-corrected standard errors, the interaction is significantly different from zero at the 5% level. Garrett's (1998) theory is hence supported by these data.

Identical estimates of the regression coefficients and almost identical estimated standard errors are obtained by using the `regress` command with the `vce(robust year)` option (output not shown).

Instead of assuming lack of longitudinal correlation when estimating the regression coefficients, we can let the total residual in model (6.1) have an AR(1) structure, $\xi_{ij} = \rho\xi_{i-1,j} + u_{ij}$, where ρ is a common autocorrelation parameter for all countries (previously, we referred to this autocorrelation parameter as α). We fit this model using the `xtpcse` command with the `correlation(ar1)` option:

```
. xtpcse gdp oild demand leftpow organ left_org, correlation(ar1)

Prais-Winsten regression, correlated panels corrected standard errors (PCSEs)

Group variable:    country                 Number of obs      =        350
Time variable:     year                    Number of groups   =         14
Panels:            correlated (balanced)   Obs per group: min =         25
Autocorrelation:   common AR(1)                           avg =         25
                                                          max =         25
Estimated covariances      =        105    R-squared          =     0.1516
Estimated autocorrelations =          1    Wald chi2(5)       =      31.55
Estimated coefficients     =          6    Prob > chi2        =     0.0000
```

gdp	Coef.	Panel-corrected Std. Err.	z	P>|z|	[95% Conf. Interval]	
oildep	-13.77227	6.587739	-2.09	0.037	-26.684	-.8605331
demand	.0060806	.0016414	3.70	0.000	.0028635	.0092977
leftpow	-1.46776	.3623476	-4.05	0.000	-2.177948	-.7575716
organ	-1.177444	.2934019	-4.01	0.000	-1.752502	-.6023872
left_org	.448846	.1112233	4.04	0.000	.2308524	.6668396
_cons	5.814019	.8076921	7.20	0.000	4.230971	7.397066
rho	.2958842					

We see that the autocorrelation parameter, called `rho` in the output, is estimated as 0.30. The estimate of the interaction is very similar to that produced by pooled OLS, and the interaction is still significant at the 5% level.

Garrett (1998) also included country-specific dummy variables in the model so that the estimated regression coefficients represent within-country effects. Instead of using an AR(1) structure to accommodate longitudinal dependence, he included the lag-1 of the response as a covariate in the fixed part of the model. However, he did not address the initial-conditions problem discussed in section 5.7.2.

When an exchangeable correlation structure is adequate both longitudinally and cross-sectionally, it is possible to specify a random intercept for occasion in addition to a random intercept for units. Such two-way error-components models or crossed random-effects models are discussed in chapter 9 (see also exercise 9.9). Asymptotics in this case rely on both the number of units and the number of occasions getting large.

6.8   Summary and further reading

In this chapter, we focused mostly on short panel data or longitudinal data; we summarize the issues concerning modeling of such data below. We also introduced methods for long panel data with few units and many occasions, where the challenge is to allow for cross-sectional correlations in addition to longitudinal correlations.

Three different marginal approaches for longitudinal data have been discussed: pooled OLS, GEE, and ML estimation. Pooled OLS corresponds to assuming uncorrelated residuals with constant variances when estimating regression coefficients and then making valid inferences regarding the coefficients based on robust standard errors. The covariance structure is not modeled but is treated as a nuisance. The original idea behind generalized estimating equations (GEE) is to do one better than pooled OLS by assuming a working correlation structure that may lead to efficiency improvements. However, due to robust standard errors invariably being used in GEE, the choice of correlation structure is not an important consideration in this approach either. In contrast, traditional ML estimation is accompanied by model-based standard errors, which explains the greater importance attached to specifying a reasonable covariance structure in ML estimation compared with GEE.

In linear models, all three approaches usually give nearly identical estimates of regression coefficients except when there are missing data and the probability of missingness depends on other responses. In the latter case, correct specification of the covariance structure is necessary for consistent estimation of regression coefficients (see exercise 6.6).

There is an overwhelming range of covariance structures available to choose from. When there are few (no more than 4) occasions with balance, an unstructured covariance matrix might be a good choice unless there are only a small number of subjects. With more occasions, a structured covariance matrix should generally be used to improve efficiency. The smaller the interval between occasions and the larger the number of occasions, the more likely it is that a decaying correlation structure is needed. However, correlations are unlikely to decay to zero, so combining random intercepts with autoregressive or Toeplitz structures may be a good idea. The combination of random intercept and exponential covariance structures is particularly attractive because it can be used with nonconstant time intervals. For balanced data and when there are not too many occasions, it may be worth checking whether the variance changes over time.

Marginal approaches are susceptible to the problem of subject-level confounding or level-2 endogeneity discussed in chapters 3 and 5, so inconsistent estimators of regression coefficients are produced if relevant covariates at the subject level are omitted.

Good books on marginal approaches to the analysis of panel and longitudinal data include Diggle et al. (2002), Hedeker and Gibbons (2006), and Fitzmaurice, Laird, and Ware (2011). Frees (2004, sec. 8.3) discusses methods for long panels (few units and many occasions). Weiss (2005) provides a particularly extensive discussion of different covariance structures for marginal modeling.

Exercises 6.1, 6.2, and 6.3 consider marginal modeling, and exercises 6.2 and 6.3 also consider random-intercept and random-coefficient models. More exercises on random-coefficient models can be found in the next chapter. Exercise 6.4 concerns the cross-sectional time-series approach to long panel data, and exercise 6.5 demonstrates the use of `xtmixed` for fitting multivariate regression models or seemingly unrelated regressions. Exercise 6.6 is a simulation study to investigate the consequences of misspecifying the covariance structure with and without missing data.

6.9 Exercises

6.1 Antisocial-behavior data

Consider the data from Allison (2005) that are described in exercise 5.2.

1. Find out if there are any missing values for the response variable `anti`.

2. Calculate a new variable, `time`, taking the values 0, 1, and 2 for the three time points.

3. Use pooled OLS to fit a linear model for `anti` with `time`, `pov`, `momage`, `female`, `childage`, `hispanic`, `black`, `momwork`, and `married` as covariates. Obtain robust standard errors that take the clustering of the data into account.

4. Now use GEE, again with robust standard errors.

   a. Use GEE with the same covariates as in step 3 but with the following correlation structures: unstructured, exchangeable, and AR(1). Store each set of estimates.

   b. Obtain the corresponding estimated correlation matrices and standard deviations. Comparing the two restricted correlation matrices with the unstructured ones, which restricted structure seems to be more reasonable?

   c. Do the estimated regression coefficients change appreciably when different working correlation structures are used? Does their significance at the 5% level change? (Hint: Use `estimates table` with the `star` option to compare the estimates.)

   d. How would you use GEE to obtain the pooled OLS estimates in step 2?

5. Use `xtmixed` to fit the models in step 4 by ML and obtain the estimated residual standard deviations and correlations.

6. Fit a random-coefficient model with the same covariates as in step 3 and with a child-specific random intercept and slope of `time`.

7. ❖ Derive the three residual variances and correlations implied by the random-coefficient model (by hand) and compare them with the estimates from step 5.

6.2 Postnatal-depression data | Solutions |

The dataset to be analyzed in this exercise comes from a clinical trial of the use of estrogen patches in the treatment of postnatal depression; full details are given in Gregoire et al. (1996).

In total, 61 women with major depression, which began within 3 months of childbirth and persisted for up to 18 months postnatally, were randomly assigned to the active treatment (an estrogen patch) or a placebo (a dummy patch). Thirty-four women received the former and the remaining 27 received the latter.

The women were assessed pretreatment and monthly for six months after treatment using the Edinburgh postnatal depression scale (EPDS), with possible scores from 0 to 30 and with a score of 10 or greater interpreted as possible depression. Noninteger depression scores result from missing questionnaire items (in this case, the average of all available items was multiplied by the total number of items).

The main research question is whether the estrogen patch is effective at reducing postnatal depression compared with the placebo.

The following variables are in the dataset `postnatal.dta` that was supplied by Rabe-Hesketh and Everitt (2007):

- `subj`: patient identifier
- `group`: treatment group (1: estrogen patch; 0: placebo patch)
- `pre`: pretreatment or baseline EPDS depression score
- `dep1` to `dep6`: EPDS depression scores for months 1 to 6

The mean structure of all models considered in this exercise should include a term for treatment group and a term for a linear time trend (where time starts at 0 for the first posttreatment visit). Note that the treatment by time interaction is not significant at the 5% level. Use `xtmixed` to fit each of the models mentioned below by ML, followed by `estimates store` to store the results.

1. Start by preparing the data for analysis.

   a. Reshape the data to long form.
   b. Missing values for the depression scores are coded as -9 in the dataset. Recode these to Stata's missing-value code. (You may want to use the `mvdecode` command.)
   c. Use the `xtdescribe` command to investigate missingness patterns. Is there any intermittent missingness?

2. Fit a model with an unstructured residual covariance matrix. Store the estimates (also store estimates for each of the models below).

3. Fit a model with an exchangeable residual covariance matrix. Use a likelihood-ratio test to compare this model with the unstructured model.

4. Fit a random-intercept model and compare it with the model with an exchangeable covariance matrix.

5. Fit a random-intercept model with AR(1) level-1 residuals. Compare this model with the ordinary random-intercept model using a likelihood-ratio test.

6. Fit a model with a Toeplitz(5) covariance structure (without a random intercept). Use likelihood-ratio tests to compare this model with each of the models fit above that are either nested within this model or in which this model is nested. (Stata may refuse to perform a test if it thinks the models are not nested. If you are sure the models are nested, use the `force` option.)

7. Fit a random-coefficient model with a random slope of time. Use a likelihood-ratio test to compare the random-intercept and random-coefficient models.

8. Specify an AR(1) process for the level-1 residuals in the random-coefficient model. Use likelihood-ratio tests to compare this model with the models you previously fit that are nested within it.

9. Use the `estimates stats` command to obtain a table including the AIC and BIC for the fitted models. Which models are best and second best according to the AIC and BIC?

6.3 Adolescent-alcohol-use data

Singer and Willett (2003) analyzed a dataset from Curran, Stice, and Chassin (1997). As part of a larger study of substance abuse, 82 adolescents were interviewed yearly from ages 14–16 and asked about their alcohol consumption during the previous year. Specifically, they were asked to report the frequency in the past 12 months of each of the following behaviors on an 8-point scale from 0 (not at all) to 7 (every day):

1. drinking wine or beer
2. drinking hard liquor
3. drinking five or more drinks in a row
4. getting drunk

Following Singer and Willett, we will use the square root of the mean of these four items as the response variable.

At age 14, the adolescents were also asked how many of their peers drank alcohol 1) occasionally and 2) regularly over the past 12 months, with each answer scored on a 6-point rating scale from 0 (none) to 5 (all). The square root of the mean of these two items was used as a covariate.

The original sample comprised 246 children of alcoholics (recruited through court records, wellness questionnaires from health maintenance organizations, and community telephone surveys) and 208 demographically matched controls. For the children of alcoholics, at least one biological and custodial parent had to have a lifetime Diagnostic Interview Schedule III (DSM-III) diagnosis of alcohol abuse or dependence.

The dataset `alcuse.dta` has the following variables:

- `id`: identifier for the adolescents

- `alcuse`: frequency of alcohol use (square root of mean on four alcohol items)
- `age_14`: age − 14, number of years since first interview (t_i)
- `coa`: dummy variable for being a child of an alcoholic (w_{1j})
- `peer`: alcohol use among peers at age 14 (square root of mean of two items) w_{2j}

In this exercise, we will consider a sequence of models with different marginal covariance structures. All models have the same mean structure, which includes the adolescent-level covariates `coa` and `peer` and their interaction with `age_14`.

1. Fit the following sequence of models using ML in `xtmixed` and store the estimates for each model:

   a. An AR(1) model for the total residuals.
   b. A random-intercept model.
   c. A random-intercept model with an AR(1) process for the level-1 residuals.
   d. A random-intercept model allowing the level-1 residuals to be heteroskedastic over years since first interview.
   e. A random-coefficient model with a random slope of number of years since first interview.
   f. A random-coefficient model with an AR(1) process for the level-1 residuals.
   g. A random-coefficient model with heteroskedastic level-1 residuals over years since first interview.
   h. A model with an unstructured residual covariance matrix.

2. Use the `estimates stats` command to obtain a table that includes the AIC and BIC for the fitted models. Which model is the best according to the AIC and BIC?

Growth-curve models are applied to this dataset in exercise 7.7.

6.4 ❖ Cigarette-consumption data

Baltagi, Griffin, and Xiong (2000) investigated how cigarette prices and disposable income affect cigarette consumption using annual data for 1963 to 1992 for 46 U.S. states.

The dataset `cigar.dta` from Baltagi (2008) includes the following variables:

- `state`: identifier for U.S. state (j)
- `name`: abbreviated name of U.S. state
- `year`: year from 1963–1992 (i)
- `price`: average retail price of pack of cigarettes in U.S. dollars
- `pop`: population size
- `pop16`: population size above 16 years
- `CPI`: consumer price index (reference value of 100 in 1983)
- `NDI`: per capita annual disposable income in U.S. dollars

- c: cigarette sales in packs per capita per year (by persons of smoking age; 14 years or older)
- pimin: minimum price per pack of cigarettes in adjoining states in U.S. dollars

In this kind of long panel data for macro units, it is often of interest to accommodate potential cross-sectional dependence between states when estimating standard errors. All models will be fit using xtpcse but with different approaches to handling cross-sectional and longitudinal correlations.

1. xtset the data.

2. The consumer price index is a weighted average of prices of a basket of consumer goods and services. In this dataset, CPI is 100 times the consumer price index in a given year divided by the consumer price index in 1983. Convert price, pimin and NDI to 1983 U.S. dollars using the consumer price index. Such real prices and real income are adjusted for inflation over years. In the analyses below, you will use the natural logarithms of real prices and real disposable income.

3. Let the response variable be the logarithm of the number of packs of cigarettes sold per person of smoking age. Consider models where the mean structure contains the following covariates: the logarithm of real cigarette price, the logarithm of real disposable income, the logarithm of real minimum price in adjoining states (a proxy for casual smuggling across states), and year (treated as continuous).

   a. Fit a regression model with estimated standard errors based on assuming that there are neither cross-sectional nor longitudinal correlations. First use the regress command and then use xtpcse with the correlation(independent) and independent options; the first option implies longitudinal independence and the second option implies cross-sectional independence. Also specify the nmk option to base standard errors on 'n minus k' (number of observations minus number of estimated parameters) as in standard linear regression. The parameter estimates and the estimated standard errors produced by the two commands should be identical.

   b. Use xtpcse with the correlation(independent) option but not the independent option to accommodate cross-sectional correlations among states when estimating standard errors. (Regression coefficients are estimated by pooled OLS.) Compare the magnitude of the estimated standard errors to those assuming no correlations.

   c. Use xtpcse with the correlation(ar1) option to fit a model with longitudinal correlation, specified as an AR(1) process with autocorrelation parameter ρ for the residuals. This accommodates cross-sectional as well as longitudinal correlations among states.

      i. Why have the estimated coefficients changed compared with step 3b above?

    ii. Interpret the estimated coefficients. When the response variable is a logarithmic transformation, the coefficients associated with covariates that are logarithmic transformations can be interpreted as elasticities (% change in the expectation of response for a 1% change in the covariate); see display 6.2.

  d. Use `xtgls` with the `panels(correlated)` and `corr(ar1)` options to fit the same model as in step 3c by FGLS.

    i. How many variance and covariance (or correlation) parameters are estimated?

    ii. Compare the estimated coefficients and their estimated standard errors with those from the previous step.

Consider a linear regression model where the response variable y_i and a covariate x_i are both log-transformed:

$$\ln y_i = \beta_1 + \beta_2 \ln x_i + \cdots + \epsilon_i$$

The logarithm of the conditional expectation of y_i given the covariates then is (see display 1.1 on page 64)

$$\ln\{E(y_i|\mathbf{x}_i)\} \;=\; \beta_1 + \sigma^2/2 + \beta_2 \ln x_i + \cdots$$

Taking derivatives with respect to x_i (using the chain rule), we find that

$$\frac{\partial \ln E(y_i|\mathbf{x}_i)}{\partial x_i} \;=\; \frac{1}{E(y_i|\mathbf{x}_i)}\frac{\partial E(y_i|\mathbf{x}_i)}{\partial x_i} \;=\; \beta_2\frac{1}{x_i}$$

so that the elasticity is

$$\frac{\partial E(y_i|\mathbf{x}_i)}{E(y_i|\mathbf{x}_i)} \bigg/ \frac{\partial x_i}{x_i} \;=\; \beta_2$$

The regression coefficient can therefore be interpreted as the relative change in the conditional expectation of y_i associated with unit relative change in x_i. Note, however, that this is the correct effect only for small changes in x_i.

Display 6.2: Elasticities

See exercise 9.9 for further analysis of this dataset.

6.5 ❖ Wages-and-fringe-benefits data

Wooldridge (2010) analyzed and provided a subset of data from the 1977 Quality of Employment Survey. The survey recruited workers aged 16 and older who were working for pay for 20 or more hours per week.

The variables in `fringe.dta` that we will use here are the following:

- `hrearn`: hourly wages in U.S. dollars

- `hrbens`: hourly fringe benefits in U.S. dollars, including value of vacation days, sick leave, employee insurance, and employee pension
- `educ`: years of schooling
- `exper`: years of work experience
- `expersq`: years of work experience squared
- `tenure`: number of years with the current employer
- `tenuresq`: number of years with the current employer squared
- `union`: dummy variable for being a union member
- `south`: dummy variable for living in the south of the U.S.
- `nrtheast`: dummy variable for living in the northeast of the U.S.
- `nrthcen`: dummy variable for living in the north central U.S.
- `married`: dummy variable for being married
- `white`: dummy variable for being white
- `male`: dummy variable for being male

Wooldridge considered a bivariate model for wages and fringe benefits, where fringe benefits are the total value of employee benefits including vacation days, sick leave, insurance, and pension. Annual wages and annual fringe benefits were divided by the total number of hours worked per year to obtain hourly wages, `hrearn`, and hourly benefits, `hrbens`. Each response variable was regressed on `educ`, `exper`, `expersq`, `tenure`, `tenuresq`, `union`, `south`, `nrtheast`, `nrthcen`, `married`, `white`, and `male`, giving two regression equations, each with its own set of regression coefficients. The residual error terms for the two regressions had different variances and were allowed to be correlated (a positive correlation would be expected because ability and other unobserved covariates that increase wages would also be expected to increase fringe benefits).

The model is a bivariate linear regression model (because of the correlation between the two response variables). Econometricians often call such models seemingly unrelated regression (SUR) models because the only aspect connecting the equations is the residual correlation (neither response variable appears as a covariate in the other equation, and there are typically no constraints across equations).

An advantage of joint instead of separate modeling of the two response variables is that hypotheses across equations can be tested. For instance, we can test the joint null hypothesis that there is no gender gap for earnings and that there is no gender gap for benefits. Using one such joint test for two or more response variables has the advantage that multiple testing (one test for each response variable) and the resulting increase in type I error are avoided.

1. Fit the model using `mvreg` with the `corr` option (see `help mvreg`). What is the estimated residual correlation? Interpret the relationship between experience and wages, and compare it with the relationship between experience and benefits.

2. Fit the model using `sureg` with the `corr` option (see `help sureg`). Are there any differences between the estimates from step 1 and step 2?

3. Fit the model using `xtmixed`.

    a. Stack `hrearn` and `hrbens` into one variable by using the `reshape` command.

    b. Form dummy variables `e_` for the observations corresponding to `hrearn` and `f_` for the observations corresponding to `hrbens`.

    c. Form interactions between all the covariates and the two dummy variables, `e_` and `f_`. (To avoid a large block of commands, you can use `foreach var of` *varlist*; see page 154 for an example.)

    d. Now fit the model by ML using `xtmixed`, treating subjects as clusters and the identifier for the two response variables as the `t()` variable in the `residuals()` option.

4. Compare the estimates from step 3 with those from steps 1 and 2. Note that in contrast to `mvreg` and `sureg`, `xtmixed` uses ML estimation. Unlike the other commands, `xtmixed` uses the response for wages if someone's response for benefits is missing and vice versa, which yields consistent estimates under the missing at random (MAR) assumption (see section 5.8).

5. Use `testparm` to test whether there is a gender gap in wages or benefits (that is, simultaneously test two hypotheses for both response variables) after controlling for the other covariates.

6. Test whether there are regional differences in wages or benefits after controlling for the other covariates.

See also exercise 8.9 for an example of a multivariate multilevel model.

6.6 Simulation study

In this exercise, we simulate data with a particular covariance structure and then fit marginal models by ML with both the correct and several incorrect covariance structures. We also use pooled OLS. We then repeat the exercise, this time creating missing data where missingness depends on the response at the first occasion. (See also section 5.8.1 for a simulation study.)

1. Simulate data for 1,000 subjects and three occasions with mean structure

$$E(y_{ij}|t_{ij}) \; = \; \beta_1 + \beta_2 t_{ij}$$

setting $\beta_1 = 0$ and $\beta_2 = 0$, and with an exchangeable residual covariance matrix with standard deviations equal to 2 and correlations equal to 0.5. You may use the following commands:

```
clear
set seed 1231214
set obs 1000

matrix define R = (1, .5, .5 \ .5, 1, .5 \ .5, .5, 1)
matrix list R

matrix define s = (2,2,2)

drawnorm y1 y2 y3, sds(s) corr(R)

generate id = _n
reshape long y, i(id) j(time)
```

Here the **drawnorm** command simulates variables—with names y1, y2, and y3—from a multivariate normal distribution with zero means (the default), standard deviations 2 (given in the row matrix s), and correlation matrix R. The **matrix define** commands are used to define the matrices R and s element by element. For the R matrix, the elements are given in the order R_{11}, R_{12}, R_{13}, R_{21}, etc., with elements separated by commas and rows separated by backslashes. We can think of this model as having the mean structure $\beta_1 + \beta_2 t_{ij}$ when both coefficients are zero because this gives zero mean at each occasion. (If you have a fast computer, you can also simulate 10,000 or more subjects to be able to better judge consistency of point estimates.)

2. Fit models with the same mean structure as the simulated data, treating β_1 and β_2 as parameters to be estimated.

   a. Use ML estimation with the correct covariance structure.
   b. Use ML estimation with a Toeplitz(1) covariance structure.
   c. Use ML estimation with an AR(1) covariance structure.
   d. Use pooled OLS with robust standard errors.

3. Compare the estimated slope of **time** with the true slope for each model. Are there important differences? Do the estimated standard errors of the slope differ substantially between models? Explain the findings.

4. Now create missing values by dropping observations at occasions 2 and 3 with probability 0.8 if the subject's response at occasion 1 is greater than 1. You may use these commands:

   ```
   generate rand = runiform()
   by id (time): generate firsty = y[1]
   drop if firsty>1 & rand<.8 & time>1
   ```

5. Investigate the missingness patterns using **xtdescribe**.

6. Produce a table of means and sample sizes at the three occasions. Explain what you find.

7. Fit models with the same mean structure as the simulated data, treating β_1 and β_2 as parameters to be estimated.

   a. Use ML estimation with the correct covariance structure.
   b. Use ML estimation with a Toeplitz(1) covariance structure.
   c. Use ML estimation with an AR(1) covariance structure.
   d. Use pooled OLS with robust standard errors.

8. Compare the estimated slope of **time** with the true slope for each model. Are there important differences? Explain the findings, including the direction of any difference between estimate and truth, and how this may have come about, taking into account the estimated correlation matrix.

9. ❖ Design your own simulation study, for instance, simulating with an AR(1) structure and fitting the models above, or changing the probability of missingness.

6.7 Variances and correlations of total residual in random-coefficient model

Consider a random-coefficient model with a random intercept and random slope of time, where time takes on the values $t_{1j} = 0$, $t_{2j} = 1$, and $t_{3j} = 2$. The covariance matrix of the intercept and slope is estimated as

$$\widehat{\Psi} = \begin{bmatrix} 4 & 1 \\ 1 & 2 \end{bmatrix}$$

and the level-1 residual variance is estimated as

$$\widehat{\theta} = 1$$

1. Calculate the estimated model-implied variance of the total residual

$$\xi_{ij} = \zeta_{1j} + \zeta_{2j} t_{ij} + \epsilon_{ij}$$

   for the three time points.
2. Calculate the estimated model-implied correlation matrix.

6.8 ❖ Covariance structure for random-intercept model with AR(1) errors

Consider a random-intercept model with level-1 errors following an AR(1) process.

1. Using the information that $\text{Var}(\epsilon_{ij}) = \frac{\sigma_e^2}{1-\alpha^2}$ and $\text{Cov}(\epsilon_{ij}, \epsilon_{i'j}) = \frac{\sigma_e^2}{1-\alpha^2}\alpha^{|i-i'|}$, derive an expression for the variance $\text{Var}(\xi_{ij})$ and covariance $\text{Cov}(\xi_{ij}, \xi_{i'j})$, where $\xi_{ij} = \zeta_j + \epsilon_{ij}$ and $\text{Var}(\zeta_j) = \psi$. ($\zeta_j$ and ϵ_{ij} are uncorrelated with each other and across subjects j and occasions i.)
2. Describe the covariance and correlation structure.
3. Calculate the standard deviations and correlation matrix for four equally spaced occasions for the case $\alpha = 0.6$, $\sigma_e^2 = 1$, and $\psi = 1$.

7 Growth-curve models

7.1 Introduction

In this chapter, we discuss perhaps the most prominent multilevel approach to longitudinal data, so-called *growth-curve models*, also sometimes referred to as *latent-trajectory models* or *latent growth-curve models*. Many people find this approach to longitudinal or panel modeling attractive because it explicitly models the shape of trajectories of individual subjects over time and how these trajectories vary, both systematically, due to occasion-level and subject-level covariates, and randomly. A somewhat flamboyant expression for this kind of modeling is to study *interindividual differences in intraindividual change*.

Growth-curve models are a special case of random-coefficient models where it is the coefficient of time that varies randomly between subjects. Random-coefficient models were discussed in chapter 4 and in sections 5.5 and 6.3.3 of the previous two chapters. That material can be viewed as sufficient background for standard linear growth-curve modeling, whereas the present chapter covers more advanced topics.

We first use growth-curve models to study children's physical growth, and we illustrate two different ways of modeling nonlinear growth, using polynomial functions and piecewise linear functions. How to handle heteroskedasticity or complex variation is also discussed both for the level-1 residuals and for random effects. Growth-curve models are subsequently applied to balanced panel data on growth in reading ability from kindergarten through third grade. In the balanced case, we show how growth-curve models can alternatively be expressed as structural equation models with latent variables and can be fit using Stata's `sem` command.

7.2 How do children grow?

The dataset considered here is on Asian children in a British community who were weighed on up to four occasions, roughly at ages 6 weeks, and then 8, 12, and 27 months. The dataset `asian.dta` is a 12% random sample, stratified by gender, from the dataset `asian.dat` available from the webpage of the Centre for Multilevel Modelling.[1] The full data were previously analyzed by Prosser, Rasbash, and Goldstein (1991).

---

1. See http://www.cmm.bristol.ac.uk/learning-training/multilevel-m-support/datasets.shtml.

The dataset `asian.dta` has the following variables:

- `id`: child identifier (j)
- `weight`: weight in kilograms (y_{ij})
- `age`: age in years (t_{ij})
- `gender`: gender (1: male; 2: female) (w_j)

The data are read in using

```
. use http://www.stata-press.com/data/mlmus3/asian
```

We want to investigate the growth trajectories of childrens' weights as they get older. Both the shape of the trajectories and the degree of variability in growth among the children are of interest.

7.2.1 Observed growth trajectories

We start by plotting the observed growth trajectories (connecting the observations by straight lines) by gender, after defining value labels for `gender` to make them appear on the graph:

```
. label define g 1 "Boy" 2 "Girl"
. label values gender g
. sort id age
. graph twoway (line weight age, connect(ascending)),
> by(gender) xtitle(Age in years) ytitle(Weight in Kg)
```

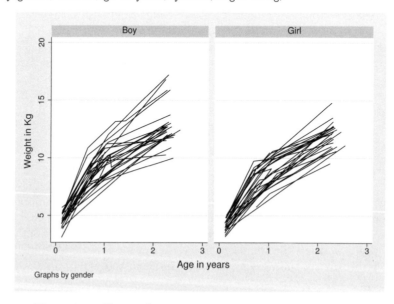

Figure 7.1: Observed growth trajectories for boys and girls

From figure 7.1, it is clear that the growth trajectories are nonlinear; growth is initially fast and then slows down. We will consider two methods of modeling this nonlinearity: polynomials and piecewise linear functions.

7.3 Models for nonlinear growth

7.3.1 Polynomial models

Growth trajectories can take a variety of shapes. A flexible approach to modeling possibly nonlinear growth in y_{ij} is to use a pth degree polynomial function of time t_{ij},

$$y_{ij} = \underbrace{\beta_1}_{\text{constant}} + \underbrace{\beta_2 t_{ij}}_{\text{linear}} + \underbrace{\beta_3 t_{ij}^2}_{\text{quadratic}} + \underbrace{\beta_4 t_{ij}^3}_{\text{cubic}} + \underbrace{\beta_5 t_{ij}^4}_{\text{quartic}} + \cdots + \beta_{p+1} t_{ij}^p + \xi_{ij}$$

where the names of the most commonly used terms are given below the braces and the random part of the model is represented by the term ξ_{ij}. A quadratic function will include a quadratic term and all terms of lower degree (a linear term and a constant), a cubic function includes a cubic term and all terms of lower degree, and so on.

An illustration of linear, quadratic, cubic, and quartic functions fit to the same data is given in figure 7.2. The largest number of extrema (maxima and minima) that can occur is $p - 1$, and they may not all occur within the range of the data. So a quadratic curve can have at most one maximum or minimum, a cubic curve can have at most one maximum and one minimum, etc. For low-degree polynomials, rapid changes in curvature are not possible. We see that these restrictions can cause artifacts in the fitted curve, such as the quadratic and cubic curves in figure 7.2 declining much sooner than the quartic curve.

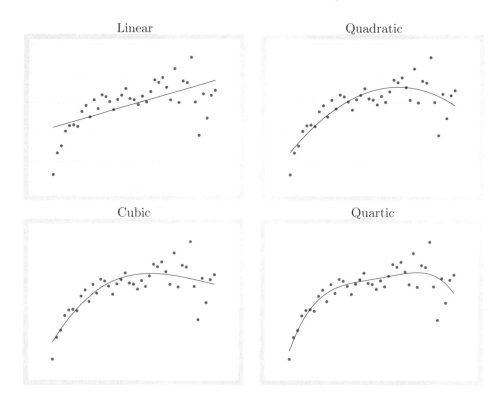

Figure 7.2: Illustration of different polynomial functions

It generally does not make sense to set the coefficients of lower-degree terms to zero while retaining higher-degree terms. This is because the implied restriction on the shape of the curve is usually not of interest and would only hold if the origin of the time scale is in a certain position. For instance, in a linear function, setting the intercept to zero implies that the line is at zero when time is zero; in a quadratic function, setting the coefficient of the linear term to zero implies that the minimum or maximum of the curve occurs when time is zero.

For balanced data with the same timing of occasions across subjects and a maximum of n occasions per subject (the number per person may vary because of missing data), a polynomial of degree $n-1$ will produce fitted means that are equal to the corresponding sample means at each occasion. It is therefore not possible to fit higher than $(n-1)$-degree polynomials in this case.

Fitting the models

We first model the nonlinearity in the growth trajectories by including a quadratic term for `age` in the model, giving a second-degree polynomial. It is also expected that boys and girls differ in their mean weight at any given age and that children vary in the initial weight and rate of growth. We therefore consider the model

$$y_{ij} = \beta_1 + \beta_2 w_j + \beta_3 t_{ij} + \beta_4 t_{ij}^2 + \zeta_{1j} + \zeta_{2j} t_{ij} + \epsilon_{ij} \qquad (7.1)$$

where y_{ij} is the weight, t_{ij} is the age of child j at occasion i, w_j is a dummy variable for being a girl, and ζ_{1j} and ζ_{2j} are a random intercept and random slope, respectively. The occasion-specific error term ϵ_{ij} allows the responses y_{ij} to deviate from the perfectly quadratic trajectories defined by the first four terms.

The usual assumptions for random-coefficient models (discussed in section 4.4.1) are made throughout this chapter. Point estimates and standard errors of regression coefficients are consistent if the mean structure (the fixed part of the model) and the covariance structure of the total residual are correctly specified.

We construct a dummy variable, `girl`, for the child being a girl:

```
. recode gender 2=1 1=0, generate(girl)
```

We also construct the variable `age2` for age squared:

```
. generate age2 = age^2
```

The model can then be fit in `xtmixed` using

```
. xtmixed weight girl age age2 || id: age, covariance(unstructured) mle
```

Mixed-effects ML regression   Number of obs   =   198
Group variable: id   Number of groups   =   68

Obs per group: min =   1
              avg =   2.9
              max =   5

Wald chi2(3)   =   1975.44
Log likelihood = -253.86692   Prob > chi2   =   0.0000

weight	Coef.	Std. Err.	z	P>|z|	[95% Conf. Interval]	
girl	-.5960093	.196369	-3.04	0.002	-.9808854	-.2111332
age	7.697967	.2382121	32.32	0.000	7.23108	8.164854
age2	-1.657843	.0880529	-18.83	0.000	-1.830423	-1.485262
_cons	3.794769	.1655053	22.93	0.000	3.470385	4.119153

Random-effects Parameters	Estimate	Std. Err.	[95% Conf. Interval]	
id: Unstructured				
sd(age)	.5097091	.0871791	.3645319	.7127041
sd(_cons)	.5947313	.128989	.3887827	.9097764
corr(age,_cons)	.1571078	.3240797	-.4564673	.6694134
sd(Residual)	.57233	.0496273	.4828786	.678352

LR test vs. linear regression: chi2(3) =   104.17   Prob > chi2 = 0.0000

Note: LR test is conservative and provided only for reference.

and we store the estimates using

```
. estimates store rc
```

We see that girls weigh on average about 0.6 kg less than boys of the same age. The coefficients of both `age` and `age2` are significant at the 5% level. The growth curve moves up at 0 (because of the positive coefficient of `age`) but the rate of growth declines (because of the negative coefficient of `age2`). There is a considerable estimated random-intercept standard deviation of 0.59 kg. The mean increase in weight per month varies with a standard deviation of 0.51 kg per month. The estimates are also shown under "Model 1: Polynomial" in table 7.1.

Note that we have included a higher-degree polynomial in the fixed part of the model than in the random part. This is often a good idea because there may be insufficient within-subject information to estimate the variability in the coefficients of higher powers of time, here the variability in the curvature of the relationships. If we also include a random slope for `age2`, which implies three extra parameters for the random part (one variance and two covariances), we obtain the following:

```
. xtmixed weight girl age age2 || id: age age2, covariance(unstructured) mle
Mixed-effects ML regression                     Number of obs     =        198
Group variable: id                              Number of groups  =         68

                                                Obs per group: min =          1
                                                               avg =        2.9
                                                               max =          5

                                                Wald chi2(3)      =    2159.56
Log likelihood = -241.01654                     Prob > chi2       =     0.0000
```

weight	Coef.	Std. Err.	z	P>|z|	[95% Conf.	Interval]
girl	-.450187	.1582819	-2.84	0.004	-.7604139	-.1399602
age	7.790089	.2585223	30.13	0.000	7.283395	8.296784
age2	-1.700954	.1017658	-16.71	0.000	-1.900412	-1.501497
_cons	3.703218	.1200624	30.84	0.000	3.4679	3.938536

Random-effects Parameters	Estimate	Std. Err.	[95% Conf.	Interval]
id: Unstructured				
sd(age)	1.350001	.3137402	.8560894	2.128898
sd(age2)	.5173013	.1234178	.3240886	.8257021
sd(_cons)	.1911551	.2846017	.0103293	3.537521
corr(age,age2)	-.9193943	.0427844	-.9719891	-.7791064
corr(age,_cons)	.7019993	1.97522	-.9999973	.9999999
corr(age2,_cons)	-.4414962	1.581981	-.9996501	.9976714
sd(Residual)	.4706363	.0497117	.3826273	.5788885

```
LR test vs. linear regression:      chi2(6) =   129.87   Prob > chi2 = 0.0000
Note: LR test is conservative and provided only for reference.
```

The model converges (after 13 iterations, not shown), but the confidence intervals for the last two correlations cover nearly the entire range. There seems to be insufficient information to reliably estimate these correlations, and we will return to the model with a quadratic fixed part and linear random part:

```
. estimates restore rc
```

In general, it is perfectly reasonable to allow only the lower-order terms of the polynomial used in the fixed part of the model to vary randomly between subjects.

Table 7.1: Maximum likelihood estimates of random-coefficient models for children's growth data (in kilograms)

	Model 1: Polynomial		Model 2: Piecewise linear	
	Est	(SE)	Est	(SE)
Fixed part				
β_1 [_cons]	3.79	(0.17)	3.37	(0.22)
β_2 [girl]	−0.60	(0.20)	-0.64	(0.19)
β_3 [age] or [ages1]	7.70	(0.24)	9.94	(1.07)
β_4 [age2] or [ages2]	−1.66	(0.09)	4.44	(0.75)
β_5 [ages3]			2.34	(0.64)
β_6 [ages4]			2.21	(0.36)
Random part				
$\sqrt{\psi_{11}}$	0.59		0.66	
$\sqrt{\psi_{22}}$ [age]	0.51		0.57	
ρ_{21}	0.16		−0.04	
$\sqrt{\theta}$	0.57		0.49	
Log likelihood	−253.87		−242.00	

Predicting the mean trajectory

Except in the linear case, it is not easy to understand the shape of a polynomial without plotting it. To visualize the mean curve for boys, we produce a graph, shown in figure 7.3:

```
. twoway (function Weight=_b[_cons]+_b[age]*x+_b[age2]*x^2,
>          range(0.1 2.6)), xtitle(Age in years) ytitle(Weight in Kg)
>          yscale(range(0 20)) ylab(0(5)20)
```

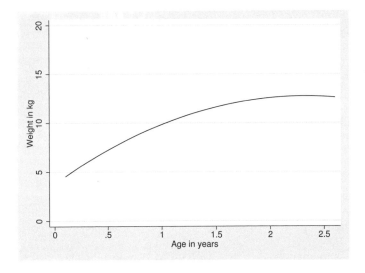

Figure 7.3: Mean trajectory for boys from quadratic model

(We have scaled the y axis from 0 to 20 to facilitate comparison with curves we will produce later.)

To visualize the variability in the growth curves implied by the model, we add the limits of the range within which 95% of the subject-specific growth curves are expected to lie. We first calculate the estimated standard deviation of $\zeta_{1j}+\zeta_{2j}t_{ij}$ which is given by $\sqrt{\widehat{\psi}_{11} + 2\widehat{\psi}_{21}t_{ij} + \widehat{\psi}_{22}t_{ij}^2}$. We can display the required variance and covariance estimates to plug into this expression by using

```
. estat recovariance
Random-effects covariance matrix for level id
```

	age	_cons
age	.2598033	
_cons	.0476257	.3537053

A more elegant approach is to access the logarithms of the standard deviations by using [lns1_1_1]_cons and [lns1_1_2]_cons and access the arc tanh of the correlation (often called Fisher's z transformation) by using [atr1_1_1_2]_cons (to find out which labels to use, run xtmixed with the estmetric option, as shown in section 2.11.2). We calculate the variances and covariance below, storing the values in the *scalars* var1, var2, and cov:

```
. scalar var1 = exp(2*[lns1_1_1]_cons)
. scalar var2 = exp(2*[lns1_1_2]_cons)
. scalar cov = tanh([atr1_1_1_2]_cons)*sqrt(var1*var2)
```

We can now use these scalars within the `twoway function` command, which produces figure 7.4 for boys:

```
. twoway (function Weight=_b[_cons]+_b[age]*x+_b[age2]*x^2,
>           range(0.1 2.6) lwidth(medium))
>         (function upper=_b[_cons]+_b[age]*x+_b[age2]*x^2
>         +1.96*sqrt(var1+2*cov*x+var2*x^2), range(0.1 2.6) lpatt(dash))
>         (function lower=_b[_cons]+_b[age]*x+_b[age2]*x^2
>         -1.96*sqrt(var1+2*cov*x+var2*x^2), range(0.1 2.6) lpatt(dash)),
>         legend(order(1 "Mean" 2 "95% range")) xtitle(Age in years)
>         ytitle(Weight in kg) yscale(range(0 20)) ylab(0(5)20)
```

To obtain the corresponding figure for girls, add `_b[girl]` to the fixed part of the model.

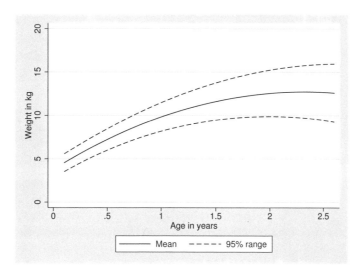

Figure 7.4: Mean trajectory and 95% range of subject-specific trajectories for boys from quadratic model

Predicting trajectories for individual children

After estimation with `xtmixed`, the predicted trajectories—based on substituting empirical Bayes (EB) predictions $\widetilde{\zeta}_{1j}$ and $\widetilde{\zeta}_{2j}$ for the random intercepts and random slopes—can be obtained using `predict` with the `fitted` option:

```
. predict traj, fitted
```

We can plot these predicted trajectories together with the observed responses using a *trellis graph*, a graph containing a separate two-way plot for each subject. This is accomplished using the by(id) option. For girls, we use the command

```
. twoway (scatter weight age) (line traj age, sort) if girl==1,
>         by(id, compact legend(off))
```

and similarly for boys. The graphs are shown in figures 7.5 and 7.6 for girls and boys, respectively, and suggest that the model fits well.

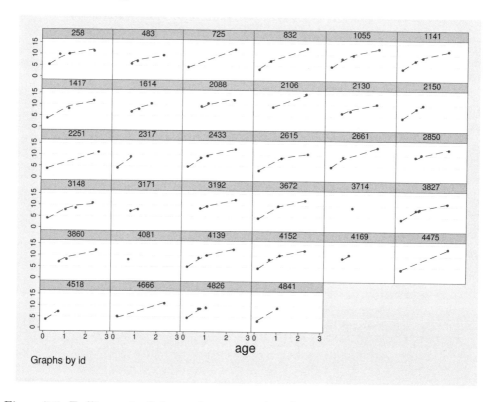

Figure 7.5: Trellis graph of observed responses (dots) and predicted trajectories (dashed lines) from quadratic model for girls

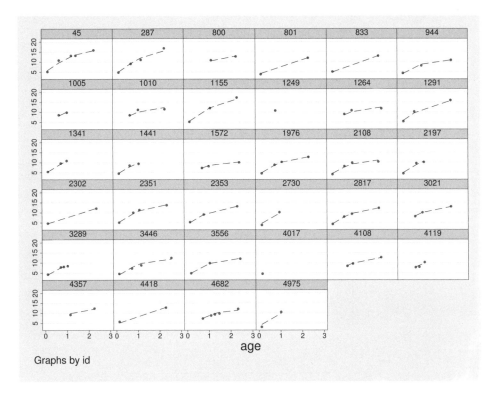

Figure 7.6: Trellis graph of observed responses (dots) and fitted trajectories (dashed lines) for boys

7.3.2 Piecewise linear models

Instead of using a polynomial to model the nonlinear relationship between weight and age, we can also split the age range into intervals, at *knots*, and fit straight line-segments between the knots, giving a piecewise linear model. Such a piecewise linear curve is an example of a *spline*, specifically a linear spline. As for polynomials, the piecewise linear function can be written as a linear model, $\beta_1 + \beta_2 z_{1ij} + \beta_3 z_{2ij} + \cdots$, where the variables z_{1ij}, z_{2ij}, etc., are called linear spline basis functions.

To see this, refer to the illustration in figure 7.7. The bottom panel shows the spline basis functions, which start at zero, increase with a slope of 1 between two knots, and then remain constant. The function in the top panel is a linear combination of these basis functions. We see that the function up to the first knot at 2 is simply $1 + z_{1ij}$. Between the knots at 2 and 6, the slope is only 0.25, so the second basis function z_{2ij} is multiplied by 0.25 before being added to $1 + z_{1ij}$. Finally, the slope in the last interval, from 6 to the maximum (here 7), is 2, so the final term in the model is $2z_{3ij}$.

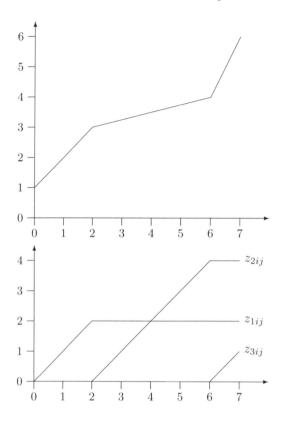

Figure 7.7: Illustration of piecewise linear function $1 + z_{1ij} + 0.25z_{2ij} + 2z_{3ij}$ with knots at 2 and 6

Fitting the models

For the children's growth data, we will arbitrarily let the knots be located at ages 0.4, 0.9, and 1.5 years. It is of course preferable to base the choice of knots on subject-matter considerations when possible (see exercise 7.6 for an example).

To fit a linear spline, we must create basis functions to include in the model as covariates. To produce nice graphs, we first add the knot values to the data so that we can later display the basis functions (generally, this is not something you would have to do).

There are 198 observations in the data:

```
. display _N
198
```

We increase the number of observations by 3 and substitute the values 0.4, 0.9, and 1.5 for age:

```
. set obs 201
obs was 198, now 201
. replace age = 0.4 in 199
(1 real change made)
. replace age = 0.9 in 200
(1 real change made)
. replace age = 1.5 in 201
(1 real change made)
```

(Note that we do not risk accidentally treating these invented data as real data during model fitting because the response variable is missing in rows 199 to 201.)

The spline basis functions are created using the Stata command `mkspline`,

```
. mkspline ages1 .4 ages2 .9 ages3 1.5 ages4 = age
```

and we can plot them using

```
. twoway (line ages1 age, sort) (line ages2 age, sort)
>        (line ages3 age, sort) (line ages4 age, sort),
>        xscale(range(0 2.5)) xlabel(0 .4 .9 1.5)
```

giving the graph in figure 7.8.

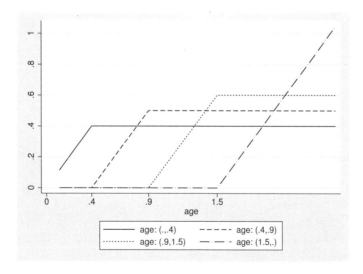

Figure 7.8: Spline basis functions for piecewise-linear model for children's growth data

We see that the first basis function (`ages1`) increases linearly from 0 (although the plotted line starts at the lowest age that occurs in the data), with a slope of 1, up to the first knot, and then stays constant. The second basis function, `ages2`, increases linearly from 0, between the first and second knots, and then stays constant, etc. We can include these basis functions as covariates, and their coefficients will represent the slopes in the corresponding intervals. We can fit the piecewise-linear random-coefficient model using

```
. xtmixed weight girl ages1 ages2 ages3 ages4 || id: age,
> covariance(unstructured) mle
```

```
Mixed-effects ML regression                     Number of obs      =        198
Group variable: id                              Number of groups   =         68

                                                Obs per group: min =          1
                                                               avg =        2.9
                                                               max =          5

                                                Wald chi2(5)       =    2075.53
Log likelihood = -242.00811                     Prob > chi2        =     0.0000
```

weight	Coef.	Std. Err.	z	P>|z|	[95% Conf. Interval]	
girl	-.6358097	.1947046	-3.27	0.001	-1.017424	-.2541958
ages1	9.936391	1.065415	9.33	0.000	7.848216	12.02457
ages2	4.441785	.7492172	5.93	0.000	2.973346	5.910224
ages3	2.344885	.6430038	3.65	0.000	1.084621	3.605149
ages4	2.209955	.3629384	6.09	0.000	1.498609	2.921301
_cons	3.374523	.2216652	15.22	0.000	2.940067	3.808979

Random-effects Parameters	Estimate	Std. Err.	[95% Conf. Interval]	
id: Unstructured				
sd(age)	.5678807	.0818571	.4281153	.7532747
sd(_cons)	.6608931	.1158728	.468697	.9319021
corr(age,_cons)	-.0431119	.2332324	-.4629951	.3925743
sd(Residual)	.4878942	.0445727	.4079079	.583565

```
LR test vs. linear regression:       chi2(3) =    121.88   Prob > chi2 = 0.0000
Note: LR test is conservative and provided only for reference.
```

Estimates for this model were shown under "Model 2: Piecewise linear" in table 7.1. The estimates of the gender difference and the random part of the model are similar to the estimates for the quadratic function. The slope in the first interval, up to 0.4 months, is estimated as 9.94 kg/month; the slope between 0.4 and 0.9 months is estimated as 4.44 kg/month; and the last two slopes are estimated as about 2.3 kg/month and 2.2 kg/month. The intercept, estimated as 3.37 kg, represents the mean initial weight (at birth), but this is an extrapolation outside the range of the data. An advantage of the piecewise linear model is that it is easier to interpret the relationship between mean birthweight and age than for the quadratic model.

The random part of the model is linear (with a random slope of age), so for a given child the same amount is added to the slope of each line segment. The estimated residual random-intercept standard deviation is 0.66 kg, and the mean increase in weight per month varies with an estimated standard deviation of 0.49 kg per month.

Predicting the mean trajectory

As for the quadratic model, we see that the slope decreases with age. We can obtain a graph of the mean trajectory for boys and the 95% range of trajectories as before. Here, it is easier to use the **predict** command to produce the mean curve. To make predictions for the three invented datapoints at the spline knots, we first set **girl** equal to 0 for these three observations:

```
. replace girl = 0 in 199/201
(3 real changes made)
. predict mean, xb
```

We produce scalars for the variances and covariance as before,

```
. scalar var1 = exp(2*[lns1_1_1]_cons)
. scalar var2 = exp(2*[lns1_1_2]_cons)
. scalar cov = tanh([atr1_1_1_2]_cons)*sqrt(var1*var2)
```

and then we generate variables **upper** and **lower** containing the limits of the 95% range:

```
. generate upper = mean + 1.96*sqrt(var1+2*cov*age+var2*age^2)
. generate lower = mean - 1.96*sqrt(var1+2*cov*age+var2*age^2)
```

We can then plot the ranges in addition to the mean by using the **twoway** command:

```
. twoway (line mean age, sort lwidth(medium))
>        (line lower age, sort lpatt(dash))
>        (line upper age, sort lpatt(dash)) if girl==0,
>        legend(order(1 "Mean" 2 "95% range")) xtitle(Age in years)
>        ytitle(Weight in kg) xtick(.4 .9 1.5, grid) ylabel(,nogrid)
>        yscale(range(0 20)) ylab(0(5)20)
```

The resulting graph in figure 7.9 clearly shows the kinks at the first two knots (helped by having made predictions at the generated ages at the spline knots).

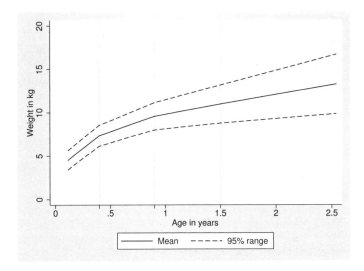

Figure 7.9: Mean trajectory and 95% range of subject-specific trajectories for boys from piecewise-linear model

Trellis graphs can be produced as shown for the quadratic model.

7.4   Two-stage model formulation

As is often done in the literature, we now express the polynomial random-coefficient model in (7.1) using the two-stage formulation described in section 4.9. The level-1 model is written as

$$y_{ij} = \pi_{0j} + \pi_{1j}t_{ij} + \pi_2 t_{ij}^2 + \epsilon_{ij}$$

where the intercept π_{0j} and slope π_{1j} are child-specific coefficients. The level-2 model has these coefficients as responses:

$$\pi_{0j} = \gamma_{00} + \gamma_{01}w_j + r_{0j}$$
$$\pi_{1j} = \gamma_{10} + r_{1j} \tag{7.2}$$

where `girl` (w_j) is a covariate only in the intercept equation. As usual, r_{0j} and r_{1j} are assumed to have a bivariate distribution with zero means and unstructured covariance matrix.

Substituting the level-2 models into the level-1 model, we obtain the reduced form

$$
\begin{aligned}
y_{ij} &= \underbrace{\gamma_{00} + \gamma_{01}w_j + r_{0j}}_{\pi_{0j}} + \underbrace{\left(\gamma_{10} + r_{1j}\right)}_{\pi_{1j}} t_{ij} + \pi_2 t_{ij}^2 + \epsilon_{ij} \\
&= \gamma_{00} + \gamma_{01}w_j + \gamma_{10}t_{ij} + \pi_2 t_{ij}^2 + r_{0j} + r_{1j}t_{ij} + \epsilon_{ij} \\
&\equiv \beta_1 + \beta_2 w_j + \beta_3 t_{ij} + \beta_4 t_{ij}^2 + \zeta_{1j} + \zeta_{2j}t_{ij} + \epsilon_{ij}
\end{aligned}
$$

where $\beta_1 \equiv \gamma_{00}$, $\beta_2 \equiv \gamma_{01}$, $\beta_3 \equiv \gamma_{10}$, $\beta_4 \equiv \pi_2$, $\zeta_{1j} \equiv r_{0j}$, and $\zeta_{2j} \equiv r_{1j}$. This reduced-form model is equivalent to model (7.1).

In the two-stage formulation, a natural extension of this model is to include `girl` as a covariate also in the level-2 model for π_{1j} in (7.2) by adding the term $\gamma_{11}w_j$ there. This results in a *cross-level interaction* $\gamma_{11}w_j t_{ij}$, between the level-2 covariate w_j and the level-1 covariate t_{ij}.

After constructing the interaction term for `age` and `girl`,

```
. generate age_girl = age*girl
```

the model can be fit using

```
. xtmixed weight girl age_girl age age2 || id: age, covariance(unstructured) mle
```

```
Mixed-effects ML regression              Number of obs      =        198
Group variable: id                       Number of groups   =         68

                                         Obs per group: min =          1
                                                        avg =        2.9
                                                        max =          5

                                         Wald chi2(4)       =    2023.53
Log likelihood = -252.99486              Prob > chi2        =     0.0000
```

weight	Coef.	Std. Err.	z	P>|z|	[95% Conf. Interval]	
girl	-.5040553	.2071801	-2.43	0.015	-.9101208	-.0979897
age_girl	-.2303089	.1731563	-1.33	0.183	-.569689	.1090712
age	7.814711	.2526441	30.93	0.000	7.319538	8.309885
age2	-1.658569	.087916	-18.87	0.000	-1.830881	-1.486257
_cons	3.748607	.1682409	22.28	0.000	3.418861	4.078353

Random-effects Parameters	Estimate	Std. Err.	[95% Conf. Interval]	
id: Unstructured				
sd(age)	.4969469	.0875711	.3518137	.7019517
sd(_cons)	.5890289	.1291188	.3833081	.9051599
corr(age,_cons)	.1870199	.3361438	-.4569593	.7023664
sd(Residual)	.5729779	.0498678	.4831206	.679548

```
LR test vs. linear regression:       chi2(3) =    104.77   Prob > chi2 = 0.0000
Note: LR test is conservative and provided only for reference.
```

The estimates are shown under "Model 3" in table 7.2, next to the previous model that had no cross-level interaction, repeated here as "Model 1". Because the interaction is not significant at the 5% level, we retain Model 1 in the next section.

Table 7.2: Maximum likelihood estimates for quadratic models for children's growth data. "Model 3" includes cross-level interaction. "Model 4" and "Model 5" allow the random part of the model at level 1 and level 2, respectively, to differ between boys (B) and girls (G).

	Model 1		Model 3		Model 4		Model 5	
	Est	(SE)	Est	(SE)	Est	(SE)	Est	(SE)
Fixed part								
β_1 [_cons]	3.79	(0.17)	3.75	(0.17)	3.83	(0.17)	3.82	(0.16)
β_2 [girl]	−0.60	(0.20)	−0.50	(0.21)	−0.60	(0.20)	−0.61	(0.20)
β_3 [girl×age]			−0.23	(0.17)				
β_4 [age]	7.70	(0.24)	7.81	(0.25)	7.63	(0.23)	7.61	(0.23)
β_5 [age2]	−1.66	(0.09)	−1.66	(0.09)	−1.64	(0.09)	−1.65	(0.09)
Random part								
							B	G
$\sqrt{\psi_{11}}$	0.59		0.59		0.63		0.54	0.70
$\sqrt{\psi_{22}}$ [age]	0.51		0.50		0.49		0.69	0.22
ρ_{21}	0.16		0.19		0.15		0.05	0.39
					B	G		
$\sqrt{\theta}$	0.57		0.57		0.64	0.49	0.57[†]	0.57
Log likelihood	−253.87		−252.99		−252.41		−249.71	

[†]Constrained equal across genders

7.5 Heteroskedasticity

7.5.1 Heteroskedasticity at level 1

In all models considered so far in this chapter, we have assumed that the random intercept, random slope, and level-1 residual all have constant variance for all children. However, it is sometimes necessary to allow variances to depend on covariates.

We first allow the level-1 residual variance θ to differ between boys and girls by introducing gender-specific parameters $\theta^{(B)}$ and $\theta^{(G)}$, respectively. This is easily accomplished in xtmixed using the residuals(independent, by(gender)) option to let the level-1 residuals be independent over occasions and have different variances for each gender:

```
. xtmixed weight girl age age2 || id: age, covariance(unstructured) mle
> residuals(independent, by(gender))
```

Mixed-effects ML regression		Number of obs	=	198
Group variable: id		Number of groups	=	68

```
                                      Obs per group: min =          1
                                                     avg =        2.9
                                                     max =          5

                                      Wald chi2(3)        =    2093.69
Log likelihood = -252.40553           Prob > chi2         =     0.0000
```

weight	Coef.	Std. Err.	z	P>|z|	[95% Conf. Interval]	
girl	-.6026741	.2011077	-3.00	0.003	-.9968378	-.2085103
age	7.629066	.2327022	32.78	0.000	7.172978	8.085154
age2	-1.635112	.0859732	-19.02	0.000	-1.803617	-1.466608
_cons	3.829123	.1731333	22.12	0.000	3.489788	4.168458

Random-effects Parameters	Estimate	Std. Err.	[95% Conf. Interval]	
id: Unstructured				
sd(age)	.4857393	.0889252	.339291	.6953993
sd(_cons)	.6271286	.1180433	.4336471	.9069362
corr(age,_cons)	.1457196	.2996523	-.4245955	.6332441
Residual: Independent, by gender				
Boy: sd(e)	.6397179	.0697704	.5165981	.7921805
Girl: sd(e)	.4937988	.0581918	.3919585	.6220996

```
LR test vs. linear regression:      chi2(4) =   107.09   Prob > chi2 = 0.0000
Note: LR test is conservative and provided only for reference.
```

The level-1 residual standard deviations $\sqrt{\theta^{(B)}}$ and $\sqrt{\theta^{(G)}}$ are estimated as 0.64 and 0.49, respectively. The point estimates look quite similar to each other, but we can use a likelihood-ratio test to formally test the null hypothesis that both residual variances (and hence standard deviations) are the same, H_0: $\theta^{(B)} = \theta^{(G)}$, using

```
. lrtest rc .
Likelihood-ratio test                       LR chi2(1)  =       2.92
(Assumption: rc nested in .)                Prob > chi2 =     0.0873
Note: The reported degrees of freedom assumes the null hypothesis is not on
      the boundary of the parameter space. If this is not true, then the
      reported test is conservative.
```

There is no strong evidence against the null hypothesis, and we retain a common level-1 variance or homoskedastic level-1 residual. (We can ignore the note in the output for the likelihood-ratio test because our null hypothesis is not on the boundary of parameter space.)

For balanced data with a sufficient number of observations at each occasion, we can also let the level-1 residuals be heteroskedastic over occasions, as will be described in section 7.6.

7.5.2   Heteroskedasticity at level 2

We can also allow the random-intercept variance ψ_{11} and random-slope variance ψ_{22}, and their covariance ψ_{21}, to differ between boys and girls, with parameters $\psi_{11}^{(B)}$, $\psi_{22}^{(B)}$, and $\psi_{21}^{(B)}$ for boys and $\psi_{11}^{(G)}$, $\psi_{22}^{(G)}$, and $\psi_{21}^{(G)}$ for girls.

This is most easily done by specifying four different random effects—a random intercept ζ_{1j} and slope ζ_{2j} for boys, and a random intercept ζ_{3j} and slope ζ_{4j} for girls—giving level-2 random parts $\zeta_{1j} + \zeta_{2j}t_{ij}$ for boys and $\zeta_{3j} + \zeta_{4j}t_{ij}$ for girls. These level-2 random parts must be multiplied by dummy variables for boys and girls, respectively, to switch them on for the appropriate gender. The model can be written as

$$
\begin{aligned}
y_{ij} &= \beta_1 + \beta_2 \texttt{girl}_j + \beta_3 t_{ij} + \beta_4 t_{ij}^2 + (\zeta_{1j} + \zeta_{2j}t_{ij})\texttt{boy}_j + (\zeta_{3j} + \zeta_{4j}t_{ij})\texttt{girl}_j + \epsilon_{ij} \\
&= \cdots + (\zeta_{1j}\texttt{boy}_j + \zeta_{2j}t_{ij}\texttt{boy}_j) + (\zeta_{3j}\texttt{girl}_j + \zeta_{4j}t_{ij}\texttt{girl}_j) + \epsilon_{ij},
\end{aligned}
$$

where the dummy variable `girl`, previously denoted w_j in equations, already exists and the dummy variable for boys, `boy`, can be created using

```
. generate boy = 1-girl
```

The model includes random coefficients of the gender dummies, as well as random coefficients of the products, $t_{ij}\texttt{boy}_j$ and $t_{ij}\texttt{girl}_j$. The latter variable already exists and is called `age_girl`. We create $t_{ij}\texttt{boy}_j$ using

```
. generate age_boy = age*boy
```

In the `xtmixed` command, we first specify random coefficients for `age_boy` and `boy`. After that, we specify a new random part, also with `id` as cluster variable, but this time with random coefficients for `age_girl` and `girl`. Specifying two random parts for the same cluster is a way of estimating two separate covariance matrices, hence achieving zero correlations between the random effects for boys and girls. Such correlations cannot be estimated because no child can be both a boy and a girl. Because the coefficients of `boy` and `girl` play the role of random intercepts, we must use the `noconstant` option for each random part:

```
. xtmixed weight girl age age2
> || id: age_boy boy, noconstant covariance(unstructured)
> || id: age_girl girl, noconstant covariance(unstructured)  mle
```

```
Mixed-effects ML regression                    Number of obs      =        198
Group variable: id                             Number of groups   =         68

                                               Obs per group: min =          1
                                                              avg =        2.9
                                                              max =          5

                                               Wald chi2(3)       =    2413.70
Log likelihood = -249.70684                    Prob > chi2        =     0.0000
```

weight	Coef.	Std. Err.	z	P>|z|	[95% Conf. Interval]	
girl	-.6066394	.1996813	-3.04	0.002	-.9980075	-.2152713
age	7.613015	.2335188	32.60	0.000	7.155326	8.070703
age2	-1.646256	.0869013	-18.94	0.000	-1.81658	-1.475933
_cons	3.820121	.157553	24.25	0.000	3.511323	4.128919

Random-effects Parameters	Estimate	Std. Err.	[95% Conf. Interval]	
id: Unstructured				
sd(age_boy)	.6891223	.1350621	.4693189	1.01187
sd(boy)	.5364599	.1807413	.2771755	1.038292
corr(age_boy,boy)	.0501319	.3995425	-.6260466	.6832774
id: Unstructured				
sd(age_girl)	.218154	.1450495	.0592657	.8030133
sd(girl)	.6987662	.1611978	.4445995	1.098234
corr(age_girl,girl)	.388837	.7897376	-.8881593	.9773197
sd(Residual)	.567544	.049112	.4790065	.6724465

```
LR test vs. linear regression:        chi2(6) =    112.49   Prob > chi2 = 0.0000
Note: LR test is conservative and provided only for reference.
```

A likelihood-ratio test can be performed to compare this model with the one that constrained the covariance matrices of the random effects to be equal between boys and girls, H_0: $\psi_{11}^{(B)} = \psi_{11}^{(G)}$, $\psi_{22}^{(B)} = \psi_{22}^{(G)}$, $\psi_{21}^{(B)} = \psi_{21}^{(G)}$, using

```
. lrtest rc .
Likelihood-ratio test                          LR chi2(3)  =       8.32
(Assumption: rc nested in .)                   Prob > chi2 =     0.0398
Note: The reported degrees of freedom assumes the null hypothesis is not on
      the boundary of the parameter space.  If this is not true, then the
      reported test is conservative.
```

The null hypothesis is not on the boundary, so the *p*-value is 0.04.

7.6   How does reading improve from kindergarten through third grade?

We now consider data from the 1986, 1988, 1990, and 1992 panel waves of the U.S. National Longitudinal Survey of Youth (NLSY) provided by Bollen and Curran (2006). The children were in kindergarten, grade 1, or grade 2 in 1986. Here the time-scale of interest is the grade the child is in (see Bollen and Curran [2006, 82]); we will consider grades 0 (kindergarten), 1, 2, and 3. The data have a considerable amount of missing values because they were collected only every other year; some children were already in year 3 at the first wave, and therefore contributed no further data in subsequent waves. The response variable used here is the reading recognition subscore of the Peabody Individual Achievement Test, scaled as the percentage of 84 items that were answered correctly.

The variables in the dataset `reading.dta` are as follows:

- `id`: child identifier
- `read0`, `read1`, `read2`, `read3`: reading scores in kindergarten (grade 0), grade 1, grade 2, and grade 3
- `age0`, `age1`, `age2`, `age3`: ages in months in kindergarten and grades 1 through 3
- `math0`, `math1`, `math2`, `math3`: math scores in kindergarten and grades 1 through 3 (not used here)
- `female`: dummy variable for being female (1: female; 0: male)
- `minority`: dummy variable for being a minority (1: minority; 0: nonminority)

Unlike the Asian child growth data, these data are balanced with constant spacing of occasions (apart from gaps due to missing data).

7.7   Growth-curve model as a structural equation model

When the data are balanced, the responses y_{ij} at different occasions i can be viewed as different variables or in other words as a multivariate response for child j. A growth-curve model can then be set up as a structural equation model (SEM) with latent (unobserved) variables. Such models are also called covariance structure models and are a special case of latent variable models. When applied to longitudinal data, SEMs are sometimes referred to as *latent growth-curve models* because the smooth growth curves for individual children (with the error ϵ_{ij} removed) are unobserved.

A linear growth-curve model for four occasions (as in the reading data) can be written as

$$
\begin{bmatrix} y_{1j} \\ y_{2j} \\ y_{3j} \\ y_{4j} \end{bmatrix} = \begin{bmatrix} 1 & 0 \\ 1 & 1 \\ 1 & 2 \\ 1 & 3 \end{bmatrix} \begin{bmatrix} \eta_{1j} \\ \eta_{2j} \end{bmatrix} + \begin{bmatrix} \epsilon_{1j} \\ \epsilon_{2j} \\ \epsilon_{3j} \\ \epsilon_{4j} \end{bmatrix} = \begin{bmatrix} \eta_{1j} + 0\eta_{2j} + \epsilon_{1j} \\ \eta_{1j} + 1\eta_{2j} + \epsilon_{2j} \\ \eta_{1j} + 2\eta_{2j} + \epsilon_{3j} \\ \eta_{1j} + 3\eta_{2j} + \epsilon_{4j} \end{bmatrix}
$$

This is just the level-1 model in a two-stage formulation (see section 7.4), written out for each occasion i, using matrix algebra and the notation $\eta_{1j} \equiv \pi_{0j}$ and $\eta_{2j} \equiv \pi_{1j}$. Here the values of t_i are 0, 1, 2, and 3 (note that the time variable does not vary between children here because the data are balanced). For a conventional SEM, it is not possible to have different sets of time points or occasions for different children.

In SEM terminology, this part of the model is called the *measurement model*. The child-specific intercept η_{1j} and slope η_{2j} are called common factors (or latent variables), and the matrix multiplying these factors on the left is the factor-loading matrix. The factor loadings (elements of the factor-loading matrix) are usually parameters to be estimated in SEM, but for linear growth-curve models they are constrained equal to fixed constants. The errors ϵ_{1j} to ϵ_{4j} have zero means and a diagonal 4×4 covariance matrix, with variances θ_{11} to θ_{44} and covariances set to zero. A diagonal covariance matrix means that the pairwise correlations are all zero. Typically, the residual variances are not constrained equal, and this is usually the only difference between SEM and multilevel approaches to growth-curve modeling.

The *structural model* is just matrix notation for the level-2 models in the two-stage formulation

$$
\begin{bmatrix} \eta_{1j} \\ \eta_{2j} \end{bmatrix} = \begin{bmatrix} \gamma_{11} \\ \gamma_{21} \end{bmatrix} + \begin{bmatrix} \zeta_{1j} \\ \zeta_{2j} \end{bmatrix} = \begin{bmatrix} \gamma_{11} + \zeta_{1j} \\ \gamma_{21} + \zeta_{2j} \end{bmatrix}
$$

(with the notation $\gamma_{11} \equiv \gamma_{00}$, $\gamma_{21} \equiv \gamma_{10}$, $\zeta_{1j} \equiv r_{0j}$, and $\zeta_{2j} \equiv r_{1j}$). The disturbances ζ_{1j} and ζ_{2j} have zero means and covariance matrix $\mathbf{\Psi}$.

A path diagram of the linear growth-curve model is shown in figure 7.10. As usual in SEM, observed variables are represented by rectangles and latent variables by circles. The four responses are regressed on the random intercept η_{1j} (as indicated by the long arrows from η_{1j} to the responses), with regression coefficients or factor loadings set to 1 (as indicated by the labels attached to the arrows). The responses are also regressed on the random slope η_{2j} with regression coefficients set equal to the time points t_i, here 0, 1, 2, and 3. The short arrows represent the occasion-specific error terms ϵ_{1j} to ϵ_{4j}. The curved double-headed arrow connecting η_{1j} and η_{2j} indicates that these latent variables (random effects), or the corresponding disturbances ζ_{1j} and ζ_{2j}, are correlated.

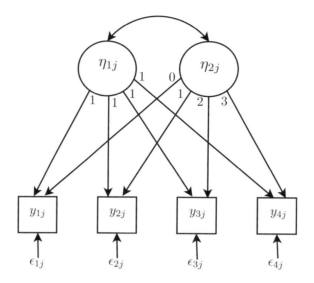

Figure 7.10: Path diagram of linear growth-curve model with four time points

Although we have written the model as an SEM, it can equivalently be fit as a multilevel model, as long as we allow the residual variances to differ between occasions. Allowing such heteroskedastic variances is possible if we have balanced data, such as those used here, but becomes unfeasible with unbalanced data if there are insufficient observations at any of the occasions.

7.7.1 Estimation using sem

The data are in wide form with responses at the four occasions represented by different variables read0, read1, read2, and read3. Because SEM is a method designed for multivariate data, the sem command (introduced in Stata 12) requires the data to be in wide form.

We start by reading in the data:

```
. use http://www.stata-press.com/data/mlmus3/reading, clear
```

In wide form, we can investigate the missingness patterns by using the misstable command (introduced in Stata 11):

```
. misstable patterns read*
```

Missing-value patterns
(1 means complete)

	Pattern			
Percent	1	2	3	4
<1%	1	1	1	1
15	1	0	0	0
14	1	0	1	0
12	0	1	0	0
10	0	0	1	0
8	1	0	0	1
8	1	1	0	0
8	0	1	0	1
7	0	0	0	1
6	0	1	1	0
5	0	0	0	0
3	1	0	1	1
2	1	1	0	1
2	0	0	1	1
<1	0	1	1	1
<1	1	1	1	0
100%				

Variables are   (1) read0   (2) read1   (3) read2   (4) read3

In this table, a "1" means that the variable is observed and a "0" represents missing. We see, for instance, that 15% of the children only have observations for the reading score at the first occasion (kindergarten) and that fewer than 1% of children have observations at the first three occasions.

We can also produce box plots of the observations at each occasion using

```
. graph box read0 read1 read2 read3, ascategory intensity(0) medtype(line)
```

with the result shown in figure 7.11.

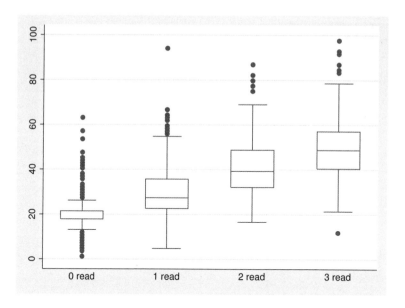

Figure 7.11: Box plots of reading scores for each grade

There does not seem to be much cause for concern. For instance, there is no sign of a ceiling effect by grade 3 (which occurs if a substantial proportion of the children get a perfect score). However, there appears to be an outlier for grade 1.

Note that traditional software for SEM, based on fitting model-implied covariance matrices to empirical covariance matrices, could only handle complete data, but modern software applies ML estimation to all available data. Hence, consistent estimates are produced if missing responses are missing at random (MAR) and the growth-curve model is correctly specified (see section 5.8.1 for more on MAR). In the `sem` command, ML estimation is obtained using the `method(mlmv)` option (`mlmv` stands for ML with missing values).

In the `sem` syntax, latent variables can be referred to using any label of our choosing as long as it begins with a capital letter. We use `L1` and `L2` for the latent variables η_{1j} and η_{2j}. For the measurement model, we want to regress the reading scores on the latent variables or, in other words, include paths from the latent variables to the response variables (see figure 7.10). Such paths are specified using arrows, `<-` or `->`. For example, the reading score in second grade is regressed on both latent variables, denoted as (`read2 <- L1 L2`) (the level-1 error ϵ_{2j} is implicit).

By default, all regression coefficients (factor loadings) and intercepts in the measurement model are free parameters, but we constrain all intercepts to zero using the `noconstant` option and constrain the factor loadings to 1 and 2, respectively, using the syntax (`read2 <- L1@1 L2@2`).

For the structural model, we want to allow the latent variables to have nonzero means (γ_{11}) and (γ_{21}). By default, sem sets means of latent variables to zero, so we relax this constraint by using the means(L1 L2) option.

Putting it all together, the sem command becomes

```
. sem (read0 <- L1@1 L2@0)
>     (read1 <- L1@1 L2@1)
>     (read2 <- L1@1 L2@2)
>     (read3 <- L1@1 L2@3), means(L1 L2) noconstant method(mlmv)
(90 all-missing observations excluded)

Endogenous variables

Measurement:  read0 read1 read2 read3

Exogenous variables

Latent:       L1 L2

Structural equation model                   Number of obs      =        1677
Estimation method  = mlmv
Log likelihood     =    -9594.51

 ( 1)   [read0]L1 = 1
 ( 2)   [read1]L1 = 1
 ( 3)   [read1]L2 = 1
 ( 4)   [read2]L1 = 1
 ( 5)   [read2]L2 = 2
 ( 6)   [read3]L1 = 1
 ( 7)   [read3]L2 = 3
 ( 8)   [read0]_cons = 0
 ( 9)   [read1]_cons = 0
 (10)   [read2]_cons = 0
 (11)   [read3]_cons = 0
```

	Coef.	OIM Std. Err.	z	P>|z|	[95% Conf. Interval]	
Measurement read0 <-						
L1	1	(constrained)				
_cons	0	(constrained)				
read1 <-						
L1	1	(constrained)				
L2	1	(constrained)				
_cons	0	(constrained)				
read2 <-						
L1	1	(constrained)				
L2	2	(constrained)				
_cons	0	(constrained)				
read3 <-						
L1	1	(constrained)				
L2	3	(constrained)				
_cons	0	(constrained)				
Mean						
L1	19.86971	.1915621	103.72	0.000	19.49426	20.24517
L2	10.25926	.1508238	68.02	0.000	9.963652	10.55487
Variance						
e.read0	18.44756	5.410867			10.38179	32.77973
e.read1	61.00465	4.689062			52.47304	70.92343
e.read2	72.7696	7.215578			59.9167	88.37963
e.read3	42.8985	11.88719			24.92148	73.84318
L1	17.98089	5.403334			9.977496	32.40417
L2	7.925603	1.933835			4.912929	12.78569
Covariance L1						
L2	2.887407	2.615503	1.10	0.270	-2.238885	8.0137

LR test model vs. saturated: chi2(5) =   71.97, Prob > chi2 = 0.0000

In the output, the constraints are listed first, followed by the parameters (here all constrained) for the regressions of each of the response variables. This is followed by the estimated means of L1 and L2, and then the residual variances of the response variables and the variances and covariance for L1 and L2.

The intercept η_{1j} therefore has an estimated mean of 19.9 and variance of 18.0, and the slope η_{2j} has an estimated mean of 10.3 and variance of 7.9. The correlation between intercepts and slopes is estimated as 0.24 ($= 2.887407/\sqrt{7.925603 \times 17.98089}$). It is therefore estimated that the reading score increases, on average, by 10.3 units per grade level, and for 95% of children, it increases between 4.7 and 15.8 units per grade level ($10.25926 \pm 1.96 \times \sqrt{7.925603}$). The estimated residual variance is much lower in kindergarten than in grades 1 and 2, but it decreases somewhat again in year 3. The estimates are also reported in table 7.3.

Table 7.3: Maximum likelihood estimates for reading data

	Est	(SE)
Fixed part		
γ_{11} [_cons]	19.87	(0.19)
γ_{21} [grade]	10.26	(0.15)
Random part		
ψ_{11}	17.98	
ψ_{22}	7.93	
ψ_{21}	2.89	
θ_{11}	18.45	
θ_{22}	61.00	
θ_{33}	72.77	
θ_{44}	42.90	
Log likelihood	-9594.51	

An alternative syntax for the same model is

```
sem (L1 -> read0@1 read1@1 read2@1 read3@1)
    (L2 -> read1@1 read2@2 read3@3), means(L1 L2) nocons method(mlmv)
```

We can obtain EB predictions of the child-specific intercepts and slopes by using

```
. predict l1 l2, latent
(latent(L1 L2) assumed)
```

These predictions are called factor scores in SEM. There are two methods for factor scoring: the regression method (used by **sem**) corresponds to EB prediction, whereas the Bartlett method corresponds to ML estimation, as described in section 2.11.1 for variance-components models and in section 4.8.1 for random-coefficient models.

7.7.2 Estimation using xtmixed

We now demonstrate how the same growth-curve model can be fit using **xtmixed**. For analysis using **xtmixed**, the data should be in long form, so we begin by reshaping the data using the **reshape** command:

```
. reshape long read math age, i(id) j(grade)
(note: j = 0 1 2 3)
```

Data		wide	->	long
Number of obs.		1767	->	7068
Number of variables		15	->	7
j variable (4 values)			->	grade
xij variables:				
	read0 read1 ... read3		->	read
	math0 math1 ... math3		->	math
	age0 age1 ... age3		->	age

In long form, we can use `xtdescribe` to explore the missing data patterns

```
. quietly xtset id grade
. drop if read >= .
(4392 observations deleted)
. xtdescribe
        id:  301, 401, ..., 1266701                      n =      1677
     grade:  0, 1, ..., 3                                T =         4
             Delta(grade) = 1 unit
             Span(grade)  = 4 periods
             (id*grade uniquely identifies each observation)
```

Distribution of T_i:	min	5%	25%	50%	75%	95%	max
	1	1	1	2	2	3	3

Freq.	Percent	Cum.	Pattern
268	15.98	15.98	1...
249	14.85	30.83	1.1.
215	12.82	43.65	.1..
170	10.14	53.79	..1.
144	8.59	62.37	1..1
136	8.11	70.48	11..
133	7.93	78.41	.1.1
119	7.10	85.51	...1
112	6.68	92.19	.11.
131	7.81	100.00	(other patterns)
1677	100.00		XXXX

Note that children with missing responses at all four occasions are not counted here, unlike in the `misstable` command. We see that at least 92.19% of children (with at least one nonmissing response) have fewer than three observations.

Because of the balanced nature of the data, we have sufficient data per occasion to plot the mean trajectory,

```
. egen mn_read = mean(read), by(grade)
. twoway (connected mn_read grade, sort), xtitle(Grade)
> ytitle(Mean reading score)
```

giving the graph in figure 7.12, where the mean growth trajectory is remarkably linear. (Using the plot type `connected` is useful because it shows where the observations occurred.)

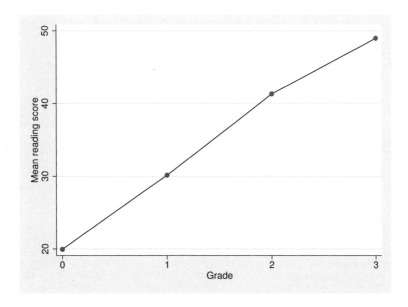

Figure 7.12: Sample mean growth trajectory for reading score

Box plots such as those shown in figure 7.11 can be produced using the command

```
graph box read, over(grade) ytitle(Reading score) intensity(0) medtype(line)
```

We can now fit a linear growth model using **grade** as the time variable and using the **residuals()** option to let the residual variance differ between occasions or, in other words, to allow heteroskedastic level-1 residuals over occasions. Specifically, we request an independence correlation structure (uncorrelated residuals) and estimate a separate variance parameter for each occasion by using the **residuals(independent, by())** option with **grade** as stratification variable:

```
. xtmixed read grade || id: grade, covariance(unstructured) mle
> variance residuals(independent, by(grade))
```

```
Mixed-effects ML regression                    Number of obs      =      2676
Group variable: id                             Number of groups   =      1677

                                               Obs per group: min =         1
                                                              avg =       1.6
                                                              max =         3

                                               Wald chi2(1)       =   4917.74
Log likelihood =   -9594.51                    Prob > chi2        =    0.0000
```

read	Coef.	Std. Err.	z	P>|z|	[95% Conf.	Interval]
grade	10.25926	.1462962	70.13	0.000	9.972526	10.546
_cons	19.86971	.1887822	105.25	0.000	19.49971	20.23972

Random-effects Parameters	Estimate	Std. Err.	[95% Conf.	Interval]
id: Unstructured				
var(grade)	7.925667	1.933749	4.913092	12.78547
var(_cons)	17.9805	5.403015	9.977499	32.40275
cov(grade,_cons)	2.887511	2.615339	-2.238458	8.013481
Residual: Independent, by grade				
0: var(e)	18.448	5.410546	10.38254	32.77895
1: var(e)	61.0048	4.689077	52.47316	70.92361
2: var(e)	72.76976	7.215601	59.91681	88.37983
3: var(e)	42.89778	11.88699	24.92107	73.84193

```
LR test vs. linear regression:        chi2(6) =    580.77   Prob > chi2 = 0.0000
Note: LR test is conservative and provided only for reference.
```

As expected, the estimates agree almost perfectly with those obtained using the sem command.

We can compare the sample mean reading scores with the estimated model-implied means graphically (shown in figure 7.13) using

```
. predict fixed, xb

. twoway (connected mn_read grade, sort lpatt(solid))
>        (connected fixed grade, sort lpatt(dash)), xtitle(Grade)
>        ytitle(Mean reading score) legend(order(1 "Raw mean" 2 "Fitted mean"))
```

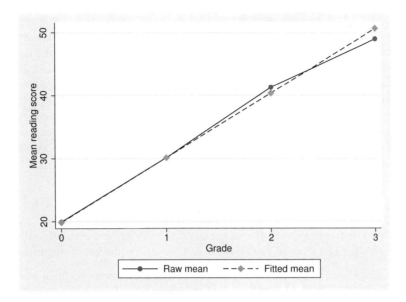

Figure 7.13: Fitted mean trajectory and sample mean trajectory for reading scores

The fitted means of the reading scores from the model are close to the sample means.

We can obtain EB predictions of the latent variables η_{1j} and η_{2j} by first predicting the corresponding disturbances,

```
. predict zeta2 zeta1, reffects
```

and then adding these to the estimated means:

```
. generate eta1 = _b[_cons] + zeta1
. generate eta2 = _b[grade] + zeta2
```

These predictions agree to at least three decimal places with the predictions l1 and l2 produced by predict after the sem command.

7.8 Summary and further reading

In this chapter, we have discussed growth-curve models, demonstrating how to model nonlinear growth using either polynomial or piecewise linear functions. We have also shown how heteroskedasticity can be modeled for both level-1 residuals and random effects. Serial correlations for the level-1 residuals can also be modeled; see section 6.4.1 and exercise 7.8. Keep in mind that adding more parameters to a model can render the model unidentified for a given data structure, as discussed for random-coefficient models in section 4.10.4. Importantly, the standard linear growth-curve model is not identified with two balanced occasions.

We also described how growth-curve models can alternatively be viewed as SEMs with latent variables and fit using the `sem` command. Whether a growth-curve model is viewed as a multilevel model or an SEM is mostly a question of tradition, but there are some differences. In practice, structural equation modelers invariably allow for occasion-specific residual variances, whereas multilevel modelers (implicitly) tend to constrain these variances to be equal. Another difference is that conventional structural equation modeling requires balanced data. A great advantage of SEM is that it is possible to include latent variables measured by multiple indicators. When the response variable in a growth-curve model is latent, such models are sometimes called curve-of-factors growth models or second-order latent growth models. Another advantage of SEM is that it is straightforward to include intervening variables or mediators in the model.

Useful introductions to linear growth-curve models include the book by Singer and Willett (2003, chap. 3–8), the articles by Bryk and Raudenbush (1987) and Willett, Singer, and Martin (1998), and the encyclopedia entry by Singer and Willett (2005). The book by Bollen and Curran (2006) takes a structural equation modeling perspective.

The exercises apply growth-curve models to problems from a range of disciplines. Whereas exercises 7.1 and 7.4 consider heteroskedasticity of the level-1 error over time, exercise 7.3 considers heteroskedasticity of both the level-1 and the level-2 random part between groups, and exercise 7.8 introduces an AR(1) process for the level-1 error (see section 6.4.1). Exercise 7.6 involves a piecewise linear spline with the knot occurring at different times for different individuals. Exercises 7.4 and 7.7 use both the two-stage model formulation and the reduced form. In exercise 7.4, the data come from a randomized experiment.

Growth-curve models for subjects nested in clusters are discussed in section 8.13 and exercises 8.1, 8.4, and 8.7 of the next chapter on higher-level models with nested random effects.

7.9   Exercises

7.1 Growth-in-math-achievement data [Solutions]

Consider the math outcome in the National Longitudinal Survey of Youth data, `reading.dta`, provided by Bollen and Curran (2006) and described in section 7.6. The math outcome, `math0` to `math3`, is the percentage of correctly answered items on a math achievement test (the same test was used in every grade).

1. Reshape the data to long form, and plot the mean math trajectory over time by minority status.

2. Fit a linear growth-curve model using `xtmixed` with `minority`, a dummy variable for being a minority, as a covariate. The fixed part should include an intercept and a slope for `grade`, and the random part should include random intercepts and random slopes of `grade`. Allow the residual variances to differ between grades.

3. By extending the model from step 2, test whether there is any evidence for a narrowing or widening of the minority gap over time.

4. Plot the mean fitted trajectories for minority and nonminority students.

5. Plot fitted and observed growth trajectories for the first 20 children (`id` less than 15900).

6. Fit the model in step 2, but without `minority` as a covariate, using `sem`.

7.2 Children's growth data

In this exercise, we revisit the Asian children's growth data used in the first part of this chapter.

1. Fit the model in (7.1) using `xtmixed`.

2. For the model and estimates from step 1, obtain EB predictions for the random intercepts and slopes.

3. Perform some residual diagnostics.

4. ❖ Also obtain ML (or OLS) estimates of ζ_{1j} and ζ_{2j} by first subtracting the predicted fixed part of the model and then using the `statsby` command. Merge these estimates into the data. Compare the ML estimates with the EB predictions using scatterplots by sex.

7.3 Jaw-growth data

In this exercise, we use the jaw-growth data `growth.dta`, previously considered in exercise 3.3.

1. Extend the model from exercise 3.3 (with an interaction between age and a dummy variable for girl) to investigate whether there is significant between-subject variability in the growth rate. Retain the simpler model if the likelihood-ratio test is not significant at the 5% level.

2. For the model chosen in step 1, investigate whether the child-level random-effects variances (and covariance if applicable) differ between boys and girls, again using a 5% level of significance.

3. For the model chosen in step 2, relax the assumption that the level-1 errors have the same variance for boys and girls, and test this assumption at the 5% level.

4. For the model selected in step 3, plot the fitted growth trajectories by sex and compare them with the corresponding observed growth trajectories.

7.4 Calcium-supplementation data

Lloyd et al. (1993) describe a study to evaluate the effect of calcium supplementation on bone acquisition in adolescent white girls. Ninety-four girls with a mean age of 11.9 years were randomized to receive 18 months of calcium supplementation (500 mg per day of calcium as calcium citrate malate) or placebo pills. Total body bone mineral density was measured approximately at six-month intervals for about 30 months using a bone absorptiometer.

The data in `calcium.dta` are provided by Vonesh and Chinchilli (1997) and Demidenko (2004), and contain the following variables:

- `id`: subject identifier
- `treat`: treatment group (1: calcium group; 0: placebo group)
- `time`: time in weeks after first intake
- y: total body bone mineral density (g/cm^2)

1. Convert the time scale to years after first intake (there are on average 52.177457 weeks per year) and call the new variable `years`.
2. Plot the observed growth trajectories by treatment group.
3. Consider a linear growth-curve model with a random intercept and random slope of `years` and with fixed effects of `years`, `treat`, and the `years` by `treat` interaction.

   a. Write down the model using a two-stage formulation.
   b. Write down the model in reduced form.
   c. Fit the model and interpret the estimated treatment effect.

4. Perform a likelihood-ratio test for the random slope of `years`, and retain it if significant at the 5% level.
5. For the model selected in step 4, relax the assumption that the level-1 residual variance is the same across occasions (create a variable, `visit`, for the occasion or visit number). Use a 5% significance level to decide whether the assumption should be rejected.
6. For the model chosen in step 5, obtain the fitted growth trajectories and plot them for comparison with the observed growth trajectories from step 1.
7. Plot the predicted mean growth trajectories for the two treatment groups.

7.5 Diffusion-of-innovations data

In some U.S. states, the introduction of new medical technology requires a certificate-of-need review to prevent unnecessary capital expenditure by health care facilities. Caudill, Ford, and Kaserman (1995) investigated whether such certificate-of-need regulation has an effect on the diffusion of innovations. Specifically, they considered the adoption of hemodialysis (blood filtering) for kidney failure in 50 U.S. states between 1977 and 1990.

Many states implemented certificate-of-need review of dialysis clinics' investments in the late 1970s and early 1980s, and many states eliminated such a review in the late 1980s. This change in policy allowed Caudill, Ford, and Kaserman to examine whether certificate-of-need regulation has slowed the rate of diffusion of hemodialysis technology.

They let L_i be the number of dialysis machines in state i in the most recent period in the sample and P_{it} be the number of dialysis machines in state i at time t. Using the transformation $\ln\{P_{it}/(L_i - P_{it})\}$, Caudill, Ford, and Kaserman specified the following model (using their notation):

$$\ln\{P_{it}/(L_i - P_{it})\} = a_i + c_iT_t + d_iT_t \times \text{Con}_{it} + \epsilon_{it}$$

where T_t is an index of the time period that begins at $T_1 = 1$ in 1977 and Con_{it} is a dummy variable indicating that certificate-of-need regulation was in effect in state i at time t.

The random intercept a_i and the random coefficients c_i and d_i are assumed to be independently distributed with means μ_a, μ_c, and μ_d, respectively, and variances σ_a^2, σ_c^2, and σ_d^2, respectively. ϵ_{it} has zero mean, variance σ_ϵ^2, and is independent of a_i, c_i, and d_i.

The dataset `data.cfk` from the *Journal of Applied Econometrics Data Archive* has the following variables:

- `state`: an identifier for the U.S. states
- `T`: the time index T_t
- `TCon`: the interaction $T_t \times \text{Con}_{it}$
- `resp`: the response variable $\ln\{P_{it}/(L_i - P_{it})\}$

1. Read the data using the `infile` command (the variables are in the same order as listed above).
2. Fit the model specified by Caudill, Ford, and Kaserman using the `xtmixed` command, noting that they specified the random effects as uncorrelated. The random effects have nonzero means, whereas `xtmixed` assumes zero means. You can accommodate the means in the fixed part of the model.
3. Do the estimates suggest that certificate-of-need legislation slows the rate of diffusion of hemodialysis technology?
4. If you can get hold of the paper, compare your estimates with the ML estimates in table 1 of Caudill, Ford, and Kaserman (1995).
5. Perform a likelihood-ratio test for the null hypothesis

$$\text{H}_0: \sigma_c^2 = \sigma_d^2 = 0$$

   Note that the "naïve" test is conservative because the null hypothesis involves two parameters on the border of the parameter space. (You could use the correct asymptotic null distribution given in display 8.1.)
6. Does the conclusion for the effect of certificate-of-need-legislation change when the restricted model (with $\sigma_c^2 = \sigma_d^2 = 0$) is used?
7. Perform residual diagnostics for the selected model, as described in section 3.9.

7.6 Fat-accretion data

Fitzmaurice, Laird, and Ware (2011) analyzed data on the influence of menarche (onset of menstruation) on body fat accretion. They used longitudinal data from a prospective study on body fat accretion in a cohort of 162 girls from the MIT Growth and Development Study (Bandini et al. 2002; Phillips et al. 2003; Naumova, Must, and Laird 2001). (Fitzmaurice, Laird, and Ware [2011] point out that the data represent a subset of the full study and should not be used to draw substantive conclusions.)

At the start of the study, all the girls were premenarcheal and nonobese, as determined by a triceps skinfold thickness less than the 85th percentile. The girls were examined annually until four years after menarche. The response variable is a measure of body fatness based on bioelectric impedance analysis from which a measure of percent body fat was derived.

The variables in the dataset `fat.dta` are the following:

- `id`: subject identifier
- `time`: time since menarche in years (negative if before menarche) (t_{ij})
- `fat`: percent body fat (y_{ij})
- `age`: age in years
- `menarche`: age at menarche (in years)

Fitzmaurice (1998) considers the following model:

$$y_{ij} = \beta_1 + \beta_2 t_{ij} + \beta_3 (t_{ij})_+ + \zeta_{1j} + \zeta_{2j} t_{ij} + \zeta_{3j} (t_{ij})_+ + \epsilon_{ij},$$

where t_{ij} is time since menarche and $(t_{ij})_+ = t_{ij}$ if $t_{ij} > 0$ and $(t_{ij})_+ = 0$ if $t_{ij} \leq 0$. The usual assumptions are made regarding the random effects and error term.

1. Interpret the following terms:
    - $\beta_1 + \zeta_{1j}$
    - $\beta_2 + \zeta_{2j}$
    - $\beta_3 + \zeta_{3j}$

2. Fit the model by ML and interpret the estimates.

3. Produce a scatterplot of the percent body fat measurements versus time, with the fitted mean curve superimposed.

7.7 Adolescent-alcohol-use data

Consider the data described in exercise 6.3. Using Singer and Willett's (2003) two-stage formulation (in their notation except that i is occasion and j is subject), consider the level-1 model

$$Y_{ij} = \pi_{0j} + \pi_{1j} t_i + \epsilon_{ij}$$

and the level-2 models

$$\pi_{0j} = \gamma_{00} + \gamma_{01} w_{1j} + \gamma_{02} w_{2j} + \zeta_{0j}$$
$$\pi_{1j} = \gamma_{10} + \gamma_{11} w_{1j} + \gamma_{12} w_{2j} + \zeta_{1j}$$

1. Substitute the level-2 models into the level-1 model to obtain the reduced-form model.
2. Interpret each of the parameters in terms of initial status at age 14, π_{0j}, and the rate of growth π_{1j}.
3. Fit the model by ML using xtmixed and interpret the estimates.
4. Separately for children of alcoholics and other children, plot the fitted trajectories together with the data using trellis graphs.
5. Obtain the estimated marginal variances and correlation matrix of the total residuals. You can use the xtmixed_corr command described in section 6.3.1. Note that xtmixed_corr stores the covariance matrix as r(V).
6. Fit latent growth-curve models using sem.

   a. Fit a standard latent growth-curve model without covariates as shown in section 7.7.1.
   b. Include covariates by specifying equations for the latent variables. For example, if the random intercept is called L1, add (L1 <- coa peer _cons). Also add cov(e.L1*e.L2) to specify that the residuals ζ_{0j} and ζ_{1j} are correlated, and remove the means() option.
   c. Why are the estimates not the same as in step 3?
   d. ❖ Modify the model so that it corresponds exactly to the model fit in step 3 (with practically identical estimates).

7.8 Unemployment-claims data

Papke (1994) analyzed panel data on Indiana's enterprise zone program which provided tax credits for areas of cities with high unemployment and high poverty levels. One of the purposes was to investigate whether inclusion in an enterprise zone would reduce the number of unemployment claims (see also exercises 5.3 and 5.4). Data were from 22 unemployment claims offices (serving a zone and the surrounding city), 6 of which were included as enterprise zones in 1984 and 4 more of which were included as zones in 1985.

The dataset ezunem.dta supplied by Wooldridge (2010) contains the following variables:

- city: unemployment claims office identifier (j)
- year: year (i)
- uclms: number of unemployment claims (y_{ij})
- ez: dummy variable for office being in an enterprise zone (x_{2ij})
- t: time 1, ..., 9 (x_{3i})

Use ML to fit the models in this exercise.

1. Fit the random-intercept model

$$\ln(y_{ij}) = \tau_i + \beta_2 x_{2ij} + \zeta_j + \epsilon_{ij}$$

where τ_i is a fixed year-specific intercept and ζ_j is an office-specific random intercept.

    a. Interpret the estimated regression coefficients and random-intercept variance.

    b. Does the enterprise zone program appear to reduce unemployment claims?

2. Fit the random-intercept model

$$\ln(y_{ij}) \;=\; \tau_i + \beta_2(x_{2ij} - \overline{x}_{2\cdot j}) + \beta_3\overline{x}_{2\cdot j} + \zeta_j + \epsilon_{ij}$$

where $\overline{x}_{2\cdot j}$ is the proportion of the panel waves when unemployment claims office j is in an enterprise zone.

    a. Interpret the estimated regression coefficients β_2 and β_3.

    b. Use the fitted model to test for level-2 endogeneity or office-level confounding.

3. Fit a random-coefficient model with a random intercept and random slope of time:

$$\ln(y_{ij}) \;=\; \tau_i + \beta_2 x_{2ij} + \zeta_{1j} + \zeta_{2j}x_{3i} + \epsilon_{ij}$$

Note that time is treated as a categorical variable in the fixed part of the model and as a continuous variable in the random part.

    a. Is the random slope required? Use a 5% level of significance.

    b. Interpret the fixed and random effects of time in this model.

    c. Instead of treating **year** as continuous in the random part of the model, consider a model where the random part consists of a random intercept and random coefficients of eight dummy variables for the years. Explain why such a model is not identified (see section 4.10.4).

4. Fit a random-coefficient model including a random intercept, a random slope of time, and an AR(1) process for the residuals

$$\ln(y_{ij}) \;=\; \tau_i + \beta_2 x_{2ij} + \zeta_{1j} + \zeta_{2j}x_{3i} + \epsilon_{ij}, \qquad \epsilon_{ij} = \alpha\epsilon_{i-1,j} + u_{ij}$$

under the usual assumptions.

    a. Which features of the random part of this model induce dependence of claims within offices over time?

    b. Check whether setting $\alpha = 0$ or $\psi_{22} = 0$ produces a significant deterioration in model fit.

Part IV

Models with nested and crossed random effects

8 Higher-level models with nested random effects

8.1 Introduction

We have until now considered two-level data where units are nested in groups or clusters. In three-level data,the clusters themselves are nested in superclusters, forming a hierarchical structure. For example, we may have repeated measurement occasions (at level 1) for patients (at level 2) who are clustered in hospitals (at level 3). This three-level design is displayed in figure 8.1. It seems reasonable to expect that measurements on different patients within the same hospital are correlated and that measurements on the same patient are even more correlated.

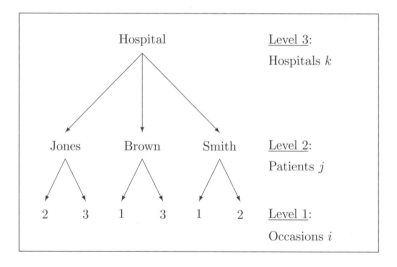

Figure 8.1: Illustration of three-level design

Other examples of three-level longitudinal designs include repeated measures on students nested in schools, and repeated measures on employees nested in firms. Examples of three-level cross-sectional designs include data on students nested in classes nested in schools, teeth nested in patients nested in dentists, and patients nested in physicians nested in hospitals.

In these examples, the level-1 units can be thought of as being grouped according to two classifications, for instance, students are grouped into both classes and schools. Nesting refers to the fact that the lower-level classification (here classes at level 2) results from subdividing units belonging to the same higher-level classification (here schools at level 3). All units belonging to the same level-2 clusters (for example, all students belonging to the same class) must also belong to the same level-3 supercluster (here school). Such data structures are referred to as *hierarchical*.

Violations of an apparently hierarchical structure are not uncommon. For instance, for patients nested in physicians nested in hospitals, some physicians may treat patients in multiple hospitals. For longitudinal data on students nested in schools, some students may switch schools. If the violations are rare, we can handle them by discarding some of the data or assigning units to the group to which they belong most often. Otherwise, models with *crossed* random effects may be necessary, as discussed in chapter 9.

A factor or classification such as school constitutes a *level* only if it is treated as random. It does not matter if students are cross-classified by ethnicity and school because ethnicity will typically be treated as fixed. Sometimes two factors are cross-classified and one factor is treated as the higher level, whereas the *interaction* between the two factors is treated as the lower level. For instance, in a study of house prices, houses may be classified by U.S. state and by whether they are in an urban or rural area. Then state may be treated as random at level 3, and the interaction between state and a dummy variable for urban area (versus rural area) treated as random at level 2. The level-2 random effect is nested in states because it takes on a different value for each combination of state and urban (versus rural) area to allow the effect of urban area to be different in different states. The model may also include a fixed main effect of a dummy variable for urban to represent the average difference in house prices between urban and rural areas across states.

8.2 Do peak-expiratory-flow measurements vary between methods within subjects?

In chapter 2, we considered test–retest data for peak-expiratory-flow measurements taken using only the Mini Wright meter. Here we will analyze the full dataset shown in table 2.1 on page 75, where two methods were used: the Mini Wright peak-flow meter and the standard Wright peak-flow meter, each on two occasions. The full dataset can be used for a method comparison study to investigate the performance of the measurement methods.

We first read in the data,

```
. use http://www.stata-press.com/data/mlmus3/pefr
```

and stack the measurements at the two occasions into one variable for each method, as shown in section 2.5:

```
. reshape long wp wm, i(id) j(occasion)
(note: j = 1 2)
Data                                    wide   ->   long

Number of obs.                            17   ->     34
Number of variables                        5   ->      4
j variable (2 values)                          ->   occasion
xij variables:
                                     wp1 wp2   ->   wp
                                     wm1 wm2   ->   wm
```

We then stack the responses wm for the Mini Wright peak-flow meter and wp for the Wright peak-flow meter into a single response, w, producing a string variable, meth, equal to the suffixes m and p. To do this using Stata's reshape long command with the string option, we must first create a new variable, i, for the i() option that takes on a different value for each observation in the wide dataset:

```
. generate i = _n
. reshape long w, i(i) j(meth) string
(note: j = m p)
Data                                    wide   ->   long

Number of obs.                            34   ->     68
Number of variables                        5   ->      5
j variable (2 values)                          ->   meth
xij variables:
                                       wm wp   ->   w
```

```
. sort id meth occasion
. list id meth occasion w in 1/8, clean noobs
      id    meth    occasion      w
       1      m           1     512
       1      m           2     525
       1      p           1     494
       1      p           2     490
       2      m           1     430
       2      m           2     415
       2      p           1     395
       2      p           2     397
```

We can convert the *string variable* meth to a numeric variable by using the encode command, which assigns successive integer values to the strings sorted in alphabetical order (here m becomes 1 and p becomes 2):

```
. encode meth, generate(method)
```

A dummy variable for method equal to 2 (the Mini Wright meter, meth equal to m) is then easy to create using the recode command:

```
. recode method 2=0
```

8.3 Inspecting sources of variability

We can plot all four peak-expiratory-flow measurements against the subject identifier using different symbols for the methods, giving the graph in figure 8.2:

```
. twoway (scatter w id if method==0, msymbol(circle))
> (scatter w id if method==1, msymbol(circle_hollow)),
> xtitle(Subject id) ytitle(Peak-expiratory-flow measurements)
> legend( order(1 "Wright" 2 "Mini Wright")) xlabel(1/17)
```

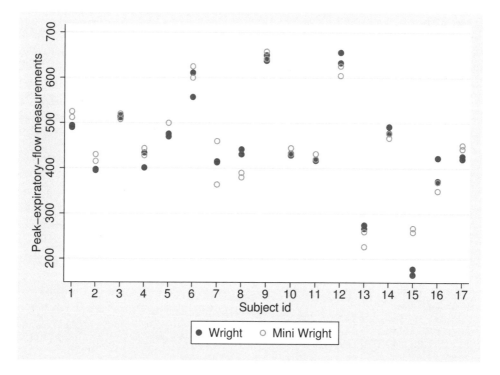

Figure 8.2: Scatterplot of peak expiratory flow measured by two methods versus subject

As would be expected, measurements on the same subjects are more similar than measurements on different subjects. This between-subject heterogeneity can be modeled by a subject-level random intercept, as we did in chapter 2. However, the figure also suggests that for a given subject, the measurements using the same method tend to resemble each other more than measurements using the other method, so that the responses for the same method are not conditionally independent given the subject-level random intercept. The phenomenon can also be described as between-method heterogeneity within subjects because the measurements made by the same instrument (or method) tend to be shifted up or down relative to the measurements made by the other instrument, with shifts that vary between subjects. For some subjects (for example, the first two), the Wright peak-flow meter measurements are lower, whereas for other subjects (for example, subjects 8 and 12) the Mini Wright meter measurements are lower. Furthermore, the difference between methods is large for some subjects (for example, subject 15) and small for others (for example, subjects 3, 9, 10, and 11). In a measurement context, this kind of method by subject interaction is sometimes referred to as subject-specific bias of the methods (see Dunn [1992]).

8.4   Three-level variance-components models

We can accommodate the between-method within-subject heterogeneity apparent in figure 8.2 by including a random intercept for each combination of method and subject in addition to a random intercept for subject. The random intercept for method is nested within subjects in the sense that it does not take on the same value for a given method across all subjects, but takes on a different value for each combination of method and subject. We will therefore think of occasions as level 1, methods as level 2, and subjects as level 3.

The three-level variance-components model can be written as

$$y_{ijk} \;=\; \beta + \zeta_{jk}^{(2)} + \zeta_k^{(3)} + \epsilon_{ijk}$$

where $\zeta_{jk}^{(2)}$ is the random intercept for method j and subject k, and $\zeta_k^{(3)}$ is the random intercept for subject k. Here the superscripts denote the levels at which the random intercepts vary.

The error components ϵ_{ijk}, $\zeta_{jk}^{(2)}$, and $\zeta_k^{(3)}$ are assumed to have zero means and to be mutually uncorrelated so that their variances add up to the total variance. The corresponding variance components are the variance θ of the level-1 residuals, the variance $\psi^{(2)}$ of the level-2 random intercepts, and the variance $\psi^{(3)}$ of the level-3 random intercepts. The level-1 variance θ can be interpreted as the between-occasion, within-method, and within-subject variance. The level-2 variance $\psi^{(2)}$ is the between-methods, within-subjects variance. And the level-3 variance $\psi^{(3)}$ is the between-subjects variance. A large between-methods within-subjects variance would mean that there is a large method by subject interaction. All three error components are uncorrelated across subjects, the level-2 random intercepts and level-1 residuals are uncorrelated across methods, and the level-1 residuals are uncorrelated across occasions.

The error components for the three-level variance-components model are shown for a subject k in figure 8.3.

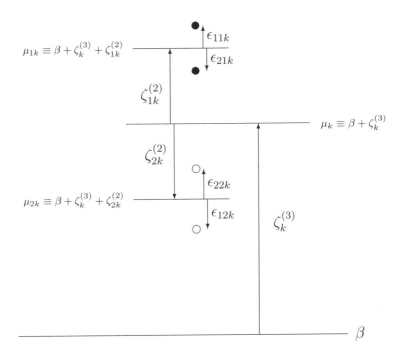

Figure 8.3: Illustration of error components for the three-level variance-components model for a subject k

The overall expectation or mean peak-expiratory-flow measurement for the population of subjects is $E(y_{ijk}) = \beta$, given as the bottom line in the figure. In the first stage, a random intercept $\zeta_k^{(3)}$ is drawn for a subject k from a distribution with mean 0 and variance $\psi^{(3)}$, resulting in a mean measurement for subject k equal to $\mu_k \equiv E(y_{ijk}|\zeta_k^{(3)}) = \beta + \zeta_k^{(3)}$. (In the measurement context, μ_k is the subject's true score.) In the second stage, a random intercept $\zeta_{1k}^{(2)}$ is drawn for the first method $j=1$ from a distribution with mean 0 and variance $\psi^{(2)}$. This produces a method-1-specific mean measurement for subject k equal to $\mu_{1k} \equiv E(y_{ijk}|\zeta_k^{(3)}, \zeta_{1k}^{(2)}) = \beta + \zeta_k^{(3)} + \zeta_{1k}^{(2)}$. For the second method $j=2$, a random intercept $\zeta_{2k}^{(2)}$ is drawn from the same distribution as for the first method, producing a method-2-specific mean measurement for subject k equal to $\mu_{2k} \equiv E(y_{ijk}|\zeta_k^{(3)}, \zeta_{2k}^{(2)}) = \beta + \zeta_k^{(3)} + \zeta_{2k}^{(2)}$. Finally, in the third stage, level-1 residuals ϵ_{11k} and ϵ_{21k} are drawn from a distribution with mean 0 and variance θ, resulting in the two observed measurements y_{11k} and y_{21k} (represented by the filled dots) for method $j=1$. Similarly, the two observed measurements y_{12k} and y_{22k} (represented by the hollow dots) for method $j=2$ are obtained by drawing level-1 residuals ϵ_{12k} and ϵ_{22k} from the same distribution.

The random part of the model is represented by a path diagram in figure 8.4. The rectangles represent observed variables, here the responses y_{11k}, y_{21k}, y_{12k}, and y_{22k} for subject k. The k subscript is implied by the label "subject k" inside the frame surrounding the diagram. The circles represent the random effects $\zeta_{1k}^{(2)}$ for method 1 and subject k, $\zeta_{2k}^{(2)}$ for method 2 and subject k, and $\zeta_k^{(3)}$ for subject k, again with k subscripts not shown. The long arrows represent regressions, here with regression coefficients set to 1, and the short arrows from below represent additive error terms ϵ_{ijk} (also with coefficients set to 1). For instance, y_{12k} (with $i=1$ and $j=2$) is regressed on $\zeta_{2k}^{(2)}$ and $\zeta_k^{(3)}$ with additive error term ϵ_{12k}

$$y_{12k} \;=\; \zeta_{2k}^{(2)} + \zeta_k^{(3)} + \epsilon_{12k} + \cdots$$

where the "$\cdots$" indicates that the fixed part of the model (here β) is not shown.

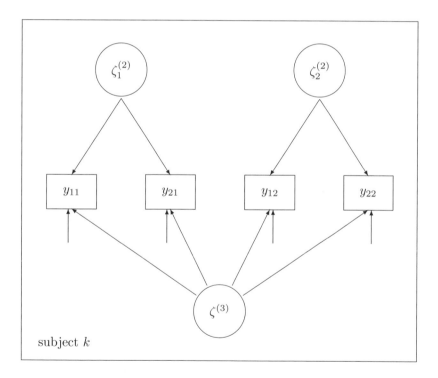

Figure 8.4: Path diagram of random part of three-level model

We see that responses (here measurements) for the same subject are correlated because they all depend on the shared level-3 random intercept $\zeta_k^{(3)}$. Conditional on $\zeta_k^{(3)}$, responses for the same method are not independent but correlated because they depend on the shared level-2 random intercepts $\zeta_{1k}^{(2)}$ or $\zeta_{2k}^{(2)}$. However, responses for different methods, such as y_{11k} and y_{12k}, are conditionally independent given the subject-level random intercept $\zeta_k^{(3)}$.

8.5   Different types of intraclass correlation

For three-level models, we can consider several types of intraclass correlations. For measurements i and i' on the same subject k, using *different* methods j and j', the intraclass correlation becomes

$$\rho(\text{subject}) \equiv \text{Cor}(y_{ijk}, y_{i'j'k}) = \frac{\psi^{(3)}}{\psi^{(2)} + \psi^{(3)} + \theta}$$

The denominator is the product of standard deviations of y_{ijk} and $y_{i'j'k}$ (each equal to the square root of the denominator), and the numerator is the covariance between these responses. We can derive this covariance by taking the expectation of the product of the random parts for each variable,

$$
\begin{aligned}
E\{(\zeta_k^{(3)} + \zeta_{jk}^{(2)} + \epsilon_{ijk})(\zeta_k^{(3)} + \zeta_{j'k}^{(2)} + \epsilon_{i'j'k})\} &= E\{(\zeta_k^{(3)})^2\} + E(\zeta_k^{(3)}\zeta_{j'k}^{(2)}) + E(\zeta_k^{(3)}\epsilon_{i'j'k}) \\
&\quad + E(\zeta_{jk}^{(2)}\zeta_k^{(3)}) + E(\zeta_{jk}^{(2)}\zeta_{j'k}^{(2)}) + E(\zeta_{jk}^{(2)}\epsilon_{i'j'k}) \\
&\quad + E(\epsilon_{ijk}\zeta_k^{(3)}) + E(\epsilon_{ijk}\zeta_{j'k}^{(2)}) + E(\epsilon_{ijk}\epsilon_{i'j'k}) \\
&= E\{(\zeta_k^{(3)})^2\} = \psi^{(3)}
\end{aligned}
$$

All terms except the first are zero because the error components in the product are uncorrelated, either because the error components are at different levels or because they are not for the same level-2 unit [in the case of $E(\zeta_{jk}^{(2)}\zeta_{j'k}^{(2)})$] or the same level-1 unit [in the case of $E(\epsilon_{ijk}\epsilon_{i'j'k})$].

For measurements on the same subject k, using the *same* method j, we get

$$\rho(\text{method}, \text{subject}) \equiv \text{Cor}(y_{ijk}, y_{i'jk}) = \frac{\psi^{(2)} + \psi^{(3)}}{\psi^{(2)} + \psi^{(3)} + \theta}$$

We can derive the numerator by substituting j for j' in the derivation above. Then we see that another nonzero term now contributes to the covariance, namely, $E(\zeta_{jk}^{(2)}\zeta_{jk}^{(2)}) = \psi^{(2)}$. This intraclass correlation can be thought of as the test–retest reliability for each method. The models considered here assume that both methods are equally reliable, an assumption we relax in exercise 8.11.

In both intraclass correlations, the numerator is equal to the variance shared by the measurements, and the denominator is just the total variance. In a proper three-level model, the variances of the random intercepts are positive, $\psi^{(2)} > 0$ and $\psi^{(3)} > 0$, and it follows that $\rho(\text{method}, \text{subject}) > \rho(\text{subject})$. This relationship makes sense because, as we saw in figure 10.2, measurements for the same subject are more correlated if they use the same method than if they use different methods.

We also saw in figure 8.4 that responses for the same method (for example, y_{11k} and y_{21k}) are connected via two paths, one through $\zeta_{1k}^{(2)}$ and the other through $\zeta_{k}^{(3)}$, whereas responses for different methods (for example, y_{21k} and y_{12k}) are connected via only one of these paths, through $\zeta_{k}^{(3)}$, making them less correlated.

8.6   Estimation using xtmixed

When there are several nested levels, we simply specify an equation for the random part at each level, starting with the highest level and working our way down.

Here we start with the random part || id: for subjects at level 3, followed by the random part || method: for methods within subject at level 2:

```
. xtmixed w || id: || method:, mle
Mixed-effects ML regression                    Number of obs      =      68
```

Group Variable	No. of Groups	Observations per Group		
		Minimum	Average	Maximum
id	17	4	4.0	4
method	34	2	2.0	2

```
                                               Wald chi2(0)       =       .
Log likelihood = -345.29005                    Prob > chi2        =       .
```

w	Coef.	Std. Err.	z	P>|z|	[95% Conf. Interval]	
_cons	450.8971	26.63839	16.93	0.000	398.6868	503.1074

Random-effects Parameters	Estimate	Std. Err.	[95% Conf. Interval]	
id: Identity				
sd(_cons)	108.6037	19.05411	77.00246	153.1739
method: Identity				
sd(_cons)	19.47623	4.829488	11.97937	31.66474
sd(Residual)	17.75859	2.153545	14.00184	22.52329

```
LR test vs. linear regression:       chi2(2) =   143.81   Prob > chi2 = 0.0000
Note: LR test is conservative and provided only for reference.
```

At the top of the output, we see that there are 17 subjects with 4 observations each and 34 methods-within-subjects (combinations of 2 methods with 17 subjects) with 2 observations each. The mean peak-expiratory-flow measurement is estimated as 450.9. In the random part of the model, the between-subject standard deviation $\sqrt{\psi^{(3)}}$ is estimated as 108.6, and the between-methods within-subjects standard deviation $\sqrt{\psi^{(2)}}$

is estimated as 19.5. Finally, the standard deviation $\sqrt{\theta}$ between occasions, within methods and subjects, is estimated as 17.8. We could obtain the corresponding variances using the `variance` option. The estimates for this variance-components model are also presented under "VC" in table 8.1.

Table 8.1: Maximum likelihood estimates for two-level and three-level variance-components (VC) and random-intercept (RI) models for peak-expiratory-flow data

	VC	RI
	Est (SE)	Est (SE)
Fixed part		
β_1	450.90 (26.6)	447.9 (26.9)
β_2		6.0 (7.8)
Random part		
$\sqrt{\psi^{(2)}}$	19.5	19.0
$\sqrt{\psi^{(3)}}$	108.6	108.6
$\sqrt{\theta}$	17.8	17.8
Log likelihood	-345.29	-345.00

The number of highest-level units, subjects, is quite small, so the subject-level variance is not very precisely estimated. Perhaps the parameters should be estimated by REML instead of ML to avoid a downward-biased variance estimate. In higher-level models, asymptotics rely on the number of highest-level units becoming large, so tests and confidence intervals may not perform well in this application.

Plugging in estimates for the variance components in the expressions for the intraclass correlations, the estimated intraclass correlation between measurements on the same subject using the same method is

$$\widehat{\rho}(\text{method}, \text{subject}) = \frac{19.476^2 + 108.604^2}{19.476^2 + 108.604^2 + 17.759^2} = 0.97$$

and the corresponding estimated intraclass correlation using different methods is

$$\widehat{\rho}(\text{subject}) = \frac{108.604^2}{19.476^2 + 108.604^2 + 17.759^2} = 0.94$$

We see that the peak-expiratory-flow measurements are extremely reliable, whether the same or different methods are used.

8.7   Empirical Bayes prediction

Empirical Bayes prediction of the random intercepts $\zeta_{jk}^{(2)}$ for the combination of methods and subjects and the random intercepts $\zeta_k^{(3)}$ for subjects is straightforward after estimation with `xtmixed`. We simply use the `predict` command with the `reffects` option, specifying as many variables as there are random effects (here 2), keeping in mind the ordering of random effects from the highest to the lowest levels:

```
. predict subj instr, reffects
```

We can list the predictions for the first seven subjects (after defining labels for the values of `method`):

```
. sort id method
. label define m 0 "Wright" 1 "Mini Wright"
. label values method m
. list id method subj instr if id<8 & occasion==1, noobs sepby(id)
```

id	method	subj	instr
1	Wright	53.14315	-8.504796
1	Mini Wright	53.14315	10.2139
2	Wright	-40.72008	-10.01413
2	Mini Wright	-40.72008	8.704561
3	Wright	61.6984	.9921208
3	Mini Wright	61.6984	.9921208
4	Wright	-23.60959	-6.91353
4	Mini Wright	-23.60959	6.154238
5	Wright	34.81049	-8.97618
5	Mini Wright	34.81049	10.0957
6	Wright	144.0732	-7.748992
6	Mini Wright	144.0732	12.38243
7	Wright	-37.05355	.1105384
7	Mini Wright	-37.05355	-1.302193

We see from the predictions $\widetilde{\zeta}_{jk}^{(2)}$ in the last column that, just as suggested by figure 8.2, for subjects 1, 2, 4, 5, and 6 the Mini Wright meter appears to be positively biased compared with the Wright peak-flow meter; this bias is reversed for subject 7. For subject 3, the empirical Bayes predictions for both methods are small because all four measurements nearly coincide.

8.8 Testing variance components

Is the between-methods within-subjects variance significantly different from zero? We can answer this question by testing

$$H_0: \psi^{(2)} = 0 \qquad \text{against} \qquad H_a: \psi^{(2)} > 0$$

using a likelihood-ratio test. We first save the estimates from the three-level model:

```
. estimates store thrlev
```

Then we fit the two-level model in which the level-2 random intercept no longer appears (equivalent to setting its variance $\psi^{(2)}$ to zero)

```
. quietly xtmixed w || id:, mle
```

where the `quietly` prefix command suppressed the display of output. We perform a likelihood-ratio test using

```
. lrtest thrlev .
Likelihood-ratio test                              LR chi2(1)  =      9.20
(Assumption: . nested in thrlev)                   Prob > chi2 =    0.0024
```

> Note: The reported degrees of freedom assumes the null hypothesis is not on
>   the boundary of the parameter space. If this is not true, then the
>   reported test is conservative.

The test is conservative because the null hypothesis is on the boundary of the parameter space. We have encountered this problem in previous chapters when testing a random-intercept variance in a random-intercept model (section 2.6.2) or when testing a random-slope variance in a model containing a random intercept and random slope (section 4.6). Although the latter situation seems similar to our current one in the sense that we are testing the variance of one random effect when another random effect is in the model, the situation is different here because the random effects are *uncorrelated*. Display 8.1 shows the asymptotic null distributions for various testing situations when all the random effects in the model are mutually uncorrelated. For testing one variance when another uncorrelated random effect is in the model, the asymptotic null distribution is $0.5\chi^2(0) + 0.5\chi^2(1)$ and we can simply divide the p-value by 2, giving 0.0012.

The asymptotic null distribution for testing the null hypothesis of k *uncorrelated* random effects (where k could be zero) against the alternative of $k + \ell$ *uncorrelated* random effects is

$$\sum_{m=0}^{\ell} \frac{1}{2^{\ell}} \binom{\ell}{m} \chi^2(m)$$

(see also Verbeke and Molenberghs [2003]).

We give the asymptotic null distributions for $\ell = 1, 2, 3$:

- The asymptotic null distribution for testing k uncorrelated random effects against the alternative of $k + 1$ uncorrelated random effects becomes

$$\sum_{m=0}^{1} \frac{1}{2} \binom{1}{m} \chi^2(m) \;=\; \frac{1}{2}\chi^2(0) + \frac{1}{2}\chi^2(1)$$

- The asymptotic null distribution for testing k uncorrelated random effects against the alternative of $k + 2$ uncorrelated random effects becomes

$$\sum_{m=0}^{2} \frac{1}{4} \binom{2}{m} \chi^2(m) \;=\; \frac{1}{4}\chi^2(0) + \frac{1}{2}\chi^2(1) + \frac{1}{4}\chi^2(2)$$

- The asymptotic null distribution for testing k uncorrelated random effects against the alternative of $k + 3$ uncorrelated random effects becomes

$$\sum_{m=0}^{3} \frac{1}{8} \binom{3}{m} \chi^2(m) \;=\; \frac{1}{8}\chi^2(0) + \frac{3}{8}\chi^2(1) + \frac{3}{8}\chi^2(2) + \frac{1}{8}\chi^2(3)$$

Display 8.1: Asymptotic null distributions for likelihood-ratio testing of variance components when random effects are uncorrelated

8.9   Crossed versus nested random effects revisited

The peak-expiratory-flow measurements are cross-classified by subject (17 individuals), method (Wright meter or Mini Wright meter), and occasion (first or second occasion for each method) according to a $17 \times 2 \times 2$ full factorial design. In such designs, it is natural to consider "main effects" of factors or categorical explanatory variables (here subjects, methods, and occasions), that is, effects that are constant across the categories of the other variables, as well as interactions between factors. If the main effects of several cross-classified factors are random, the model is said to include crossed random effects. Models with crossed random effects are the topic of chapter 9.

The three-level variance-components model considered above has a random main effect for only *one* of the factors, namely, subjects. There is another random effect for methods within subjects, or in other words, the method by subject interaction. This interaction (each of the 34 combinations of subjects and methods) is nested within subjects, so crossed random effects were not required.

The model does not include a main effect for methods and therefore assumes that the mean response over subjects and occasions is the same for both methods. We can relax this assumption by including a dummy variable x_j for the Mini Wright meter:

$$y_{ijk} = \beta_1 + \beta_2 x_j + \zeta_{jk}^{(2)} + \zeta_k^{(3)} + \epsilon_{ijk}$$

Occasion is ignored here, treating the two measurements for a given subject–method combination as exchangeable replicates.

This model can be fit by including `method` as a covariate in the fixed part of the `xtmixed` command:

```
. xtmixed w method || id: || method:, mle
Mixed-effects ML regression                    Number of obs      =        68
```

Group Variable	No. of Groups	Observations per Group Minimum	Average	Maximum
id	17	4	4.0	4
method	34	2	2.0	2

```
                                        Wald chi2(1)       =       0.60
Log likelihood = -344.99736             Prob > chi2        =      0.4403
```

w	Coef.	Std. Err.	z	P>|z|	[95% Conf. Interval]
method	6.029412	7.812739	0.77	0.440	-9.283275 21.3421
_cons	447.8824	26.92329	16.64	0.000	395.1137 500.651

Random-effects Parameters	Estimate	Std. Err.	[95% Conf. Interval]
id: Identity			
sd(_cons)	108.6455	19.04646	77.05295 153.1915
method: Identity			
sd(_cons)	19.00386	4.789038	11.59675 31.14208
sd(Residual)	17.75859	2.153545	14.00184 22.52329

```
LR test vs. linear regression:       chi2(2) =    144.35   Prob > chi2 = 0.0000
Note: LR test is conservative and provided only for reference.
```

The main effect of method is not significant at the 5% level ($z = 0.77$, $p = 0.44$). The estimates for this three-level random-intercept model were also presented under "RI" in table 8.1 on page 394.

Using ANOVA terminology, the above model includes main effects for methods and subjects and the method by subject interaction. Method is a fixed factor, whereas subject is a random factor. An interaction between a fixed and a random factor is automatically random. The model is called a two-way mixed-effects ANOVA model.

8.10   Does nutrition affect cognitive development of Kenyan children?

Whaley et al. (2003) describe a cluster-randomized intervention study from rural Kenya that was designed to investigate the impact of three different diets on the cognitive development of school children.

The primary dish consumed in the study area, the Embu district on the southeastern slopes of Mount Kenya, is githeri, a vegetable dish composed of maize, beans, vegetable oil, and some greens. Children begin regularly attending school around age 6, when they enter "Standard 1" (grade 1). The school year comprises three 3-month terms, with 1-month breaks between them (April, August, and December). In Standards 1 and 2, the school day lasts from 8 a.m. to 1 p.m. with a 30-minute playground break, and in Standards 3–8, school lasts until 4 p.m. with a 1-hour lunch break. Before the study, no meals were provided at any of the schools, and children infrequently brought a snack or lunch to school.

Twelve schools were selected for the study in 1998. Using a cluster-randomized design, these schools were randomly assigned to one of four nutrition interventions:

1. Githeri and meat (Meat group)

2. Githeri and milk (Milk group)

3. Githeri with additional oil (Calorie group)

4. No feeding (Control group)

All three supplements initially provided 240 kcal. After one year, the amount of supplement was increased to 313 kcal as the children grew. The study continued for 7 terms until the end of Standard 3, for a total of 21 months.

Cognitive assessments were carried out on all children at baseline and in terms 1, 2, 4, and 6. Here we will focus on Raven's colored progressive matrices assessment (Raven's score, for short), where children are presented with a matrix-like arrangement of symbols and are asked to complete the matrix by selecting the appropriate missing symbol from a group of symbols.

Variables in the dataset `kenya.dta` provided by Weiss (2005) include

- `id`: child identifier (j)
- `schoolid`: school identifier (k)
- `rn`: observation number (1, 2, 3, 4, 5) (i)
- `ravens`: Raven's score (Raven's colored progressive matrices assessment) (y_{ijk})
- `relyear`: time in years from baseline (t_{ijk})
- `treatment`: intervention group (1=meat; 2=milk; 3=calorie; 4=control)
- `age_at_time0`: age at baseline
- `gender`: gender of child (1=boy; 2=girl)

We read in the data using

```
. use http://www.stata-press.com/data/mlmus3/kenya
```

8.11   Describing and plotting three-level data

8.11.1   Data structure and missing data

Let us first investigate the missing-data patterns using xtdescribe. In some datasets, the level-2 identifier may not be unique across level-3 units, for example, children may be labeled from 1 to n_k for each school, where n_k is the number of children in school k. Although not necessary for the Kenya data, it is good practice to define a child identifier taking a unique value for each combination of id and schoolid (instead of relying on the values being unique already):

```
. egen child = group(id schoolid)
```

Now we look at missing-data patterns for the children:

```
. quietly xtset child rn
. xtdescribe if ravens<.
   child:  1, 2, ..., 546                              n =        546
      rn:  1, 2, ..., 5                                T =          5
            Delta(rn) = 1 unit
            Span(rn)  = 5 periods
            (child*rn uniquely identifies each observation)
  Distribution of T_i:   min      5%     25%      50%      75%     95%      max
                           1       3       5        5        5       5        5

      Freq.  Percent   Cum. |  Pattern
                            |
        475    87.00  87.00 |  11111
         19     3.48  90.48 |  1111.
         12     2.20  92.67 |  11...
         11     2.01  94.69 |  111..
          8     1.47  96.15 |  1.111
          6     1.10  97.25 |  .1111
          5     0.92  98.17 |  1....
          2     0.37  98.53 |  1..11
          2     0.37  98.90 |  111.1
          6     1.10 100.00 |  (other patterns)
                            |
        546   100.00        |  XXXXX
```

We see that 87% of children have complete data and that more than 8.6% (3.48% + 2.20% + 2.01% + 0.92%) of children have monotone missingness patterns—they drop out and do not return after missing an assessment.

We will now investigate how many children there are per school and what proportion of children per school have complete data. To do this, we first count the number of observations per child where the response variable is not missing:

```
. egen numobs2 = count(ravens), by(child)
```

We then define dummy variables `compl` for the child having complete data and `any` for the child having any data:

```
. generate compl = numobs2==5
. generate any = numobs2>0
```

Now we are ready to count the number of children with complete data and any data per school. We cannot simply add up the dummy variable `compl` within school because children with complete data would count five times. We therefore create a dummy variable, `pick_child`, taking the value one exactly once per child and zero otherwise and multiply `compl` and `any` by this variable before summing:

```
. egen pick_child = tag(child)
. egen numcomp3 = total(compl*pick_child), by(schoolid)
. egen numany3 = total(any*pick_child), by(schoolid)
```

We also calculate the proportion of children with complete data in each school:

```
. generate rate = numcomp3/numany3
```

We can now summarize these school-level variables by creating a dummy variable, `pick_school`, to pick out one observation per school:

```
. egen pick_school = tag(schoolid)
. summarize numany3 numcomp3 rate if pick_school==1
```

Variable	Obs	Mean	Std. Dev.	Min	Max
numany3	12	45.5	21.28807	12	91
numcomp3	12	39.58333	18.69593	9	80
rate	12	.8646263	.0560276	.75	.95

The number of children per school ranges from 12 to 91. Between 9 and 80 children per school have complete data, corresponding to a range of proportions of children with complete data from 0.75 to 0.95.

8.11.2 Level-1 variables

We have three kinds of variables—namely, level-1, level-2, and level-3 variables—that vary at the respective levels. `relyear` and `ravens` are level-1 (time-varying) variables, `age_at_time0` and `gender` are level-2 (child-specific) variables, and `treatment` is a level-3 (school-specific) variable. If there were a school-specific, time-varying variable, such as percentage of students with low income, this would typically be treated as level 1.

We can summarize the variability of time-varying variables within and between children, ignoring schools, using the `xtsum` command:

```
. quietly xtset child

. xtsum ravens relyear
```

Variable		Mean	Std. Dev.	Min	Max	Observations
ravens	overall	18.24904	2.984872	0	31	N = 2598
	between		1.946549	6	25.6	n = 546
	within		2.276076	7.999038	27.64904	T-bar = 4.75824
relyear	overall	.6496536	.7074833	-.23	2.01	N = 2598
	between		.1655664	-.15	1.4	n = 546
	within		.6975168	-.4403464	1.993654	T-bar = 4.75824

We see that the Raven's score varies nearly as much between children as within children and that the timing of the assessments varies between children.

To see how much these variables vary between schools, we can find the child means,

```
. egen mn_raven = mean(ravens), by(child)

. egen mn_relyr = mean(relyear), by(child)
```

and then use the xtsum command again, after declaring schoolid as the cluster identifier using xtset, as described below.

8.11.3 Level-2 variables

Before summarizing the child-level variables, we form a dummy variable for boys:

```
. quietly tabulate gender, generate(g)

. rename g1 boy
```

We now find the means and within- and between-school standard deviations of all child-level variables, being sure to use only one observation per child by specifying if pick_child==1.

```
. quietly xtset schoolid

. xtsum mn_raven mn_relyr boy age_at_time0 if pick_child==1
```

Variable		Mean	Std. Dev.	Min	Max	Observations
mn_raven	overall	18.25577	1.946549	6	25.6	N = 546
	between		.5474184	17.3875	19.33889	n = 12
	within		1.899147	6.270604	25.47102	T = 45.5
mn_relyr	overall	.6305443	.1655664	-.15	1.4	N = 546
	between		.0303738	.5888706	.6893792	n = 12
	within		.163019	-.1376712	1.402116	T = 45.5
boy	overall	.5201465	.5000521	0	1	N = 546
	between		.0929989	.3333333	.65	n = 12
	within		.4948453	-.1298535	1.186813	T = 45.5
age_at~0	overall	7.630406	1.414575	4.84	15.18	N = 542
	between		.3993327	7.017241	8.487359	n = 12
	within		1.356501	5.05585	14.42585	T = 45.1667

There is some between-school variability in the Raven's sore, but there is negligible between-school variability in the timing of assessments. From the `Min` and `Max` values for `boy`, we see that the percentage of boys per school varies between 33% and 65%.

8.11.4 Level-3 variables

We can summarize the level-3 variable, `treatments`, at the school level (treating school as the unit of analysis) by reporting the number of schools in each treatment group,

```
. tabulate treatment if pick_school==1
```

	Freq.	Percent	Cum.
meat	3	25.00	25.00
milk	3	25.00	50.00
calorie	3	25.00	75.00
control	3	25.00	100.00
Total	12	100.00	

or at the child level (treating child as the unit of analysis) by reporting the number of children in each treatment group:

```
. tabulate treatment if pick_child==1
```

	Freq.	Percent	Cum.
meat	131	23.99	23.99
milk	142	26.01	50.00
calorie	146	26.74	76.74
control	127	23.26	100.00
Total	546	100.00	

There are three schools per group with a total of 131 children in the meat group, 142 children in the milk group, 146 children in the calorie group, and 127 children in the control group.

For later use, we create dummy variables for the intervention groups and give them descriptive names:

```
. tabulate treatment, generate(treat)
```

	Freq.	Percent	Cum.
meat	655	23.99	23.99
milk	710	26.01	50.00
calorie	730	26.74	76.74
control	635	23.26	100.00
Total	2,730	100.00	

```
. rename treat1 meat
. rename treat2 milk
. rename treat3 calorie
```

8.11.5   Plotting growth trajectories

We now combine the ideas of spaghetti and trellis plots by displaying spaghetti plots for the children within panels for schools. It is important to first sort the data by `relyear` within child so that the `connect(ascending)` option can be used:

```
. sort schoolid id relyear
. twoway (line ravens relyear, connect(ascending)), by(schoolid, compact)
>         xtitle(Time in years) ytitle(Raven's score)
```

The graph is shown in figure 8.5.

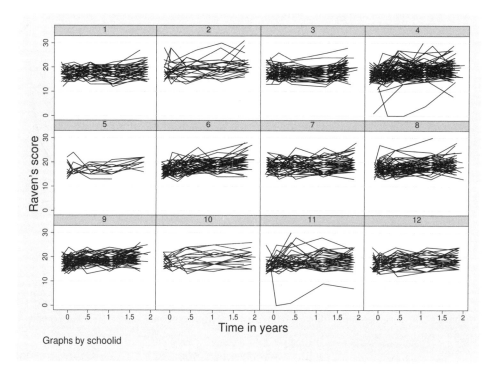

Figure 8.5: Trellis of spaghetti plots for schools in Kenyan nutrition study, showing observed growth trajectories

Schools 4 and 11 each have a child who scores very low at several occasions, and these children may merit further investigation.

8.12 Three-level random-intercept model

8.12.1 Model specification: Reduced form

For occasion i, child j, and school k, we will initially consider the three-level random-intercept model

$$
\begin{aligned}
y_{ijk} &= \beta_1 + \beta_2 w_{1k} + \beta_3 w_{2k} + \beta_4 w_{3k} + \beta_5 t_{ijk} + \beta_6 w_{1k} t_{ijk} + \beta_7 w_{2k} t_{ijk} + \beta_8 w_{3k} t_{ijk} \\
&\quad + \beta_9 x_{1jk} + \beta_{10} x_{2jk} + \zeta_{jk}^{(2)} + \zeta_k^{(3)} + \epsilon_{ijk}
\end{aligned} \tag{8.1}
$$

Here w_{1k}, w_{2k}, and w_{3k} are dummy variables for the three interventions meat, milk, and calorie, respectively. Note that these intervention variables are only indexed by k because the intervention was randomized at the school level. t_{ijk} is the time since baseline at occasion i for child jk, x_{1jk} is the child's age at baseline, and x_{2jk} is a dummy variable for being a boy. The fixed part of the model assumes that children's growth trajectories are linear with intervention-group-specific slopes of time.

We let all observed covariates in school k be denoted $\mathbf{X}_k$. It is assumed that the school-level random intercept $\zeta_k^{(3)}$ has zero mean and variance $\psi^{(3)}$, given the covariates $\mathbf{X}_k$; that the student-level random intercept $\zeta_{jk}^{(2)}$ has zero mean and variance $\psi^{(2)}$, given $\zeta_k^{(3)}$ and $\mathbf{X}_k$; and that the level-1 error term ϵ_{ijk} has zero mean and variance θ, given $\zeta_k^{(3)}$, $\zeta_{jk}^{(2)}$, and $\mathbf{X}_k$. The error terms in the model are assumed to be uncorrelated across levels.

8.12.2 Model specification: Three-stage formulation

The model can also be defined via a three-stage formulation using the notation of Raudenbush and Bryk (2002). The level-1 model for occasion i, child j, and school k is a linear regression on time and can be written as

$$
y_{ijk} = \pi_{0jk} + \pi_{1jk} t_{ijk} + \epsilon_{ijk}
$$

The intercept π_{0jk} and slope π_{1jk} in the level-1 model vary between children according to the following level-2 models:

$$
\begin{aligned}
\pi_{0jk} &= \beta_{00k} + \beta_{01} x_{1jk} + \beta_{02} x_{2jk} + r_{0jk} \\
\pi_{1jk} &= \beta_{10k}
\end{aligned} \tag{8.2}
$$

The intercepts in the level-2 models vary between schools according to the following level-3 models:

$$
\begin{aligned}
\beta_{00k} &= \gamma_{000} + \gamma_{001} w_{1k} + \gamma_{002} w_{2k} + \gamma_{003} w_{3k} + u_{0k} \\
\beta_{10k} &= \gamma_{100} + \gamma_{101} w_{1k} + \gamma_{102} w_{2k} + \gamma_{103} w_{3k}
\end{aligned} \tag{8.3}
$$

All Greek letters without i, j, or k subscripts are fixed parameters. u_{0k} is a level-3 random intercept, r_{0jk} is a level-2 random intercept, and ϵ_{ijk} is the level-1 residual as before. We see that the slopes of time are determined by the treatment group dummy variables. There is no random slope at level 2 or level 3.

Substituting the level-3 models into the level-2 models, we obtain

$$
\begin{aligned}
\pi_{0jk} &= \gamma_{000} + \gamma_{001}w_{1k} + \gamma_{002}w_{2k} + \gamma_{003}w_{3k} + u_{0k} + \beta_{01}x_{1jk} + \beta_{02}x_{2jk} + r_{0jk} \\
\pi_{1jk} &= \gamma_{100} + \gamma_{101}w_{1k} + \gamma_{102}w_{2k} + \gamma_{103}w_{3k}
\end{aligned}
$$

and substituting the level-2 models into the level-1 model gives the reduced form

$$
\begin{aligned}
y_{ijk} &= (\gamma_{000} + \gamma_{001}w_{1k} + \gamma_{002}w_{2k} + \gamma_{003}w_{3k} + u_{0k} + \beta_{01}x_{1jk} + \beta_{02}x_{2jk} + r_{0jk}) \\
&\quad + (\gamma_{100} + \gamma_{101}w_{1k} + \gamma_{102}w_{2k} + \gamma_{103}w_{3k})t_{ijk} + \epsilon_{ijk} \\
&= \gamma_{000} + \gamma_{001}w_{1k} + \gamma_{002}w_{2k} + \gamma_{003}w_{3k} \\
&\quad + \gamma_{100}t_{ijk} + \gamma_{101}w_{1k}t_{ijk} + \gamma_{102}w_{2k}t_{ijk} + \gamma_{103}w_{3k}t_{ijk} \\
&\quad + \beta_{01}x_{1jk} + \beta_{02}x_{2jk} + r_{0jk} + u_{0k} + \epsilon_{ijk}
\end{aligned}
$$

which is equivalent to the model in (8.1).

8.12.3 Estimation using xtmixed

Before fitting the three-level random-intercept model, we create the interactions between the intervention dummy variables and time:

```
. generate meat_year = meat*relyear
. generate milk_year = milk*relyear
. generate calorie_year = calorie*relyear
```

The three-level random-intercept model can now be fit using the following `xtmixed` command:

```
. xtmixed ravens meat milk calorie relyear meat_year milk_year calorie_year
> age_at_time0 boy || schoolid: || id:, mle
Mixed-effects ML regression                     Number of obs      =      2593
```

Group Variable	No. of Groups	Observations per Group		
		Minimum	Average	Maximum
schoolid	12	57	216.1	434
id	542	1	4.8	5

```
                                                Wald chi2(9)       =    271.77
Log likelihood = -6255.892                      Prob > chi2        =    0.0000
```

ravens	Coef.	Std. Err.	z	P>|z|	[95% Conf. Interval]	
meat	-.255788	.3306285	-0.77	0.439	-.903808	.392232
milk	-.4792384	.3224394	-1.49	0.137	-1.111208	.1527312
calorie	-.3714558	.3290351	-1.13	0.259	-1.016353	.2734412
relyear	.9131069	.1406314	6.49	0.000	.6374745	1.188739
meat_year	.5111626	.1967277	2.60	0.009	.1255833	.8967418
milk_year	-.1057318	.1914511	-0.55	0.581	-.480969	.2695055
calorie_year	.122682	.193161	0.64	0.525	-.2559065	.5012705
age_at_time0	.1163285	.0614812	1.89	0.058	-.0041724	.2368295
boy	.5372871	.1657506	3.24	0.001	.2124219	.8621523
_cons	16.70638	.5077426	32.90	0.000	15.71122	17.70153

Random-effects Parameters	Estimate	Std. Err.	[95% Conf. Interval]	
schoolid: Identity				
sd(_cons)	.2209702	.1334575	.0676458	.7218166
id: Identity				
sd(_cons)	1.532073	.0728596	1.395724	1.681742
sd(Residual)	2.412956	.0376427	2.340294	2.487874

```
LR test vs. linear regression:        chi2(2) =     306.40    Prob > chi2 = 0.0000
Note: LR test is conservative and provided only for reference.
. estimates store mod1
```

From the first three coefficients, we see that there are no significant differences in the mean Raven's scores between the intervention groups at baseline, as would be expected in a well-executed randomized experiment. (All three coefficients could be tested simultaneously using testparm.) After controlling for the other variables, the mean slope of time in the control group is estimated as 0.91 units per year. The meat group grows an extra 0.51 units per year ($z = 2.60$, $p = 0.009$), but there is little evidence of a difference in growth between the other nutritional supplements and the control group. After controlling for treatment group, time, and gender, each extra year of age is associated with a mean increase in the Raven's score of 0.12 units, but this is not significant at the 5% level. Boys score on average 0.54 points higher than girls ($z=3.24$, $p = 0.001$) at a given time within a given intervention group and for a given age.

We can fit the model without explicitly creating dummy variables by using factor variables as follows:

```
xtmixed ravens ib4.treatment##c.relyear age_at_time0 boy || schoolid: || id:, mle
```

Because there are only 12 schools, it might be advisable to use REML for this application.

The estimates from the `xtmixed` output shown above are also reported under RI(2) & RI(3) in table 8.2.

Table 8.2: Maximum likelihood estimates for Kenyan nutrition data. Models with random intercept at both child and school levels [RI(2) & RI(3)], random coefficient at child level and random intercept at school level [RC(2) & RI(3)], and random coefficients at both child and school levels [RC(2) & RC(3)].

Parameter	RI(2) & RI(3) Est	(SE)	RC(2) & RI(3) Est	(SE)	RC(2) & RC(3) Est	(SE)
Fixed part						
β_1 [_cons]	16.71	(0.51)	16.74	(0.51)	16.77	(0.52)
β_2 [meat]	-0.26	(0.33)	-0.24	(0.34)	-0.24	(0.37)
β_3 [milk]	-0.48	(0.32)	-0.48	(0.33)	-0.50	(0.37)
β_4 [calorie]	-0.37	(0.33)	-0.37	(0.34)	-0.37	(0.38)
β_5 [relyear]	0.91	(0.14)	0.92	(0.16)	0.92	(0.17)
β_6 [meat_year]	0.51	(0.20)	0.50	(0.22)	0.48	(0.24)
β_7 [milk_year]	-0.11	(0.19)	-0.12	(0.21)	-0.11	(0.23)
β_8 [calorie_year]	0.12	(0.19)	0.12	(0.21)	0.11	(0.24)
β_9 [age_at_time0]	0.12	(0.06)	0.11	(0.06)	0.11	(0.06)
β_{10} [boy]	0.54	(0.17)	0.49	(0.16)	0.49	(0.16)
Random part						
$\sqrt{\psi_{11}^{(2)}}$	1.53		1.46		1.45	
$\sqrt{\psi_{22}^{(2)}}$			0.86		0.86	
$\psi_{21}^{(2)}/\sqrt{\psi_{11}^{(2)}\psi_{22}^{(2)}}$			-0.00		0.01	
$\sqrt{\psi_{11}^{(3)}}$	0.22		0.25		0.32	
$\sqrt{\psi_{22}^{(3)}}$					0.11	
$\psi_{21}^{(3)}/\sqrt{\psi_{11}^{(3)}\psi_{22}^{(3)}}$					-1.00	
$\sqrt{\theta}$	2.41		2.32		2.32	
Log likelihood	$-6,255.89$		$-6,241.34$		$-6,240.65$	

8.13   Three-level random-coefficient models

8.13.1   Random coefficient at the child level

Within intervention groups, children may well vary in their individual rate of cognitive growth. Such variability can be modeled by adding the term $\zeta_{2jk}^{(2)} t_{ij}$ to the reduced form model (8.1) and renaming $\zeta_{jk}^{(2)}$ to $\zeta_{1jk}^{(2)}$, giving the random-coefficient model

$$
\begin{aligned}
y_{ijk} = {} & \beta_1 + \beta_2 w_{1k} + \beta_3 w_{2k} + \beta_4 w_{3k} + \beta_5 t_{ijk} + \beta_6 w_{1k} t_{ijk} + \beta_7 w_{2k} t_{ijk} + \beta_8 w_{3k} t_{ijk} \\
& + \beta_9 x_{1jk} + \beta_{10} x_{2jk} + \zeta_{1jk}^{(2)} + \zeta_{2jk}^{(2)} t_{ij} + \zeta_k^{(3)} + \epsilon_{ijk}
\end{aligned}
$$

or by adding the term r_{1jk} to the level-2 model for π_{1jk} in the three-stage formulation.

Given the covariates $\mathbf{X}_k$, the random intercept $\zeta_{1jk}^{(2)}$ and random slope $\zeta_{2jk}^{(2)}$ at the child level have a bivariate distribution, assumed to have zero mean and covariance matrix

$$
\mathbf{\Psi}^{(2)} = \left[\begin{array}{cc} \psi_{11}^{(2)} & \psi_{12}^{(2)} \\ \psi_{21}^{(2)} & \psi_{22}^{(2)} \end{array} \right] \equiv \left[\begin{array}{cc} \mathrm{Var}(\zeta_{1jk}^{(2)} | \mathbf{X}_k, \zeta_k^{(3)}) & \mathrm{Cov}(\zeta_{1jk}^{(2)}, \zeta_{2jk}^{(2)} | \mathbf{X}_k, \zeta_k^{(3)}) \\ \mathrm{Cov}(\zeta_{2jk}^{(2)}, \zeta_{1jk}^{(2)} | \mathbf{X}_k, \zeta_k^{(3)}) & \mathrm{Var}(\zeta_{2jk}^{(2)} | \mathbf{X}_k, \zeta_k^{(3)}) \end{array} \right]
$$

where $\psi_{21}^{(2)} = \psi_{12}^{(2)}$. (You may want to refer back to section 4.4.1 on specification of two-level random-coefficient models.)

The xtmixed syntax for fitting a three-level random-coefficient model with a random coefficient at the child level for year since baseline is

```
. xtmixed ravens meat milk calorie relyear meat_year milk_year calorie_year
> age_at_time0 boy || schoolid: || id: relyear, covariance(unstructured) mle
```

Mixed-effects ML regression   Number of obs   =   2593

Group Variable	No. of Groups	Observations per Group		
		Minimum	Average	Maximum
schoolid	12	57	216.1	434
id	542	1	4.8	5

		Wald chi2(9)	=	220.66
Log likelihood = -6241.3378		Prob > chi2	=	0.0000

ravens	Coef.	Std. Err.	z	P>|z|	[95% Conf. Interval]	
meat	-.2371712	.3380466	-0.70	0.483	-.8997303	.425388
milk	-.4782935	.3303219	-1.45	0.148	-1.125712	.1691255
calorie	-.3686388	.3384154	-1.09	0.276	-1.031921	.2946432
relyear	.9235916	.156704	5.89	0.000	.6164574	1.230726
meat_year	.4990658	.2194377	2.27	0.023	.0689758	.9291558
milk_year	-.1155923	.2139539	-0.54	0.589	-.5349343	.3037497
calorie_year	.1169333	.2149049	0.54	0.586	-.3042726	.5381392
age_at_time0	.1146185	.0607949	1.89	0.059	-.0045373	.2337743
boy	.4902904	.1640913	2.99	0.003	.1686773	.8119034
_cons	16.74299	.5059629	33.09	0.000	15.75132	17.73466

Random-effects Parameters	Estimate	Std. Err.	[95% Conf. Interval]	
schoolid: Identity				
sd(_cons)	.2546548	.1291915	.0942151	.6883089
id: Unstructured				
sd(relyear)	.864055	.11576	.6645129	1.123516
sd(_cons)	1.45822	.0936564	1.28574	1.653837
corr(relyear,_cons)	-.0010829	.1403704	-.2693884	.2673786
sd(Residual)	2.315432	.0419338	2.234685	2.399097

LR test vs. linear regression: chi2(4) =   335.51   Prob > chi2 = 0.0000

Note: LR test is conservative and provided only for reference.

```
. estimates store mod2
```

Here relyear has been added in the second random part to specify a child-level random slope for relyear. The covariance(unstructured) option was used to estimate the covariance matrix freely, that is, estimate $\psi_{11}^{(2)}$, $\psi_{21}^{(2)}$, and $\psi_{22}^{(2)}$ without constraints. (By default, xtmixed sets the covariance $\psi_{21}^{(2)}$ to zero.) The estimates from the above xtmixed command were also shown under RC(2) & RI(3) in table 8.2.

To test whether the random slope is needed, we formulate the hypotheses

$$H_0 \colon \psi_{22}^{(2)} = 0 \qquad \text{against} \qquad H_a \colon \psi_{22}^{(2)} > 0$$

and perform a likelihood-ratio test using

```
. lrtest mod1 mod2
Likelihood-ratio test                            LR chi2(2)  =      29.11
(Assumption: mod1 nested in mod2)                Prob > chi2 =     0.0000
Note: The reported degrees of freedom assumes the null hypothesis is not on
      the boundary of the parameter space.  If this is not true, then the
      reported test is conservative.
```

Although conservative, the naïve p-value is tiny, so we can reject the null hypothesis. We do not know the asymptotic null distribution in this case, which would have been useful had the naïve test not been significant.

8.13.2   Random coefficient at the child and school levels

The previous model assumes that the school mean slopes of time do not vary around the treatment-group-specific means. We can relax this assumption by adding the term $\zeta_{2k}^{(3)} t_{ijk}$ to the reduced-form model (and renaming $\zeta_k^{(3)}$ to $\zeta_{1k}^{(3)}$), giving

$$
\begin{aligned}
y_{ijk} &= \beta_1 + \beta_2 w_{1k} + \beta_3 w_{2k} + \beta_4 w_{3k} + \beta_5 t_{ijk} + \beta_6 w_{1k} t_{ijk} + \beta_7 w_{2k} t_{ijk} + \beta_8 w_{3k} t_{ijk} \\
&\quad + \beta_9 x_{1jk} + \beta_{10} x_{2jk} + \zeta_{1jk}^{(2)} + \zeta_{1k}^{(3)} + (\zeta_{2jk}^{(2)} + \zeta_{2k}^{(3)}) t_{ij} + \epsilon_{ijk}
\end{aligned}
$$

or by adding the term u_{1k} to the level-3 model (8.3) for the school mean slope of time, to allow random slopes of time at the school level:

$$\beta_{10k} = \gamma_{100} + \gamma_{101} w_{1k} + \gamma_{102} w_{2k} + \gamma_{103} w_{3k} + u_{1k}$$

Given the covariates $\mathbf{X}_k$, the random intercept $\zeta_{1k}^{(3)}$ and random slope $\zeta_{2k}^{(3)}$ at the school level have a bivariate distribution, assumed to have zero mean and covariance matrix

$$
\mathbf{\Psi}^{(3)} = \begin{bmatrix} \psi_{11}^{(3)} & \psi_{12}^{(3)} \\ \psi_{21}^{(3)} & \psi_{22}^{(3)} \end{bmatrix} \equiv \begin{bmatrix} \text{Var}(\zeta_{1k}^{(3)}|\mathbf{X}_k) & \text{Cov}(\zeta_{1k}^{(3)}, \zeta_{2k}^{(3)}|\mathbf{X}_k) \\ \text{Cov}(\zeta_{2k}^{(3)}, \zeta_{1k}^{(3)}|\mathbf{X}_k) & \text{Var}(\zeta_{2k}^{(3)}|\mathbf{X}_k) \end{bmatrix}
$$

where $\psi_{21}^{(3)} = \psi_{12}^{(3)}$. Noting that `relyear` has a random slope in the random parts for schools and children and that the `covariance(unstructured)` option in `xtmixed` must be specified separately for each level. The three-level random-coefficient model with random coefficients at both the child and school levels can be fit using

```
. xtmixed ravens meat milk calorie relyear meat_year milk_year calorie_year
> age_at_time0 boy || schoolid: relyear, covariance(unstructured)
> || id: relyear, covariance(unstructured) mle
```

Mixed-effects ML regression   Number of obs   =   2593

Group Variable	No. of Groups	Observations per Group		
		Minimum	Average	Maximum
schoolid	12	57	216.1	434
id	542	1	4.8	5

	Wald chi2(9)	=	186.94
Log likelihood = -6240.6504	Prob > chi2	=	0.0000

ravens	Coef.	Std. Err.	z	P>|z|	[95% Conf. Interval]	
meat	-.2418509	.3729209	-0.65	0.517	-.9727624	.4890606
milk	-.5016431	.3655257	-1.37	0.170	-1.21806	.2147741
calorie	-.3713378	.3763629	-0.99	0.324	-1.108995	.3663199
relyear	.9186612	.1699896	5.40	0.000	.5854877	1.251835
meat_year	.4801711	.2380819	2.02	0.044	.0135392	.9468029
milk_year	-.106055	.2327757	-0.46	0.649	-.562287	.3501769
calorie_year	.1111154	.2359755	0.47	0.638	-.351388	.5736189
age_at_time0	.1127028	.0607339	1.86	0.063	-.0063335	.231739
boy	.4913901	.1639159	3.00	0.003	.1701209	.8126594
_cons	16.77486	.5180617	32.38	0.000	15.75948	17.79024

Random-effects Parameters	Estimate	Std. Err.	[95% Conf. Interval]	
schoolid: Unstructured				
sd(relyear)	.1110955	.0944235	.0210007	.5877048
sd(_cons)	.3172716	.1371533	.135977	.7402816
corr(relyear,_cons)	-.9999998	.0002197	-1	1
id: Unstructured				
sd(relyear)	.8583323	.1162664	.6581957	1.119324
sd(_cons)	1.452461	.0933819	1.280497	1.647517
corr(relyear,_cons)	.0107735	.1427232	-.2626855	.2826305
sd(Residual)	2.31525	.041923	2.234523	2.398893

LR test vs. linear regression: chi2(6) =   336.88   Prob > chi2 = 0.0000
Note: LR test is conservative and provided only for reference.

These estimates were also shown under RC(2) & RC(3) in table 8.2.

Unfortunately, the correlation between the school-level random intercepts and slopes is estimated as −1 with a confidence interval from −1 to 1, essentially implying that the data provide no information on this correlation. The model appears to be too ambitious given the small number of schools, and we return to the previous model.

```
. estimates restore mod2
```

8.14   Residual diagnostics and predictions

The retained model includes random intercepts $\zeta_{1k}^{(3)}$ at the school level, random intercepts $\zeta_{1jk}^{(2)}$ and random slopes $\zeta_{2jk}^{(2)}$ at the child level, and level-1 residuals ϵ_{ijk}. We now predict all of these random terms to perform residual diagnostics, keeping in mind that the predictions are based on normality assumptions and will appear more normal than they are when normality is violated.

Empirical Bayes predictions or BLUPs $\widetilde{\zeta}_{1jk}^{(2)}$, $\widetilde{\zeta}_{2jk}^{(2)}$, and $\widetilde{\zeta}_{1k}^{(3)}$ for the random effects can be obtained using `predict` with the `reffects` option. Here it is important to remember that the highest-level random effects come first, and within a given level, the random intercept comes last (the `predict` command defines useful labels for the variables it creates). Denoting $\widetilde{\zeta}_{1k}^{(3)}$ as `ri3`, $\widetilde{\zeta}_{2jk}^{(2)}$ as `rc2`, and $\widetilde{\zeta}_{1jk}^{(2)}$ as `ri2`, the syntax is

```
. predict ri3 rc2 ri2, reffects
```

As always, the level-1 residuals are predicted using

```
. predict res, residuals
```

Because the random intercepts at the different levels and the level-1 residuals are all on the same scale, we plot the distributions on the same graph using box plots (the random slopes are on a different scale). However, we must be careful to use only one observation per school for the box plot of `ri3` and only one observation per student for the box plot of `ri2`. The easiest way of discarding the unnecessary observations is to replace them with missing values.

```
. replace ri3=. if pick_school!=1
. replace ri2 =. if rn!=1
. graph box ri3 ri2 res, ascategory box(1, bstyle(outline))
> yvaroptions(relabel(1 "School" 2 "Child" 3 "Occasion"))
> medline(lcolor(black))
```

The resulting graph in figure 8.6 reveals that there is much more variability within schools than between schools and that there appear to be some outlying children with very low intercepts (as we saw in figure 8.5).

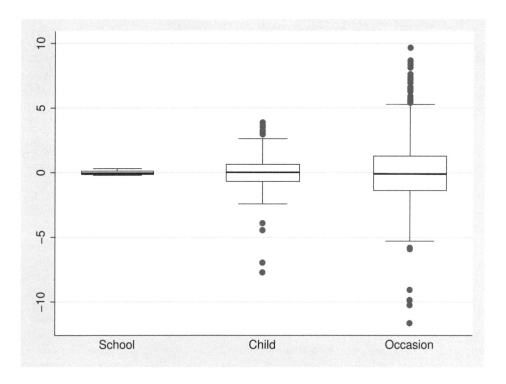

Figure 8.6: Box plots of empirical Bayes predictions for random intercepts at the school level $\widetilde{\zeta}_{1k}^{(3)}$, random intercepts at the child level $\widetilde{\zeta}_{1jk}^{(2)}$, and level-1 residuals $\widetilde{\epsilon}_{ijk}$ at the occasion level

A nice display of the bivariate distribution of the predicted child-level random intercepts and slopes, together with the univariate marginal distributions can be produced as follows (see page 205 for an explanation of the commands):

```
. scatter rc2 ri2 if rn==1, saving(yx, replace)
> xtitle("Random intercept") ytitle("Random slope")

. histogram rc2, freq horiz saving(hy, replace)
> yscale(alt) ytitle(" ") fxsize(35) normal

. histogram ri2, freq saving(hx, replace)
> xscale(alt) xtitle(" ") fysize(35) normal

. graph combine hx.gph yx.gph hy.gph, hole(2) imargin(0 0 0 0)
```

These commands produce the graph in figure 8.7.

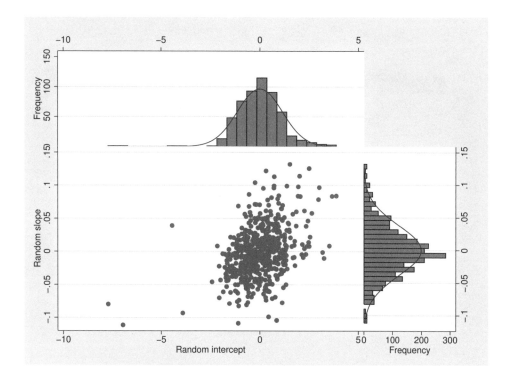

Figure 8.7: Bivariate and univariate distributions of empirical Bayes predictions for random intercepts $\widetilde{\zeta}_{1jk}^{(2)}$ and random slopes $\widetilde{\zeta}_{2jk}^{(2)}$ at the child level

We again see that there are some children with very low intercepts, whereas the slopes are more symmetrically distributed.

We can also plot the predicted child-specific growth trajectories in a trellis of spaghetti plots, analogous to the graph for the observed growth trajectories in figure 8.5. First, we use the `fitted` option in the `predict` command to obtain predicted growth trajectories,

```
. predict predtraj, fitted
```

and then we plot them using

```
. sort schoolid id relyear
. twoway (line predtraj relyear, connect(ascending)), by(schoolid, compact)
> xtitle(Time in years) ytitle(Raven's score)
```

The resulting graph is shown in figure 8.8.

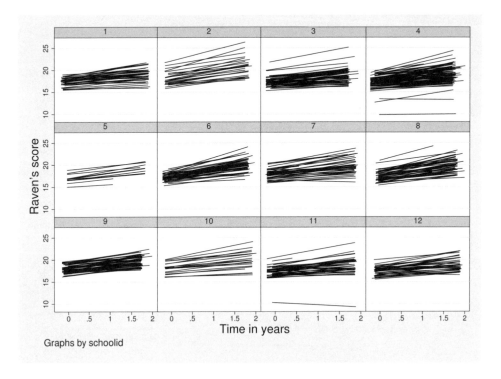

Figure 8.8: Trellis of spaghetti plots for schools in Kenyan nutrition study, showing predicted growth trajectories for children

To visualize the estimated treatment effects, it is useful to plot the model-implied mean Raven's scores as a function of time for the three intervention groups. To do this, we must first set the other covariates in the model, here `age_at_time0` (x_{1jk}) and `boy` (x_{2jk}), to constant values. We set age at baseline to its mean $\overline{x}_{1\cdot\cdot}$, and we set the dummy variable for boys to 1 to obtain predictions for boys of average age at baseline.

$$
\begin{aligned}
\widehat{E}(y_{ijk}|w_{1k}, w_{2k}, w_{3k}, t_{ijk}, x_{1jk} &= \overline{x}_{1\cdot\cdot}, x_{2jk} = 1) \ = \\
&\widehat{\beta}_1 + \widehat{\beta}_2 w_{1k} + \widehat{\beta}_3 w_{2k} + \widehat{\beta}_4 w_{3k} + \widehat{\beta}_5 t_{ijk} + \widehat{\beta}_6 w_{1k} t_{ijk} + \widehat{\beta}_7 w_{2k} t_{ijk} \\
&+ \widehat{\beta}_8 w_{3k} t_{ijk} + \widehat{\beta}_9 \overline{x}_{1\cdot\cdot} + \widehat{\beta}_{10}
\end{aligned}
$$

To obtain the predictions using `predict`, we set the values of `age_at_time0` and `boy` using

```
. summarize age_at_time0 if rn==1

    Variable |      Obs        Mean    Std. Dev.       Min        Max
-------------+------------------------------------------------------
age_at_time0 |      542    7.630406    1.414575       4.84      15.18
. replace age_at_time0 = r(mean)
. replace boy = 1
```

and then use `predict` with the `xb` option:

```
. predict means, xb
. twoway (line means relyear if meat==1, sort lpatt(solid))
>        (line means relyear if milk==1, sort lpatt(dash))
>        (line means relyear if calorie==1, sort lpatt(shortdash))
>        (line means relyear if treatment==4, sort lpatt(longdash_dot)),
>        legend(order(1 "Meat" 2 "Milk" 3 "Calorie" 4 "Control"))
>        xtitle(Time in years) ytitle(Predicted mean Raven's score)
```

The graph in figure 8.9 shows that the estimated slope is considerably steeper for the meat group than for the other three intervention groups.

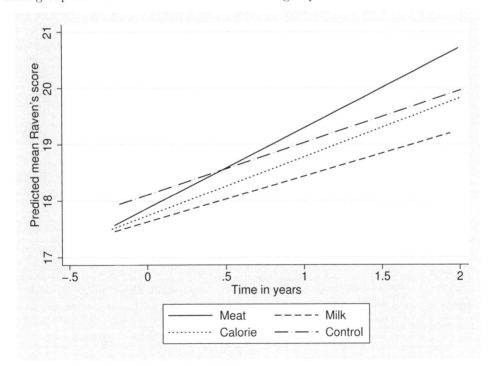

Figure 8.9: Predicted mean Raven's scores over time for the four intervention groups among boys whose age at baseline was average

8.15   Summary and further reading

We have introduced the idea of nested random effects for three-level datasets where there are two nested levels of clustering. A three-level random-intercept model includes error components for nested clusters to allow the total variance to be decomposed into the level-1 variance within clusters, the level-2 variance between clusters within super-clusters, and the level-3 variance between superclusters. We discussed several intraclass correlations for this situation. We have seen that random slopes can be included to allow the effect of level-1 covariates to vary at level 2, at level 3, or at both levels. It is also possible to allow the effect of level-2 covariates to vary at level 3. Recall that a random coefficient should vary at a higher level than the associated covariate (unless used as a device for introducing heteroskedasticity). All of these ideas extend naturally to higher-level models (see exercise 8.8 for a four-level example).

The models discussed in this chapter can be extended by relaxing the homoskedasticity assumptions for the random effects (see section 7.5.2) or the level-1 residuals (see sections 6.4.2, 6.4.3, and 7.5.1). If the data are longitudinal, we may want to allow level-1 residuals to be correlated, for instance, with an AR(1) covariance structure, as discussed for two-level models in section 6.4.1.

Good references on linear multilevel models with several levels of nested random effects include Raudenbush and Bryk (2002, chap. 8), Goldstein (2011, chap. 3), and Snijders and Bosker (2012, sec. 4.9, 5.5). A review of linear two- and three-level models with an application to political science is given by Steenbergen and Jones (2002). Raudenbush (1989) discusses two types of multilevel longitudinal designs in education: repeated measurements on the same students nested in schools (see exercises 8.1 and 8.6) and repeated measurements for the same schools, but with changing cohorts of students (see exercise 8.5). Longitudinal data from economics and medicine are considered in exercises 8.3 and 8.4. Cross-sectional examples from education are considered in exercises 8.2 and 8.7.

As we have seen in the nutrition example, three-level data sometimes arise from randomized trials. The nutrition example is a cluster-randomized trial in which schools (level 3) were randomly assigned to treatments. In exercise 8.4, we consider a multisite hypertension study, where patients (level 2) are randomly assigned to treatments *within centers* (level 3). Similarly, in the Tennessee class-size experiment used in exercise 8.6, interventions were applied to classes *within schools*. Interestingly, the classes were not preexisting but were formed by randomly allocating both students and teachers to classes within schools. Some models not explicitly covered in this chapter that can be expressed as three-level models are introduced in exercises 8.9 (multivariate multilevel model) and 8.10 (biometrical genetic model for twin data).

8.16 Exercises

8.1 Math-achievement data [Solutions]

Raudenbush and Bryk (2002) and Raudenbush et al. (2004) discuss data from a longitudinal study of children's academic growth in the six primary school years. The data have a three-level structure with repeated observations on 1,721 students from 60 urban public primary schools.

The dataset `achievement.dta` has the following variables:

- Level 1 (occasion)
    - `math`: math test score derived from an item response model
    - `year`: year of study minus 3.5 (values -2.5, -1.5, -0.5, 0.5, 1.5, 2.5) (a_{1ijk})
    - `retained`: dummy variable for child being retained in grade (1: retained; 0: not retained)
- Level 2 (child)
    - `child`: child identifier
    - `female`: dummy variable for being female
    - `black`: dummy variable for being African American (X_{1jk})
    - `hispanic`: dummy variable for being Hispanic (X_{2jk})
- Level 3 (school)
    - `school`: school identifier
    - `size`: number of students enrolled in the school
    - `lowinc`: percentage of students from low-income families (W_{1k})
    - `mobility`: percentage of students moving during the course of a school year

Raudenbush et al. (2004) specify a three-level model in three stages. The level-1 model is

$$Y_{ijk} = \pi_{0jk} + \pi_{1jk}a_{1ijk} + e_{ijk}, \quad e_{ijk} \sim N(0, \sigma^2)$$

where Y_{ijk} is the jth child's math achievement at occasion i in school k and a_{1ijk} is `year` at that occasion. The level-2 models are

$$\pi_{pjk} = \beta_{p0k} + \beta_{p1}X_{1jk} + \beta_{p2}X_{2jk} + r_{pjk}, \quad p = 0, 1 \quad (r_{0jk}, r_{1jk})' \sim N(\mathbf{0}, \mathbf{T}_\pi)$$

where X_{1jk} is `black`, X_{2jk} is `hispanic`, and r_{pjk} is a random effect (intercept if $p = 0$, slope if $p = 1$) at level 2. The covariance matrix of the level-2 random effects is defined as

$$\mathbf{T}_\pi = \begin{bmatrix} \tau_{\pi 00} & \tau_{\pi 01} \\ \tau_{\pi 10} & \tau_{\pi 11} \end{bmatrix}, \quad \tau_{\pi 10} = \tau_{\pi 01}$$

Finally, the level-3 model is

$$\beta_{p0k} = \gamma_{p00} + \gamma_{p01}W_{1k} + u_{p0k}, \quad p = 0, 1 \quad u_{p0k} \sim N(0, T_\beta)$$

where W_{1k} is `lowinc` and u_{p0k} is a random intercept at level 3.

1. Substitute the level-3 models into the level-2 models and then the resulting level-2 models into the level-1 model. Rewrite the final reduced-form model using the notation of this book.

2. Fit the model using `xtmixed` and interpret the estimates.

3. Include some of the other covariates in the model and interpret the estimates.

8.2 Instructional-improvement data

West, Welch, and Galecki (2007) analyzed a dataset on first-grade students from the Study of Instructional Improvement by Hill, Rowan, and Ball (2005).

The question of interest is how teachers' experience, mathematics preparation, and mathematics content knowledge affect students' gain in mathematics achievement scores from kindergarten to first grade.

The variables in the dataset `instruction.dta` used here are

- Level 1 (students)
  - `childid`: student identifier
  - `mathgain`: gain in math achievement score from spring of kindergarten to spring of first grade
  - `mathkind`: match achievement score in spring of the kindergarten year
  - `girl`: dummy variable for being a girl
  - `minority`: dummy variable for being a minority student
  - `ses`: socioeconomic status

- Level 2 (class)
  - `classid`: class identifier
  - `yearstea`: first-grade teacher's years of teaching experience
  - `mathprep`: first-grade teacher's mathematics preparation (score based on number of mathematics content and methods courses)
  - `mathknow`: first-grade teacher's mathematics content knowledge, based on a 30-item scale with higher values indicating better knowledge

- Level 3 (school)
  - `schoolid`: school identifier
  - `housepov`: percentage of households in the neighborhood of the school below the poverty level

1. Treating `mathgain` as the response variable, write down the "unconditional" (without covariates) three-level random-intercept model for children nested in classes nested in schools, using
   a. the three-stage formulation
   b. the reduced form

2. Fit the model in Stata, obtain the estimated intraclass correlations, and interpret them.

3. For the three-stage formulation in step 1, write down an extended level-1 model that includes the four student-level variables, `mathkind`, `girl`, `minority`, and `ses`, as covariates. Similarly, write down an extended level-2 model by including the three class-level covariates, and an extended level-3 model by including the school-level covariate.

4. Fit the model from step 3. Interpret the estimated coefficients of the class-level covariates.

5. Fit an extended model that also includes the school means of `ses`, `minority`, and `mathkind`. Is there evidence that the within-school effects of these variables differ from the between-school effects?

6. Obtain empirical Bayes predictions of the random effects, and produce graphs to assess their distribution.

8.3 U.S. production data

In economics, a production function expresses the relationship between the output or production (the monetary value of all goods produced) of an economic unit (such as a country) and different inputs. A Cobb–Douglas production function expresses production P as a log-linear model of input, such as capital K and labor L,

$$P_i = AK_i^{\beta_2} L_i^{\beta_3} e^{\epsilon_i}$$

so that

$$\ln(P_i) = \ln(A) + \beta_2 \ln(K_i) + \beta_3 \ln(L_i) + \epsilon_i$$

Thus after taking logarithms of the output and all input variables, the production function can be estimated using linear regression. The research question concerns how public spending (on highways and streets, water and sewer facilities, and other buildings and structures) affects private production.

Baltagi, Song, and Jung (2001) analyzed data from Munnell (1990) on state productivity for 48 U.S. states from nine regions over the period 1970–1986. They estimate a Cobb–Douglas production function with error components for region and state.

The variables in `productivity.dta` are

- `state`: state identifier
- `region`: region identifier
- `year`: year 1970–1986
- `private`: logarithm of private capital stock.
- `hwy`: logarithm of highway component of public stock
- `water`: logarithm of water component of public stock
- `other`: logarithm of building and other components of public stock
- `unemp`: state unemployment rate

1. Fit a three-level model for the logarithm of private capital stock, `private`, with covariates `hwy`, `water`, `other`, and `unemp` and with random intercepts for `state` and `region`. Use `xtmixed` with both the `mle` and the `reml` options.

2. Which components of public capital have a positive effect on private output?

3. Interpret the sizes of the estimated residual variance components. Also comment on any differences between the ML and the REML estimates.

See also exercise 9.3 for further analyses of these data.

8.4 Multicenter hypertension-trial data

Hall et al. (1991) describe a randomized double-blind multicenter trial of treatments for hypertension (high blood pressure). The data were made available by Brown and Prescott (2006).

Three hundred eleven patients from 29 centers met eligibility criteria. In the initial phase of the trial, the patients received a placebo treatment and were then reassessed for eligibility one week later (visit 2). The 288 patients who still met eligibility criteria were then randomized *within* each center to receive one of the three treatments (A=Carvedilol, B=Nifedipine, C=Atenolol). The patients were followed up every other week for four visits. Diastolic blood pressure (the pressure in the bloodstream when the heart fills with blood) was the primary endpoint (response) and will be considered here.

The variables in the dataset `bp.dta` that we will use here are

- `center`: center identifier
- `id`: patient identifier
- `time`: visit number 3, 4, 5, 6 (postrandomization)
- `bp`: diastolic blood pressure in mmHg
- `bp_base`: diastolic blood pressure during second visit, prior to randomization
- `treat`: treatment group A, B, C (a string variable)

1. Transform `time` so that it takes the value 0 at the first postrandomization visit (visit 3) and increases 2 units between visits (representing the number of weeks since first postrandomization visit).

2. Produce a trellis graph, with one panel per center, where each panel is a spaghetti plot of the patients' observed trajectories.

3. Fit a three-level random-intercept model for `bp` with random intercepts for patients and centers and with `bp_base`, `time`, and dummy variables for treatments B and C as covariates.

4. Use a significance level of 5% to consider the following additions to this model, keeping each significant addition in all subsequent models:

   a. a quadratic term for `time`
   b. interactions between the treatment dummies and `time`

     c. a random slope of `time` at the patient level

     d. a random slope of `time` at the center level

5. Write down and describe the chosen model, defining all the notation you are using, including subscripts and variances.

6. Interpret the estimated treatment effect.

7. Produce a graph similar to the graph in step 2 but with the predicted patient-specific growth trajectories instead of the observed trajectories.

8.5 School-effects data

The dataset considered here is a 50% random subset of data described in Nuttall et al. (1989) and Goldstein (1991) and is made available by the Centre for Multilevel Modelling at the University of Bristol.

Examination results are available for three successive cohorts of year 11 (age 16) students from 140 schools from the Inner London Education Authority in 1985, 1986, and 1987. Prior to entry to secondary school (at age 11), each child was assigned to one of three academic achievement bands, largely on the basis of a verbal reasoning test. Band 1 contains the highest 25%, band 2 contains the middle 50%, and band 3 contains the lowest 25%. The response variable is a score derived from grades obtained for ordinary level (O-level) and graduate certificate of secondary education (GCSE) exams taken at age 16.

The dataset is an example of the second type of longitudinal multilevel data discussed in Raudenbush (1989), where schools are followed over time, but at each time point a different cohort of students is considered (in a given grade). In contrast, the data in exercises 8.1 and 8.7 consider the same cohort of students over time as they progress through the grades. The present data can be used to assess the stability of school effects over time.

The variables in `schooleffects.dta` are

- `year`: year when test was taken (1: 1985; 2: 1986; 3: 1987), defining the cohort and calendar time (period)
- `school`: school identifier
- `score`: score, based on O-level and GSCE results
- `pfsm`: percentage of students in the school who are eligible for free school meals
- `pvr1`: percentage of students in the school in verbal reasoning band 1
- `female`: dummy variable for being female (1: female; 0: male)
- `vr`: verbal reasoning band of student (values 1, 2, 3)
- `ethnic`: ethnic group of student (11 groups; see value labels)
- `schgend`: school gender (1: mixed; 2: male; 3: female)
- `schden`: school denomination (1: no denomination; 2: Church of England; 3: Roman Catholic)

1. Goldstein (1991) considers a model for `score` with random intercepts for school (level 3) and cohort within school (level 2). Fit this model, also allowing the mean score to change linearly over calendar time/cohort (coded 0, 1, 2).

2. Raudenbush (1989) considers a model that is identical to the model in step 1 except that the random intercept for cohort within school is replaced by a school-level random slope of time (coded 0, 1, 2). Fit this model.

3. For the models in steps 1 and 2, write down the random part of the model for time 0, 1, and 2 (six expressions in total). Discuss how the models differ.

4. Extend the model from step 2 by adding a level-2 random effect for cohorts nested in schools (as in step 1).

   a. Write down the model.
   b. Fit the model.
   c. Interpret the estimates.
   d. Compare the model with the models in steps 1 and 2 using likelihood-ratio tests.

5. For the model from step 4b, produce a trellis graph of spaghetti plots for the predicted cohort and school-specific mean score against time, where the trellis panels are for combinations of `schgend` and `schden` (you can use the `by(schden schgen)` option). Describe the graph.

8.6 STAR data I

The Tennessee class-size reduction experiment, known as Project STAR (Student-Teacher Achievement Ratio), was a four-year experiment designed to evaluate the effect of small class sizes on learning. Three different class types were compared for kindergarten through third grade (K–3):

1. a small class size, with 13–17 students per class and no teacher's aide
2. a regular class size with 22–25 students per class and no teacher's aide
3. a regular class size with 22–25 students per class and a teacher's aide

Each of the 79 participating schools had at least one class of each type, and teachers of participating schools were randomly assigned to the classes within their schools. Students entering a participating school in kindergarten in 1985 or first grade in 1986 (kindergarten was optional) were also randomly assigned to the classes. In this exercise, we restrict analysis to the kindergarten year.

The data were made available by Heros Inc. and are documented in Finn et al. (2007). The variables in `star1.dta` that we will use here are

- `schid`: school identifier
- `class`: class identifier
- `stdntid`: student identifier
- `grade`: grade, coded 0 for kindergarten and 1, 2, 3 for grades 1, 2, and 3

- `tmaths`: total scale score for Stanford achievement test (SAT) in math (ranging from 0 to 1400)
- `classt`: class type, coded as shown above

1. Keep only the kindergarten data (grade 0).
2. Obtain the frequency distribution for the number of classes per school.
3. What are the minimum, mean, and maximum number of students per class for small and regular class sizes (not distinguishing between regular class sizes with and without an aide)?
4. Fit a three-level random-intercept model for the SAT math score, with random intercepts for classes and schools.
5. Add class type as an explanatory variable and discuss whether small class size appears to be beneficial. Also discuss the change in the estimated variance components if any.
6. The STAR study has been described by Frederick Mosteller as "one of the most important educational investigations ever carried out" (Mosteller 1995), presumably because both students and teachers were randomly assigned to classes within schools. However, Krueger (1999, 510) points out that independence between class size and other (possibly omitted) variables holds only within schools. There could be omitted school-level variables that are correlated with class size and student outcomes (that is, confounders). Repeat the analysis from step 5, but use fixed effects for schools instead of random effects to avoid school-level endogeneity or confounding. Do the conclusions regarding class size change?

8.7 STAR data II

In the previous exercise, analysis was restricted to the kindergarten year. Here we consider students' growth in math achievement from kindergarten through third grade. Students were assessed each year using the Stanford achievement test in math. Scale scores were derived from item response models to make them comparable across grades, though the test becomes more difficult over time.

As a result of mobility, school membership could change over time for some students. For these students, we found the school with the largest number of observations and resolved ties by a random choice of school. The corresponding school assignment is given in the variable `school`. For simplicity, we ignore class membership in this exercise.

Some of the variables we will use here are described in the previous exercise. The additional variables used here are

- `school`: school to which student belongs most often
- `frlnch`: percentage of students in school eligible for a free school lunch that grade

- `freelu`: student eligible for a free school lunch (1: yes; 2: no)
- `gender`: student's gender (1: male; 2: female)

1. What percentage of students change schools during the study? (You may want to use the `egen` function `sd()` to find the standard deviation of `schid` within students, keeping in mind that the standard deviation is not defined for students who were only observed once.)

2. Delete observations where `schid` differs from `school`.

3. Use `xtdescribe` to explore the patterns of missing values of `tmaths`.

4. Fit a three-level random-intercept model for `tmaths`, with random intercepts for students and schools.

5. Add grade as a covariate and allow the effect of grade to vary randomly between students and schools.

   a. Write down the model using a three-stage formulation and derive the reduced form.
   b. Fit the model.
   c. Interpret the estimates.

6. Add `classt`, `frlnch`, `freelu`, and `gender` as explanatory variables, using appropriate dummy variables for categorical variables.

7. Interpret the estimated coefficients.

8. Check whether there is any evidence that the mean annual growth in math scores differs between class types.

See also exercise 9.10 for further analyses of data from the STAR experiment.

8.8 Dairy-cow data

Dohoo et al. (2001) and Dohoo, Martin, and Stryhn (2010) analyzed data on dairy cows nested in herds and regions of Reunion Island. One outcome considered was the time interval between calving (giving birth to a calf) and first service (attempt to inseminate the cow again). This outcome was available for several lactations (calvings) per cow.

The variables in the dataset `dairy.dta` used here are

- `cow`: cow identifier
- `herd`: herd identifier
- `region`: geographic region
- `lncfs`: log of calving to first service interval (in log days)
- `fscr`: first service conception risk (dummy variable for cow becoming pregnant)
- `ai`: dummy variable for artificial insemination being used (versus natural) at first service
- `heifer`: dummy variable for being a young cow that has given birth only once

1. Fit a four-level random-intercept model with `lncfs` as the response variable and with random intercepts for cows, herds, and geographic regions. Do not include any covariates. Use restricted maximum likelihood (REML) estimation. There are only five geographic regions, so it is arguable that region should be treated as fixed.

2. Obtain the estimated intraclass correlations for 1) two observations for the same cow, 2) observations for two different cows from the same herd, and 3) observations for two different cows from different herds in the same region.

3. Use REML to fit a three-level model for lactations nested in cows nested in herds, including dummy variables for the five geographic regions and omitting the constant. Compare the estimates for this model with the estimates using a four-level model.

8.9 ❖ Exam-and-coursework data

In this exercise, we consider a so-called multivariate multilevel model—that is, a regression model with several response variables—where units are nested in clusters. A multivariate regression (also known as seemingly unrelated regressions) for single-level data was considered in exercise 6.5.

The data are on students in England who took the General Certificate of Secondary Education (GCSE) exam in science (at age 16) in 1989. In addition to a written exam, the students also completed coursework that included projects undertaken during the school year and graded by each student's teacher. Both the written exam and coursework component scores have been scaled so that the maximum score is 100. The students are nested in schools. The data are described in Rasbash et al. (2009) and Goldstein (2011, chap. 6) and made available by the Centre for Multilevel Modelling at the University of Bristol.

The variables in `coursework.dta` are

- `schid`: school identifier
- `stid`: student identifier
- `girl`: dummy variable for being a girl
- `written`: written exam score
- `coursework`: coursework score

The two component scores are expected to be correlated, possibly with different correlations at the student and school levels. Denoting the written exam score for student i in school j as w_{ij} and the corresponding coursework score as c_{ij}, an appropriate model, with a dummy variable g_{ij} for girl as a covariate, is therefore

$$w_{ij} = \beta_{w1} + \beta_{w2}g_{ij} + \zeta_{wj} + \epsilon_{wij}$$
$$c_{ij} = \beta_{c1} + \beta_{c2}g_{ij} + \zeta_{cj} + \epsilon_{cij}$$

Here β_{w1} and β_{w2} are the intercept and slope of g_{ij} for the written exam, and β_{c1} and β_{c2} are the intercept and slope of g_{ij} for the coursework.

Given gender, the school-level random intercepts ζ_{wj} and ζ_{cj} are assumed to follow a bivariate normal distribution,

$$\begin{bmatrix} \zeta_{wj} \\ \zeta_{cj} \end{bmatrix} \sim N\left(\begin{bmatrix} 0 \\ 0 \end{bmatrix}, \begin{bmatrix} \psi_{ww} & \psi_{wc} \\ \psi_{wc} & \psi_{cc} \end{bmatrix} \right)$$

as do the level-1 errors,

$$\begin{bmatrix} \epsilon_{wij} \\ \epsilon_{cij} \end{bmatrix} \sim N\left(\begin{bmatrix} 0 \\ 0 \end{bmatrix}, \begin{bmatrix} \theta_{ww} & \theta_{wc} \\ \theta_{wc} & \theta_{cc} \end{bmatrix} \right)$$

The model can be fit using `xtmixed` by changing the data to long form, stacking the written exam and coursework exams into one variable. The data can then be thought of as three-level data with response variables (w_{ij} and c_{ij}) at level 1, students at level 2, and schools at level 3. We will use the notation y_{rjk}, with $y_{1jk} \equiv w_{ij}$ and $y_{2jk} \equiv c_{ij}$. To obtain different intercepts for the two response variables, we can use dummy variables w_r and c_r, where w_r is one when $r = 1$ and zero otherwise, whereas c_r is one when $r = 2$ and zero otherwise. The slopes β_{w2} and β_{c2} of g_{ij} correspond to the coefficients of the interactions $g_{ij}w_r$ and $g_{ij}c_r$, respectively.

The Stata dataset should look like this:

```
. sort schid stid variable
. list schid stid y variable w c girl_c girl_w in 1/6, sepby(stid) noobs
```

schid	stid	y	variable	w	c	girl_w	girl_c
20920	16	23	1	1	0	0	0
20920	16	.	2	0	1	0	0
20920	25	.	1	1	0	1	0
20920	25	71.2	2	0	1	0	1
20920	27	39	1	1	0	1	0
20920	27	76.8	2	0	1	0	1

There are two rows of data per student, the variable `y` contains the two responses, the variable `variable` keeps track of which response is which (denoted r above), `w` and `c` are dummy variables for the written exam and coursework scores, respectively, and `girl_w` and `girl_c` are interactions between these dummy variables and `girl`.

The random intercepts can be specified as random coefficients of the dummy variables `w` and `c` (with an unstructured covariance matrix). Finally, an unstructured covariance matrix for the level-1 residuals can be specified using the `residuals(unstructured, t(variable))` option, as shown in section 6.3.1.

1. Reshape the data to long form, creating the variable `variable` to keep track of which row of data corresponds to which response variable.

2. Missing responses are coded -1. Replace these with missing values (.) using the `mvdecode` command.

3. Create the dummy variables `w` and `c` and the interactions `girl_w` and `girl_c`.

4. Fit the model by ML using `xtmixed`.

5. Interpret the estimates.

8.10 ❖ Twin-neuroticism data

Here we consider the twin data used in exercise 2.3. The neuroticism scores for twins i and i' in the same twin-pair j are expected to be correlated because the twins share genes and aspects of the environment. A biometric model for the effects of genes and environment can be written as

$$y_{ij} = \mu + A_{ij} + D_{ij} + C_{ij} + \epsilon_{ij}$$

where A_{ij} are additive genetic effects, D_{ij} are dominance genetic effects, C_{ij} is a common environment effect, and ϵ_{ij} is a unique environment effect. The terms are uncorrelated with each other and have variances σ_A^2, σ_D^2, σ_C^2 and σ_e^2, respectively. The genetic effects A_{ij} and D_{ij} are perfectly correlated between members of the same twin-pair for MZ (identical) twins and have correlations $1/2$ and $1/4$, respectively, for DZ (fraternal) twins. The shared environment effect C_{ij} is perfectly correlated for both types of twins, whereas ϵ_{ij} is uncorrelated for both types of twins. The full model is not identifiable using twin data, so the ACE or ADE models are usually fit (omitting either D_{ij} or C_{ij} from the model).

Rabe-Hesketh, Skrondal, and Gjessing (2008) show that the covariance structure implied by the biometrical model can be induced using the following three-level model:

$$y_{ikj} = \beta_1 + \zeta_{kj}^{(2)} + \zeta_j^{(3)} + \epsilon_{ikj} \tag{8.4}$$

where k is an artificially created identifier that equals the twin-pair identifier j for MZ twins and the person identifier i for DZ twins. The variance ψ_3 of $\zeta_j^{(3)}$ is then shared by both members of the twin-pair for both MZ and DZ twins, whereas the variance ψ_2 of $\zeta_{kj}^{(2)}$ is shared by members of the twin-pair only for MZ twins. The covariances are therefore

$$\text{Cov}(y_{ij}, y_{i'j}) = \begin{cases} \psi_2 + \psi_3 = \sigma_A^2 + \sigma_D^2 + \sigma_C^2 & \text{for MZ twins} \\ \psi_3 = \sigma_A^2/2 + \sigma_D^2/4 + \sigma_C^2 & \text{for DZ twins} \end{cases}$$

For the ACE model ($\sigma_D^2 = 0$), we have

$$\sigma_A^2 = 2\psi_2 \qquad \sigma_C^2 = \psi_3 - \psi_2 \qquad \sigma_e^2 = \theta$$

and for the ADE model ($\sigma_C^2 = 0$), we have

$$\sigma_A^2 = 3\psi_3 - \psi_2 \qquad \sigma_D^2 = 2(\psi_2 - \psi_3) \qquad \sigma_e^2 = \theta$$

1. The data are in collapsed or aggregated form with `num2` representing the number of twin-pairs having a given pair of neuroticism scores. Expand the data using `expand num2`.

2. Create an identifier for twin-pairs (for example, using `generate pair = _n`) and reshape the data to long form, stacking the neuroticism scores into one variable.

3. Create the artificial identifier k and fit the model in (8.4).

4. Obtain the estimated variance components for the ACE and ADE models. Which model would you choose?

8.11 ❖ Peak-expiratory-flow data I

The three-level model for the peak-expiratory-flow data from Bland and Altman (1986) considered in this chapter can be written as

$$
\begin{aligned}
y_{ijk} &= \pi_{0jk} + \epsilon_{ijk} \\
\pi_{0jk} &= \underbrace{\gamma_{000} + u_{0k}}_{\beta_{0k}} + r_{0jk}
\end{aligned}
$$

where we have used the notation common in the three-stage formulation but with the level-3 model for β_{0k} substituted into the level-2 model for π_{0jk}. This model can be represented by the path diagram in the left panel of figure 8.10.

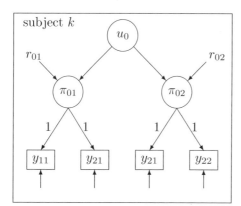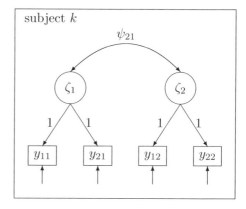

Figure 8.10: Path diagrams of equivalent models (left panel: three-stage formulation of three-level model; right panel: correlated random effects)

1. Express the correlation between π_{01k} and π_{02k} for the model represented in the left panel of figure 8.10 in terms of the variances $\psi^{(2)}$ at level 2 and $\psi^{(3)}$ at level 3.

2. The right panel in figure 8.10 represents a model with two method-specific correlated random intercepts, ζ_{1k} for method 1 and ζ_{2k} for method 2. Defining dummy variables d_{1j} for method 1 and d_{2j} for method 2, the model can be written as a two-level random-coefficient model:

$$
y_{ijk} = \beta_1 + \zeta_{1k}d_{1j} + \zeta_{2k}d_{2j} + \epsilon_{ijk}
$$

If the variances of the two random coefficients are constrained equal and if the covariance is positive, this model is equivalent to the previous three-level model. Verify this by fitting the above model by ML using xtmixed. You can use the covariance(exchangeable) option to constrain the variances of the random coefficients to be equal.

3. For the model in step 2, relax the assumption of equal variances of the random coefficients by using the covariance(unstructured) option. Compare this model with the model in step 2 using a likelihood-ratio test.

4. For the more complex model from step 3, obtain the estimated method-specific intraclass correlations, $\widehat{\rho}(\text{subject}|\text{method}=1)$ and $\widehat{\rho}(\text{subject}|\text{method}=2)$.

5. Extend the model further (still using xtmixed) by relaxing the assumption of equal measurement error variances θ. Is there any evidence that the measurement error variances differ between the methods? Again obtain the estimated method-specific intraclass correlations based on the more complex model.

8.12 ❖ Peak-expiratory-flow data II

This exercise is a useful transition to the next chapter, where we sometimes treat random intercepts for a factor as random coefficients of dummy variables for the levels of the factor.

For the peak-expiratory-flow data, we require a random intercept for method nested in subject. This can be achieved by having uncorrelated random coefficients at the subject level for a dummy variable d_{1j} for method 1 and d_{2j} for method 2. We also retain the random intercept at the subject level (now called ζ_{3k}):

$$y_{ijk} = \beta_1 + \zeta_{1k}d_{1j} + \zeta_{2k}d_{2j} + \zeta_{3k} + \epsilon_{ijk}$$

All three random effects should be mutually uncorrelated, but only the first two should have the same variance. This can be achieved by using two random parts for the same level: || id: d1 d2, covariance(identity) noconstant || id:, where the covariance(identity) option specifies equal variances and no covariance for the random coefficients of d1 and d2 (the dummy variables) and the noconstant option suppresses the random intercept because this is already specified by the next equation, || id:. Random effects in different random part specifications (separated by ||) are assumed to be uncorrelated. Also note that the noconstant option is not necessary because xtmixed puts the constant only in the last equation for a given level by default.

Fit this model using xtmixed, and show that it is equivalent to the three-level random-intercept model specified using || id: || method:.

9 Crossed random effects

9.1 Introduction

In the previous chapter, we discussed higher-level hierarchical models where units are classified by some factor (for instance, school) into top-level clusters. The units in each top-level cluster are then (sub)classified by a further factor (for instance, classroom) into clusters at a lower level, etc. The factors defining the classifications are nested in the sense that a lower-level cluster can only belong to *one* higher-level cluster (for instance, a classroom can only belong to one school).

We now discuss nonhierarchical models where units are *cross-classified* by two or more factors, with each unit potentially belonging to any combination of values of the different factors. If the *main* effects of these cross-classifications are represented by *random* effects, this leads to models with crossed random effects. Cross-classifications do of course frequently occur in the fixed part of the model where they can easily be handled by including dummy variables and interaction terms.

A prominent example of crossed data is students who are cross-classified by elementary schools and middle schools. Children from the same elementary school can go to different middle schools and children in the same middle school can come from different elementary schools. We would expect that the middle school attended by the child would have a main effect on, for instance, achievement in addition to the main effect of the elementary school attended. Usually, the factors elementary school and middle school are both treated as random. The random effects are then crossed, which produces a crossed random effects model. An example where Scottish primary schools are crossed with secondary schools is considered in section 9.4.

Longitudinal or panel data is another example of cross-classified data where the factor subject (or country or firm, etc.) is crossed with another factor, occasion (see section 9.2 for such an example). So far in this book, we have treated occasions as nested within subjects when considering longitudinal data. This means that the random part of the model has mean zero at each occasion across subjects. Occasion was represented in the random part by the level-1 error, taking a different value for each subject–occasion combination. There was hence no random main effect of occasion taking the same value across subjects. However, if all subjects are affected similarly by some events or characteristics associated with the occasions—such as weather conditions, strikes, new legislation, etc.—it seems reasonable to consider a random main effect of occasion. If the factors subject and occasion are both treated as random, the random effects are crossed and econometricians call the model a *two-way error-components model*. Such a model is considered in section 9.3.

9.2   How does investment depend on expected profit and capital stock?

Grunfeld (1958) and Boot and de Wit (1960), among others, analyzed investment data on 10 large American corporations collected annually from 1935 to 1954. This is a dataset with long panels as discussed in section 6.7.

The variables in the file `grunfeld.dta` provided by Baltagi (2008) are

- `fn`: firm identifier (i)

- `firmname`: firm name

- `yr`: year (j)

- `I`: annual gross investment (in \$1,000,000) defined as amount spent on plant and equipment plus maintenance and repairs (y_{ij})

- `F`: market value of firm (in \$1,000,000) defined as value of all shares plus book value of all debts outstanding at the beginning of the year (x_{2ij})

- `C`: real value of capital stock (in \$1,000,000) defined as the deviation of stock of plant and equipment from stock in 1933 (x_{3ij})

We read in the data using

```
. use http://www.stata-press.com/data/mlmus3/grunfeld
```

Grunfeld argued that the investment of firms depends on expected profit (measured as market value) and capital stock (see display 9.1 if you are interested in a brief summary of Grunfeld's investment theory). The theory implies an investment equation of the form

$$y_{ij} = \beta_1 + \beta_2 x_{2ij} + \beta_3 x_{3ij}$$

where we have denoted the gross investment `I` for firm i in year j as y_{ij}, the market value `F` as x_{2ij}, and the value of capital stock `C` as x_{3ij}.

Boot and de Wit (1960) summarize Grunfeld's (1958) investment theory as follows: Observed profits are rejected as an explanation of investment, and expected profits, measured as the market value of the firm $F(t)$ at time t, are used instead. The desired capital stock $C^*(t)$ is assumed to be a linear function of the market value:

$$C^*(t) = c_1 + c_2 F(t)$$

The desired net investment is the difference between desired capital stock $C^*(t)$ and the existing capital stock $C(t)$: $C^*(t) - C(t)$. Assuming that a constant fraction q_1 of the desired net investment is made between t and $t+1$, the net investment for that year becomes

$$q_1\{C^*(t) - C(t)\} = q_1 c_1 + q_1 c_2 F(t) - q_1 C(t)$$

Assuming that replacement investment plus maintenance and repairs equals a constant fraction q_2 of the existing capital stock $C(t)$, the gross investment in the following year $I(t+1)$ becomes

$$
\begin{aligned}
I(t+1) &= q_1\{C^*(t) - C(t)\} + q_2 C(t) \\
&= q_1 c_1 + q_1 c_2 F(t) + (q_2 - q_1) C(t)
\end{aligned}
$$

Denoting the gross investment I for firm i in year j as y_{ij}, the market value F as x_{2ij}, and the value of capital stock C as x_{3ij}, the investment equation can finally be written as

$$y_{ij} = \beta_1 + \beta_2 x_{2ij} + \beta_3 x_{3ij}$$

where $\beta_1 \equiv q_1 c_1$, $\beta_2 \equiv q_1 c_2$, and $\beta_3 \equiv q_2 - q_1$. Both β_2 and β_3 are expected to be positive. However, the intercept β_1 has limited meaning because capital stock has been measured as a deviation from the stock in 1933.

Display 9.1: Brief summary of Grunfeld's (1958) investment theory

9.3   A two-way error-components model

9.3.1   Model specification

Because the investment behavior of corporations is surely not deterministic, statistical models including error terms have invariably been specified. Baltagi (2008) allows the effects of both firms and years on gross investment y_{ij} to vary by specifying the following two-way error-components model:

$$y_{ij} = \beta_1 + \beta_2 x_{2ij} + \beta_3 x_{3ij} + \zeta_{1i} + \zeta_{2j} + \epsilon_{ij} \tag{9.1}$$

Here x_{2ij} and x_{3ij} represent the market value and capital stock of firm i in year j. ζ_{1i} and ζ_{2j} are random intercepts for firms i and years j, respectively, and ϵ_{ij} is a residual error term.

Given the covariates x_{2ij} and x_{3ij}, the random intercepts have zero means and are uncorrelated with each other, ζ_{1i} has variance ψ_1 and is uncorrelated across firms, and

ζ_{2j} has variance ψ_2 and is uncorrelated across years. The random intercepts ζ_{1i} and ζ_{2j} are also uncorrelated with ϵ_{ij}. The residual ϵ_{ij} has zero mean and variance θ, given the covariates and random intercepts, and the residuals are uncorrelated across firms and years.

This model differs from the models considered so far because the two random intercepts represent factors that are crossed instead of nested. The random intercept for firm ζ_{1i} is shared across all years for a given firm i, whereas the random intercept for year ζ_{2j} is shared by all firms in a given year j. The residual error ϵ_{ij} comprises both the interaction between year and firm and any other effect specific to firm i in year j. An interaction between firm and year could be due to some events occurring in some years being more beneficial (or detrimental) to some firms than others.

9.3.2   Residual variances, covariances, and intraclass correlations

It follows from the assumptions that the variance for a response given the covariates becomes

$$\mathrm{Var}(y_{ij}|x_{2ij}, x_{3ij}) \; = \; \psi_1 + \psi_2 + \theta$$

Longitudinal correlations

Given the covariates, the covariance between responses for the same firm i at different years j and j' is

$$\mathrm{Cov}(y_{ij}, y_{ij'}|x_{2ij}, x_{3ij}, x_{2ij'}, x_{3ij'}) \; = \; \psi_1$$

As in section 8.5, this covariance can be derived by taking the expectations of the product of the random parts in y_{ij} and $y_{ij'}$. The only product term with a nonzero expectation under the model assumptions is the square of the random intercept for firms, whose expectation is $E(\zeta_{1i}^2) = \mathrm{Var}(\zeta_{1i}) \equiv \psi_1$. The corresponding longitudinal (between-year, within-firm) intraclass correlation is

$$\rho(\mathrm{firm}) \equiv \mathrm{Cor}(y_{ij}, y_{ij'}|x_{2ij}, x_{3ij}, x_{2ij'}, x_{3ij'}) \; = \; \frac{\psi_1}{\psi_1 + \psi_2 + \theta}$$

Cross-sectional correlations

The covariance between responses for different firms i and i' in the same year j is

$$\mathrm{Cov}(y_{ij}, y_{i'j}|x_{2ij}, x_{3ij}, x_{2i'j}, x_{3i'j}) \; = \; \psi_2$$

This can be obtained by taking the expectations of the product of the random parts in y_{ij} and $y_{i'j}$, and here the only product term with a nonzero expectation is the square of the random intercept for year. The cross-sectional (between-firm, within-year) intraclass correlation becomes

$$\rho(\mathrm{year}) \equiv \mathrm{Cor}(y_{ij}, y_{i'j}|x_{2ij}, x_{3ij}, x_{2i'j}, x_{3i'j}) \; = \; \frac{\psi_2}{\psi_1 + \psi_2 + \theta}$$

9.3.3 Estimation using xtmixed

The `xtmixed` command is primarily designed for multilevel models with nested random effects. To fit models with crossed effects, we therefore use the following trick described by Goldstein (1987):

- Consider the entire dataset as an artificial level-3 unit a within which both firms and years are nested.

- Treat either years or firms as level-2 units j, and specify a random intercept $u_{ja}^{(2)}$ for them. It is best to choose the factor with more levels, that is, years.

- For the other factor, here firm, specify a level-3 random intercept for each firm, $u_{pa}^{(3)}$, $(p = 1, \ldots, 10)$. This can be constructed by treating $u_{pa}^{(3)}$ as the random coefficient of the dummy variable d_{pi} for firm p, where

$$d_{pi} = \left\{ \begin{array}{ll} 1 & \text{if } p = i \\ 0 & \text{otherwise} \end{array} \right.$$

  The 10 random coefficients are then specified as having equal variance ψ_1 and being uncorrelated.

Here we have used the notation u for the random effects to avoid confusion between the different formulations. Model (9.1) can then be written as

$$
\begin{aligned}
y_{ija} &= \beta_1 + \beta_2 x_{2ij} + \beta_3 x_{3ij} + u_{ja}^{(2)} + \sum_p u_{pa}^{(3)} d_{pi} + \epsilon_{ija} \\
&= \beta_1 + \beta_2 x_{2ij} + \beta_3 x_{3ij} + \underbrace{u_{ja}^{(2)}}_{\zeta_{2j}} + \underbrace{u_{ia}^{(3)}}_{\zeta_{1i}} + \underbrace{\epsilon_{ija}}_{\epsilon_{ij}}
\end{aligned}
$$

where $u_{ja}^{(2)}$ and $u_{ia}^{(3)}$ are uncorrelated because they are specified at different levels.

The basic idea is to treat the 10 different realizations (or observations) of one random intercept $\zeta_i (i = 1, \ldots, 10)$ as realizations of 10 different random coefficients (or variables) $u_{ia}^{(3)}$ (similar to transforming data from long form to wide form where observations of one variable become different variables). Each random coefficient $u_{ia}^{(3)}$ takes on only one value for the entire dataset, and the dummy variable for firm i ensures that $u_{ia}^{(3)}$ contributes to the model only for firm i.

Recall that the firm identifier is `fn` and the year identifier is `yr`. Fortunately, `xtmixed` makes it easy to specify a random intercept $u_{pa}^{(3)}$ for each level of a factor (here each firm) or, as specified above, random coefficients for the corresponding dummy variables. The syntax `R.fn` in the random part accomplishes this and also automatically sets all variances equal and all correlations to zero as required. This covariance structure is called `identity` in `xtmixed` because the covariance matrix is proportional to the 10×10 identity matrix (a matrix with ones on the diagonal and zeros elsewhere). We also do not need to create an artificial level-3 identifier because `xtmixed` accepts the cluster name `_all` for this purpose.

The first random part is therefore specified as `|| _all: R.fn`. Following this, we specify `|| yr:` to let year have a random intercept at level 2. The random effects for time and year are uncorrelated as required because they are specified in different random parts (separated by `||`).

We fit the two-way error-components model in `xtmixed` by maximum likelihood (ML) using

```
. xtmixed I F C || _all: R.fn || yr:, mle
Mixed-effects ML regression                    Number of obs      =        200
```

Group Variable	No. of Groups	Observations per Group		
		Minimum	Average	Maximum
_all	1	200	200.0	200
yr	20	10	10.0	10

```
                                               Wald chi2(2)       =     661.06
Log likelihood = -1095.2485                    Prob > chi2        =     0.0000
```

I	Coef.	Std. Err.	z	P>|z|	[95% Conf. Interval]	
F	.1099012	.010378	10.59	0.000	.0895607	.1302418
C	.3092288	.0172182	17.96	0.000	.2754818	.3429758
_cons	-58.27229	27.76304	-2.10	0.036	-112.6869	-3.857725

Random-effects Parameters	Estimate	Std. Err.	[95% Conf. Interval]	
_all: Identity				
sd(R.fn)	80.4124	18.42495	51.32	125.9968
yr: Identity				
sd(_cons)	3.864611	15.26661	.0016771	8905.534
sd(Residual)	52.34725	2.903932	46.95415	58.3598

```
LR test vs. linear regression:       chi2(2) =    193.11   Prob > chi2 = 0.0000
Note: LR test is conservative and provided only for reference
```

The market value (expected profit) and stock value (capital stock) of a firm both have positive effects on investment as expected. According to the fitted model, an increase in the market value of $1,000,000 increases the mean investment by about $110,000, controlling for stock value. An increase in the stock value of $1,000,000 increases the mean investment by $310,000, controlling for market value. The estimated residual standard deviation between firms is $80.41 million, and the estimated residual standard deviation between years is only $3.86 million. The remaining residual standard deviation, not due to additive effects of firms and years, is estimated as $52.35 million. The estimates are also given under ML in table 9.1.

Table 9.1: Maximum likelihood (ML) and restricted maximum likelihood (REML) estimates of two-way error-components model for Grunfeld (1958) data

	ML		REML	
	Est	(SE)	Est	(SE)
Fixed part				
β_1	-58.27	(27.76)	-58.84	(29.51)
β_2 [F]	0.11	(0.01)	0.11	(0.01)
β_3 [C]	0.31	(0.02)	0.31	(0.02)
Random part				
$\sqrt{\psi_1}$ [Firm]	80.41		86.07	
$\sqrt{\psi_2}$ [Year]	3.86		5.40	
$\sqrt{\theta}$	52.35		52.47	
Log likelihood	$-1,095.25$		$-1,097.90^{\dagger}$	

†Restricted log likelihood

The reason for choosing year to be at level 3 and to have a random intercept for each firm at level 2 is to minimize the computational burden. With this formulation, the model has 10 random effects at level 3 (one for each firm) and 1 random effect at level 2. If we instead had chosen firm to be at level 2 and to have a separate random intercept for each year at level 3, we would have required 20 random effects at level 3 (one for each year) and 1 random effect at level 2. The syntax would be

```
xtmixed I F C || _all: R.yr || fn:, mle
```

Although more computationally demanding, this setup produces identical estimates to the command used above.

We have used the same estimation method (maximum likelihood) as for models with nested random effects, but it should be noted that asymptotics now rely on *both* the number of firms and the number of occasions going to infinity. Because we only have data on 10 firms over 20 years here, we cannot expect asymptotic results to hold and should treat them as approximate. For this reason, we do not test hypotheses regarding variance parameters for these data (see section 9.7.4 for such tests).

Another problem with a small number of clusters is that ML estimates of variance components are downward biased (see section 2.10.2). For balanced data, such as the Grunfeld (1958) investment data, restricted maximum likelihood (REML) yields unbiased estimates (if the estimates are allowed to be negative).

```
. xtmixed I F C || _all: R.fn || yr:, reml
Mixed-effects REML regression                    Number of obs    =       200
```

Group Variable	No. of Groups	Observations per Group Minimum	Average	Maximum
_all	1	200	200.0	200
yr	20	10	10.0	10

```
                                                Wald chi2(2)     =     643.60
Log restricted-likelihood = -1097.8951          Prob > chi2      =     0.0000
```

I	Coef.	Std. Err.	z	P>|z|	[95% Conf. Interval]	
F	.1100654	.0106036	10.38	0.000	.0892828	.130848
C	.3106316	.0174474	17.80	0.000	.2764353	.344828
_cons	-58.83708	29.50687	-1.99	0.046	-116.6695	-1.00468

Random-effects Parameters	Estimate	Std. Err.	[95% Conf. Interval]	
_all: Identity				
sd(R.fn)	86.06838	20.78351	53.61647	138.1621
yr: Identity				
sd(_cons)	5.399614	11.48764	.0834521	349.3723
sd(Residual)	52.46645	2.910242	47.0616	58.49202

```
LR test vs. linear regression:      chi2(2) =   196.43   Prob > chi2 = 0.0000
Note: LR test is conservative and provided only for reference.
```

The estimates were also given under REML in table 9.1. We see that only the random-intercept standard deviations differ appreciably between REML and ML and are larger for REML as expected.

Based on the REML estimates, the residual longitudinal intraclass correlation within firms is estimated as

$$\widehat{\rho}(\text{firm}) \ = \ \frac{86.06838^2}{86.06838^2 + 5.399614^2 + 52.46645^2} = 0.73$$

and the residual cross-sectional intraclass correlation between firms within years is estimated as

$$\widehat{\rho}(\text{year}) \ = \ \frac{5.399614^2}{86.06838^2 + 5.399614^2 + 52.46645^2} = 0.003$$

Hence, there is a high correlation over years within firms and a negligible correlation over firms within years, given the covariates.

Instead of typing in the estimates, we can refer to them by the equation and column names, which we can find by displaying the matrix of estimates (or running xtmixed with the estmetric option):

```
. matrix list e(b)
e(b)[1,6]
             I:         I:         I:    lns1_1_1:   lns2_1_1:    lnsig_e:
             F          C       _cons       _cons       _cons       _cons
  y1   .11006541  .31063164  -58.837081  4.4551421   1.6863275   3.9601739
```

The last three elements are the log-standard deviations of the random effects and level-1 residuals in the same order as given in the output. We can therefore obtain the estimated within-firm intraclass correlation using

```
. display exp(2*[lns1_1_1]_cons)/(exp(2*[lns1_1_1]_cons) + exp(2*[lns2_1_1]_cons)
> + exp(2*[lnsig_e]_cons))
.72698922
```

9.3.4 Prediction

Having fit the model using REML, it is easy to obtain various predictions. For instance, we can obtain empirical Bayes predictions or best linear unbiased predictions (BLUPs) $\widetilde{\zeta}_{1i}$ and $\widetilde{\zeta}_{2j}$ of the random effects of firm and year using predict with the reffects option,

```
. predict firm year, reffects
```

and list these predictions for four of the firms for the first three years:

```
. sort fn yr
. list fn firmname yr firm year if yr<1938&fn<5, sepby(fn) noobs
```

fn	firmname	yr	firm	year
1	General Motors	1935	−11.36264	3.290692
1	General Motors	1936	−11.36264	1.778602
1	General Motors	1937	−11.36264	.0498858
2	US Steel	1935	157.7599	3.290692
2	US Steel	1936	157.7599	1.778602
2	US Steel	1937	157.7599	.0498858
3	General Electric	1935	−173.6221	3.290692
3	General Electric	1936	−173.6221	1.778602
3	General Electric	1937	−173.6221	.0498858
4	Chrysler	1935	30.43414	3.290692
4	Chrysler	1936	30.43414	1.778602
4	Chrysler	1937	30.43414	.0498858

The predictions for the firms have all conveniently been placed into 1 variable, firm, not into 10 variables as might have been expected. We see that the prediction for firm 1 (General Motors) is −11.36 and the prediction for 1935 is 3.29. After controlling for market and stock value, General Motors was a firm with lower investment than the average across firms, and 1935 was a year with higher investment than the average over the 20-year interval.

We can visualize these effects by plotting the sum of the predicted random effects $\widetilde{\zeta}_{1i} + \widetilde{\zeta}_{2j}$ versus occasion j (yr) with a separate line for each firm i (fn):

```
. generate reffpart = firm + year
. twoway (line reffpart yr, connect(ascending))
> (scatter reffpart yr if yr==1954, msymbol(none) mlabel(firmnam) mlabpos(3)),
> xtitle(Year) ytitle(Predicted random effects of firm and year)
> xscale(range(1935 1958)) legend(off)
```

It is clear from the resulting figure 9.1 that the between-firm variability is much more considerable than the between-year variability.

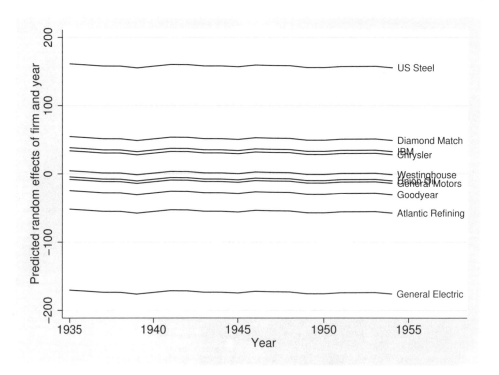

Figure 9.1: Sum of the predicted random effects $\widetilde{\zeta}_{1i} + \widetilde{\zeta}_{2j}$ versus time for 10 firms

The effect of year is hardly visible in figure 9.1 because it is negligible compared with the effect of firm. We therefore produce another graph just showing the effect of year on its own,

```
. twoway (line year yr if fn==1), xtitle(Year)
> ytitle(Predicted random effect for year)
```

which is given in figure 9.2. (Here we plotted the data for the first firm by using if fn==1 but would have obtained an identical graph if we had picked another firm.)

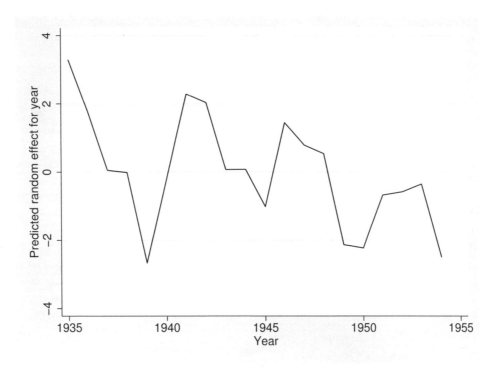

Figure 9.2: Predicted random effect of year $\widetilde{\zeta}_{2j}$

9.4 How much do primary and secondary schools affect attainment at age 16?

We now consider data described by Paterson (1991) on pupils (students) from Fife, Scotland, who are cross-classified by 148 primary schools (elementary schools) and 19 secondary schools (middle/high schools).

The dataset `fife.dta` is distributed with the MLwiN program (Rasbash et al. 2009) and has previously been analyzed by Goldstein (2011) and Rasbash (2005). The dataset has the following variables:

- `attain`: attainment score at age 16 [summary of passes in the Scottish Certificate of Education (SCE), a school exit examination] (y_{ijk})

- `pid`: identifier for primary school (up to age 12) (k)

- `sid`: identifier for secondary school (from age 12) (j)

- `vrq`: verbal reasoning score from test taken in the last year of primary school

- `sex`: gender (1: female; 0: male)

Here we investigate to what extent educational attainment at age 16 depends on the primary school attended up to age 12 and the secondary school attended thereafter.

We read in the data using

```
. use http://www.stata-press.com/data/mlmus3/fife, clear
```

9.5   Data structure

The structure of the Fife data is different than the Grunfeld investment data considered earlier in two regards: First, not every combination of primary and secondary school exists. Second, many combinations of primary and secondary school occur multiple times.

We first explore this crossed structure in more detail. For this, it is useful to define a dummy variable taking the value 1 for exactly one observation for each combination of the primary-school and secondary-school identifiers. The egen function tag() is designed for creating such dummy variables:

```
. egen pick_comb = tag(pid sid)
```

Now we can count the number of unique values of sid in each primary school (with identifier pid) using the egen function total():

```
. egen numsid = total(pick_comb), by(pid)
```

We can list the unique secondary school identifiers for the first 10 primary schools using

```
. sort pid sid
. list pid sid numsid if pick_comb==1 & pid<10, sepby(pid) noobs
```

pid	sid	numsid
1	1	3
1	9	3
1	18	3
2	7	1
3	5	1
4	6	2
4	9	2
5	1	1
6	1	4
6	3	4
6	5	4
6	11	4
7	4	4
7	6	4
7	18	4
7	19	4
8	1	3
8	9	3
8	19	3
9	1	6
9	3	6
9	6	6
9	17	6
9	18	6
9	19	6

We see that, for instance, students in this sample who attended primary school 1 ended up in three (numsid=3) secondary schools, 1, 9, and 18.

To obtain the frequency distribution of numsid, the number of secondary schools per primary school, we must first define a dummy variable equal to 1 for one student per primary school (so we do not count primary schools more than once),

```
. egen pick_pid = tag(pid)
```

and can subsequently use the `tabulate` command:

```
. tabulate numsid if pick_pid
   numsid |      Freq.     Percent        Cum.
----------+-----------------------------------
        1 |         57       38.51       38.51
        2 |         50       33.78       72.30
        3 |         26       17.57       89.86
        4 |         10        6.76       96.62
        5 |          2        1.35       97.97
        6 |          3        2.03      100.00
----------+-----------------------------------
    Total |        148      100.00
```

There are at most six secondary schools per primary school, and for about 90% of the primary schools there are at most three secondary schools per primary school.

Repeating the same commands as above, but with `sid` and `pid` interchanged, we obtain the frequency table of the number of primary schools per secondary school:

```
. egen numpid = total(pick_comb), by(sid)
. egen pick_sid = tag(sid)
. tabulate numpid if pick_sid==1
   numpid |      Freq.     Percent        Cum.
----------+-----------------------------------
        7 |          1        5.26        5.26
       10 |          2       10.53       15.79
       12 |          1        5.26       21.05
       13 |          2       10.53       31.58
       14 |          4       21.05       52.63
       15 |          1        5.26       57.89
       16 |          1        5.26       63.16
       17 |          2       10.53       73.68
       18 |          2       10.53       84.21
       23 |          1        5.26       89.47
       26 |          1        5.26       94.74
       32 |          1        5.26      100.00
----------+-----------------------------------
    Total |         19      100.00
```

There are between 7 and 32 primary schools per secondary school, the median being between 13 and 14.

9.6 Additive crossed random-effects model

9.6.1 Specification

We first consider the following model for the attainment score y_{ijk} at age 16 for student i who went to secondary school j and primary school k:

$$y_{ijk} = \beta_1 + \zeta_{1j} + \zeta_{2k} + \epsilon_{ijk} \tag{9.2}$$

The random part of this model has exactly the same structure as in the investment application with additive (and uncorrelated) random effects ζ_{1j} and ζ_{2k} of the two cross-classified factors, secondary school and primary school, respectively, plus a residual error term ϵ_{ijk}. The error components ζ_{1j} and ζ_{2i} have zero means and variances ψ_1 and ψ_2, respectively. Given ζ_{1j} and ζ_{2i}, the residual error ϵ_{ijk} has mean zero and variance θ.

9.6.2 Estimation using xtmixed

Because there are only 19 secondary schools compared with 148 primary schools, we use 19 random effects for the secondary schools at level 3 and treat primary schools as level-2 units.

The `xtmixed` command for fitting model (9.2) by ML is

```
. xtmixed attain || _all: R.sid || pid:, mle
Mixed-effects ML regression                    Number of obs    =      3435
```

Group Variable	No. of Groups	Observations per Group Minimum	Average	Maximum
_all	1	3435	3435.0	3435
pid	148	1	23.2	72

```
                                          Wald chi2(0)     =        .
Log likelihood = -8574.5655               Prob > chi2      =        .
```

attain	Coef.	Std. Err.	z	P>|z|	[95% Conf. Interval]	
_cons	5.504009	.1749325	31.46	0.000	5.161148	5.846871

Random-effects Parameters	Estimate	Std. Err.	[95% Conf. Interval]	
_all: Identity				
sd(R.sid)	.5900565	.1371156	.3741915	.9304505
pid: Identity				
sd(_cons)	1.060359	.0971078	.886135	1.268838
sd(Residual)	2.848065	.0351956	2.779912	2.91789

```
LR test vs. linear regression:        chi2(2) =    278.13   Prob > chi2 = 0.0000
Note: LR test is conservative and provided only for reference
. estimates store additive
```

The estimates are presented under "Additive" in table 9.2. We see that the estimated standard deviation $\sqrt{\widehat{\psi}_2}$ of the primary school random effect is 1.06, which is considerably larger than the estimated standard deviation $\sqrt{\widehat{\psi}_1}$ of the secondary school random effect, given by 0.59. Therefore, elementary schools appear to have greater effects or to be more variable in their effects on attainment than secondary schools. However, neither of these standard deviation estimates is very precise.

The estimated standard deviation $\sqrt{\hat{\theta}}$ of ϵ_{ijk} is 2.85. This number reflects any interactions between primary and secondary schools (deviations of the means for the combinations of primary and secondary schools from the means implied by the additive effects) as well as variability within the groups of children belonging to the same combination of primary and secondary school.

9.7   Crossed random-effects model with random interaction

9.7.1   Model specification

For many combinations of primary and secondary school, we have several observations because more than one student attended that combination of schools. We can therefore include a random interaction term ζ_{3jk} between secondary schools j and primary schools k in the model

$$y_{ijk} = \beta_1 + \zeta_{1j} + \zeta_{2k} + \zeta_{3jk} + \epsilon_{ijk} \tag{9.3}$$

The interaction term takes on a different value for each combination of secondary and primary school to allow the assumption of additive (random) effects to be relaxed. For instance, some secondary schools may be more beneficial for students who attended particular elementary schools, perhaps because of similar instructional practices.

The random intercept ζ_{3jk} has zero mean and variance ψ_3, is uncorrelated with the other random terms (ζ_{1j}, ζ_{2k}, and ϵ_{ijk}), and is uncorrelated across combinations of primary and secondary school. The residual ϵ_{ijk} represents the deviation of an individual student's response from the mean for secondary school j and primary school k. For given random effects, ϵ_{ijk} has zero mean and variance θ.

Note that we could not include an interaction in the investment application because we had no replicates for any of the firm and year combinations, so the interaction would be completely confounded with the level-1 residual.

9.7.2   Intraclass correlations

We can consider several intraclass correlations for the crossed random-effects model that includes a random interaction. For students i and i' from the same primary school k but different secondary schools j and j', we obtain

$$\rho(\text{primary}) \equiv \text{Cor}(y_{ijk}, y_{i'j'k}) = \frac{\psi_2}{\psi_1 + \psi_2 + \psi_3 + \theta}$$

where $\psi_3 = 0$ if there is no interaction. For students from the same secondary school j but different primary schools k and k', the correlation is

$$\rho(\text{secondary}) \equiv \text{Cor}(y_{ijk}, y_{i'jk'}) = \frac{\psi_1}{\psi_1 + \psi_2 + \psi_3 + \theta}$$

Finally, for students from both the same secondary school j and the same primary school k, we have

$$\rho(\text{secondary}, \text{primary}) \equiv \text{Cor}(y_{ijk}, y_{i'jk}) = \frac{\psi_1 + \psi_2 + \psi_3}{\psi_1 + \psi_2 + \psi_3 + \theta}$$

We could also condition on secondary school and consider the intraclass correlation among children from the same secondary school due to being in the same primary school:

$$\rho(\text{primary}|\text{secondary}) \equiv \text{Cor}(y_{ijk}, y_{i'jk}|\zeta_{1j}) = \frac{\psi_2 + \psi_3}{\psi_2 + \psi_3 + \theta}$$

where the between-secondary-school variance ψ_1 vanishes because secondary school is held constant. The analogous expression for the intraclass correlation due to secondary school for a given primary school is

$$\rho(\text{secondary}|\text{primary}) \equiv \text{Cor}(y_{ijk}, y_{i'jk}|\zeta_{2k}) = \frac{\psi_1 + \psi_3}{\psi_1 + \psi_3 + \theta}$$

The covariances can be derived as in section 8.5 by taking the expectation of the product of the random parts. When conditioning on a particular random term, such as ζ_{2k}, this term is simply omitted from the random part (see also exercise 9.11).

9.7.3 Estimation using xtmixed

The crossed random-effects model with a random interaction (9.3) can be fit in `xtmixed` by augmenting the random-part specification of the previous command for the additive model (9.2) given by

```
xtmixed attain || _all: R.sid || pid:, mle
```

We require a random intercept taking on distinct values for each combination of primary and secondary school. We could achieve this by defining an identifier variable for the combinations using `egen` with the `group()` function,

```
. egen comb = group(sid pid)
```

and specifying a third random part as `|| comb:`. This random intercept would be treated as nested within `pid`, the cluster identifier for the previous equation.

A more convenient setup, not requiring the variable `comb`, is to specify the last equation as `|| sid:`. Because `sid` is treated as nested in `pid`, the cluster identifier in the preceding random part, the random intercept will take on a different value for each unique secondary school, `sid`, *within* each primary school, `pid`, that is, for each combination of primary and secondary school. The syntax becomes

```
. xtmixed attain || _all: R.sid || pid: || sid:, mle

Performing EM optimization:

Performing gradient-based optimization:

Computing standard errors:

Mixed-effects ML regression                        Number of obs     =      3435
```

Group Variable	No. of Groups	Observations per Group Minimum	Average	Maximum
_all	1	3435	3435.0	3435
pid	148	1	23.2	72
sid	303	1	11.3	72

```
                                                   Wald chi2(0)      =          .
Log likelihood = -8573.9826                        Prob > chi2       =          .
```

attain	Coef.	Std. Err.	z	P>|z|	[95% Conf. Interval]	
_cons	5.501309	.1690433	32.54	0.000	5.16999	5.832627

Random-effects Parameters	Estimate	Std. Err.	[95% Conf. Interval]	
_all: Identity				
sd(R.sid)	.5595826	.1431873	.3388884	.9239993
pid: Identity				
sd(_cons)	.9502642	.1547838	.690552	1.307652
sid: Identity				
sd(_cons)	.4904358	.253612	.177007	1.3513
sd(Residual)	2.84385	.0353697	2.775364	2.914025

```
LR test vs. linear regression:     chi2(3) =    279.29   Prob > chi2 = 0.0000
Note: LR test is conservative and provided only for reference

. estimates store interaction
```

The estimates are presented under "Interaction" in table 9.2. We see that estimated standard deviations for the random effects of primary schools (with identifier pid) and secondary schools (with identifier sid) have decreased somewhat. The estimated standard deviation of the random interaction is of a similar magnitude to the secondary school standard deviation.

Table 9.2: Maximum likelihood estimates for crossed random-effects models for Fife data

	Additive		Interaction	
	Est	(SE)	Est	(SE)
Fixed part				
β_1	5.50	(0.17)	5.50	(0.17)
Random part				
$\sqrt{\psi_1}$ [Secondary]	0.59		0.56	
$\sqrt{\psi_2}$ [Primary]	1.06		0.95	
$\sqrt{\psi_3}$ [Primary×Secondary]			0.49	
$\sqrt{\theta}$	2.85		2.84	
Log likelihood	$-8,574.57$		$-8,573.98$	

The estimated intraclass correlations for the models with and without the random interaction are given in table 9.3, where we see that they generally do not change much after including the random interaction. The estimated intraclass correlation for students having attended the same primary school (but different secondary schools) is considerably higher than the correlation for students having attended the same secondary school (but different primary schools). Having attended the same primary school and secondary school increases the intraclass correlation somewhat compared with only having attended the same primary school. The estimated intraclass correlation due to having attended the same primary school given that a particular secondary school was attended is considerably higher than the correlation due to having attended the same secondary school given that a particular primary school was attended.

Table 9.3: Estimated intraclass correlations for Fife data

	Additive	Interaction
$\rho(\text{primary})$	0.12	0.09
$\rho(\text{secondary})$	0.04	0.03
$\rho(\text{secondary, primary})$	0.15	0.15
$\rho(\text{primary}|\text{secondary})$	0.12	0.12
$\rho(\text{secondary}|\text{primary})$	0.04	0.06

9.7.4   Testing variance components

A natural question to ask at this point is whether the random effects are needed in the model. We therefore first consider the joint null hypothesis that all variance components ψ_1, ψ_2, and ψ_3 are zero against the alternative that at least one of them is greater than zero:

$$H_0: \quad \psi_1 = \psi_2 = \psi_3 = 0 \qquad \text{against} \qquad H_a: \quad \psi_1 > 0 \quad \text{or} \quad \psi_2 > 0 \quad \text{or} \quad \psi_3 > 0$$

A test statistic of 279.29 is given as `LR test vs. linear regression:` at the bottom of the `xtmixed` output for the model with random effects for primary schools, secondary schools, and their interaction. The asymptotic null distribution of the likelihood-ratio statistic is $1/8\,\chi^2(0) + 3/8\,\chi^2(1) + 3/8\,\chi^2(2) + 1/8\,\chi^2(3)$ (see display 8.1 on page 397). The p-value becomes

```
. display 3/8*chi2tail(1,279.29) + 3/8*chi2tail(2,279.29) + 1/8*chi2tail(3,279.29)
4.656e-61
```

and we therefore reject the null hypothesis that all variance components are zero.

Having rejected the null hypothesis that no random effects are needed, we can proceed by testing whether the random interaction term ζ_{3jk} is required (in addition to the additive random effects). In other words, we consider the null hypothesis that the variance component ψ_3 for the random interaction is zero against the one-sided alternative that it is greater than zero:

$$H_0: \psi_3 = 0 \qquad \text{against} \qquad H_a: \psi_3 > 0$$

A likelihood-ratio test can be performed by using the `lrtest` command:

```
. lrtest additive interaction
Likelihood-ratio test                                LR chi2(1)  =      1.17
(Assumption: additive nested in interaction)         Prob > chi2 =    0.2802
Note: The reported degrees of freedom assumes the null hypothesis is not on
      the boundary of the parameter space.  If this is not true, then the
      reported test is conservative.
```

The asymptotic null distribution of the likelihood-ratio test statistic is $1/2\,\chi^2(0) + 1/2\,\chi^2(1)$, and the correct p-value can be obtained by simply dividing the naïve p-value reported in the output by two. Here the p-value becomes $0.28/2 = 0.14$, and the random interaction is hence not significant at the 5% level.

Assuming that there is no random interaction, we may want to test the joint null hypothesis that both additive variance components are zero against the alternative that at least one of the components is positive:

$$H_0: \quad \psi_1 = \psi_2 = 0 \qquad \text{against} \qquad H_a: \quad \psi_1 > 0 \quad \text{or} \quad \psi_2 > 0$$

The likelihood-ratio statistic is given as 278.13 under `LR test vs. linear regression:` in the output for the additive crossed random-effects model on page 447. In this case, the asymptotic null distribution of the likelihood-ratio test is $1/4\,\chi^2(0) + 1/2\,\chi^2(1) + 1/4\,\chi^2(2)$. The p-value becomes

```
. display 1/2*chi2tail(1,278.13) + 1/4*chi2tail(2,278.13)
1.102e-61
```

and the null hypothesis of no random effects of both primary and secondary school is therefore rejected.

We can also test the null hypothesis that the variance component for primary school ψ_2 is zero against the one-sided alternative that it is greater than zero:

$$H_0\!: \psi_2 = 0 \qquad \text{against} \qquad H_a\!: \psi_2 > 0$$

To perform a likelihood-ratio test of this hypothesis, we fit the model under H_0, where there is no random effect of primary school (only random effects of secondary school), and store the estimates:

```
. quietly xtmixed attain || sid:, mle
. estimates store secondary
. lrtest secondary additive
Likelihood-ratio test                            LR chi2(1)  =     182.92
(Assumption: secondary nested in additive)       Prob > chi2 =     0.0000
Note: The reported degrees of freedom assumes the null hypothesis is not on
      the boundary of the parameter space.  If this is not true, then the
      reported test is conservative.
```

The asymptotic null distribution of the likelihood-ratio test statistic is $1/2\,\chi^2(0) + 1/2\,\chi^2(1)$. The correct p-value can be obtained by simply dividing the naïve p-value based on the $\chi^2(1)$ by two. We see that the null hypothesis of no random effect of primary school is rejected.

In an analogous manner, we can test the null hypothesis that the variance component for secondary school ψ_1 is zero against the one-sided alternative that it is greater than zero:

$$H_0\!: \psi_1 = 0 \qquad \text{against} \qquad H_a\!: \psi_1 > 0$$

In this case, we get a likelihood-ratio statistic of 22.82 (output not shown), leading to rejection of the null hypothesis.

The null distributions used for the above tests of variance components are asymptotic and rely on the number of primary and secondary schools being large. They might not be unreasonable in the current example where there are 148 primary schools and 19 secondary schools. However, for the Grunfeld (1958) investment data, the tests would have been based on data for only 10 firms and 20 years, and the true null distributions are likely to be quite different from the asymptotic distributions. We therefore did not test variance components in that application. Note that it is generally not necessary to perform an extensive sequence of tests for variance components, but we did so here to show how to test different kinds of hypotheses.

9.7.5   Some diagnostics

Because the hypothesis tests conducted above suggested that a random interaction was not required, we return to the crossed random-effects model without an interaction, (9.2):

```
. estimates restore additive
(results additive are active now)
```

We can obtain empirical Bayes predictions or best linear unbiased predictions of both the secondary and the primary school random effects. If the random effects and the level-1 residual are assumed to have normal distributions, these predictions should have normal distributions. (This is true only for linear models.) We now use `predict` to obtain these predictions to check for outliers and assess normality by using a normal Q–Q plot:

```
. predict secondary primary, reffects
. qnorm secondary if pick_sid, xtitle(Quantiles of normal distribution)
> ytitle(Quantiles of empirical Bayes predictions)
. qnorm primary if pick_pid, xtitle(Quantiles of normal distribution)
> ytitle(Quantiles of empirical Bayes predictions)
```

Here we used the previously defined dummy variable `pick_sid` to choose one observation per secondary school, and similarly for primary school. If this were not done, the `qnorm` command would compute the quantiles for the sample of *all* students, which could be different from the quantiles required if the number of students per school is not constant.

The graph of the predictions for secondary schools is given in figure 9.3, and the graph for primary schools is given in figure 9.4. The figures suggest that the predictions have distributions that are reasonably close to normal, although there appears to be an outlying secondary school.

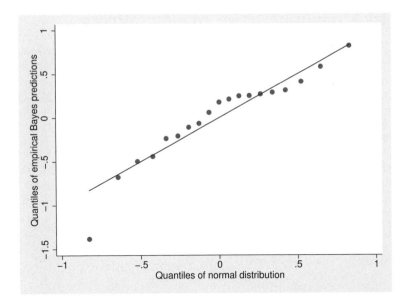

Figure 9.3: Normal Q–Q plot for secondary school predictions $\widetilde{\zeta}_{1j}$

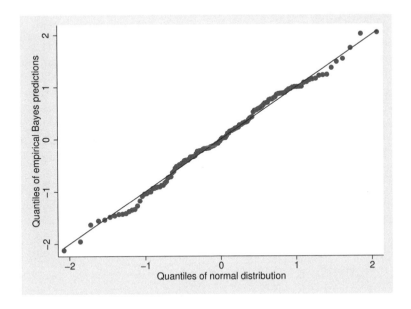

Figure 9.4: Normal Q–Q plot for primary school predictions $\widetilde{\zeta}_{2k}$

9.8 ❖ A trick requiring fewer random effects

We may have used more random effects than necessary for the primary and secondary school example. Imagine that both primary and secondary schools could be nested within regions. This would be the case if children attend both primary and secondary schools within the region in which they live and never move to a different region.

Suppose that no region has more than three secondary schools. In this case, we could arbitrarily number the secondary schools in each region from 1 to at most 3 in a variable, sec, and specify three corresponding random intercepts at the region level (with identifier `region`). The `xtmixed` syntax would be `|| region: R.sec`. Importantly, schools in different regions that happen to have the same value in sec would not have the same value of the corresponding random intercept because the intercept varies between regions. The random part for primary schools could then be specified as before, giving the `xtmixed` command (for the additive model):

```
xtmixed attain || region: R.sec || pid:, mle
```

This specification would require only three random intercepts at level 3 instead of 19.

In practice, there are no regions with insurpassable boundaries, but we could produce an identifier for a virtual level 3 within which both primary and secondary schools happen to be nested in the data. This approach becomes important if neither of the cross-classified factors has only a small number of levels and in the generalized linear mixed models discussed in volume 2 where computation may become prohibitive if there are many random effects.

The setup is shown in figure 9.5, where the students, shown as short vertical lines, can be viewed as nested in the primary schools, represented by vertical lines to their left. The students are also connected to the secondary schools to which they belong, shown as vertical lines to the right. The lines connecting students to their secondary school cross each other because primary and secondary schools are crossed. To the left of primary schools, we see the virtual level-3 units to which they belong, shown as even longer vertical lines. Both primary and secondary schools are nested within these virtual units. The random effect for primary school can be modeled by one random intercept, nested within the virtual level 3. For secondary school, the model requires 3 random coefficients of dummy variables d_{i1}, d_{i2}, and d_{i3} for the (at most) three secondary schools (numbered 1, 2, 3 in the figure) per virtual level-3 unit. The bullets show where these dummy variables take the value 1.

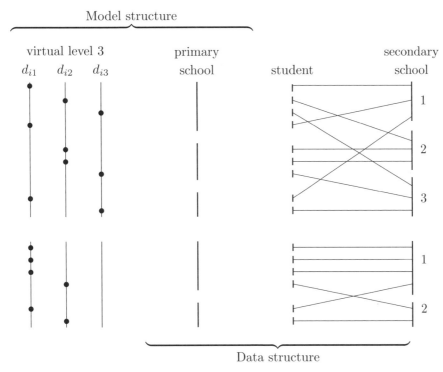

Figure 9.5: Model structure and data structure for students in primary schools crossed with secondary schools (*Source:* Skrondal and Rabe-Hesketh [2004a])

The virtual level-3 clustering variable can be produced using the command `supclust` developed by Ben Jann and available from the Statistical Software Component (SSC) archive maintained by Christopher F. (Kit) Baum. The command can be downloaded using

```
. ssc install supclust, replace
```

The syntax for generating a new variable, `region`, to serve as level-3 identifier is

```
. supclust pid sid, generate(region)
1 clusters in 3435 observations
```

Here one supercluster is found and thus the problem cannot be simplified. It is not possible to subdivide the sample into clusters within which both primary and secondary schools are nested.

To illustrate the trick, we will therefore delete primary and secondary school combinations that occur fewer than three times, which is also done in the MLwiN manual (Rasbash et al. 2009):

```
. egen num = count(attain), by(pid sid)
. drop if num<3
(168 observations deleted)
```

Now we will again try creating a virtual level-3 identifier:

```
. drop region
. supclust pid sid, gen(region)
6 clusters in 3267 observations
```

The command has identified six regions within which both `sid` and `pid` are nested. We create a new identifier for secondary schools, taking the values 1 to n_k within each region k:

```
. by region sid, sort: generate f = _n==1
. by region: generate sec = sum(f)
. table sec
```

sec	Freq.
1	1,020
2	769
3	267
4	292
5	467
6	99
7	249
8	104

There are at most 8 secondary schools per region, reducing the number of random effects required to 8 for secondary schools (compared with 19 previously) and 1 for primary schools.

In the `xtmixed` command, we simply replace `_all` with `region` and `sid` with `sec`:

```
. xtmixed attain || region: R.sec || pid:, mle
Mixed-effects ML regression                     Number of obs      =      3267
```

Group Variable	No. of Groups	Observations per Group		
		Minimum	Average	Maximum
region	6	78	544.5	1330
pid	135	3	24.2	72

```
                                        Wald chi2(0)       =         .
Log likelihood = -8153.6587             Prob > chi2        =         .
```

attain	Coef.	Std. Err.	z	P>|z|	[95% Conf. Interval]	
_cons	5.581939	.1812378	30.80	0.000	5.22672	5.937159

Random-effects Parameters	Estimate	Std. Err.	[95% Conf. Interval]	
region: Identity				
sd(R.sec)	.6196652	.1546558	.3799396	1.010647
pid: Identity				
sd(_cons)	1.050124	.0982812	.8741311	1.261551
sd(Residual)	2.84682	.0360606	2.777012	2.918382

```
LR test vs. linear regression:      chi2(2) =    263.22   Prob > chi2 = 0.0000
Note: LR test is conservative and provided only for reference
```

The estimates are not identical to the previous estimates because we dropped a small proportion of the observations. We would not generally recommend dropping any observations but did so here to illustrate the trick.

9.9   Summary and further reading

In this chapter, we have discussed crossed random-effects models for data with two crossed clustering variables, or cross-classified random factors. We have described models with additive random effects of the factors, as well as a model including a random interaction between factors. The latter is identified only if there are several observations for at least some of the combinations of the categories of the factors.

Books on multilevel models with useful chapters on crossed random-effects models include Snijders and Bosker (2012, chap. 13), Raudenbush and Bryk (2002, chap. 12), and Goldstein (2011, chap. 12). Baltagi (2008, chap. 3) provides an econometric perspective. We also recommend the excellent articles and chapter by Raudenbush (1993), Browne, Goldstein, and Rasbash (2001), and Rasbash and Browne (2001).

We have considered two typical examples requiring crossed random effects: longitudinal or panel data with random occasion or time effects and data on individuals nested in two types of institutions. Further longitudinal exercises are considered in exercises 9.2, 9.3, and 9.9. For longitudinal data, time-series–cross-sectional analysis for long panels (discussed in section 6.7) can also be used to relax the assumption that units are uncorrelated at a given occasion (see exercise 9.9). The Fife data are revisited in exercise 9.1. Exercise 9.5 considers children nested in neighborhoods and schools, and exercise 9.6 uses data from an agricultural experiment.

Another typical example where several raters (the first random factor) rate each of several objects (the second random factor) is discussed in exercises 9.4 and 9.7. The latter exercise introduces some of the basic ideas of generalizability theory. Slightly more elaborate versions of models with crossed random effects are often applied to social-network data, where each individual may rate how much they like every other individual. The crossed factors are then the senders and the receivers of the ratings, where each person is both a sender and a receiver.

There were only two random factors in the examples considered in this chapter. In exercise 9.3, we consider a problem with three random factors: occasions, states, and regions. Although states are nested in regions, both states and regions are cross-classified with occasions. In exercise 9.8, state of birth and state of residence are crossed, and mothers are nested within the cross-classifications.

Models with crossed random effects must be distinguished from *multiple membership models*. Both types of models share the common feature that the classifications represented by random effects are not nested. In multiple membership models, a unit is a member of several clusters from the *same* classification with known weights designating the degree of membership. For instance, a student may have been taught by several teachers with weights representing the time spent with each teacher (see exercise 9.10). Browne, Goldstein, and Rasbash (2001), Rasbash and Browne (2001), and Goldstein (2011, chap. 13) discuss multiple membership models.

9.10   Exercises

9.1 Fife school data

Here we revisit the Fife school data analyzed in this chapter and described on page 443. In addition to the variables listed there, the dataset `fife.dta` contains a verbal reasoning score, `vrq`, which was considered a measure of ability by Paterson (1991).

1. Fit the model with the same random part as in (9.2) but with covariates `sex` and `vrq`. Use ML estimation.
2. Interpret the estimates, and discuss the change in the estimated variance components.

9.2 Airline cost data

Greene (2012) provides data on the annual total costs and output, fuel price, and load factor for six U.S. airlines over 15 years from 1970 to 1984. The data were provided to Greene by Professor Moshe Kim.

The variables in `airlines.dta` are

- `airline`: airline identifier
- `year`: year number (1–15)
- `cost`: total annual cost in U.S. $1,000
- `output`: annual output, in revenue passenger miles, index number
- `fuelprice`: fuel price
- `loadf`: load factor, the average capacity utilization of the fleet

Here the fuel price differs between airlines in a given year because different airlines use different mixes of types of planes and because there are regional differences in supply characteristics.

1. Write down a regression model for the log cost regressed on the log output, log fuel price, and load factor with random effects for airlines and years.
2. Fit the model using `xtmixed` with the `mle` and `reml` options.
3. Compare the estimates and explain why they differ.
4. Based on the REML estimates, interpret the effect of fuel price as an elasticity (see display 6.2).
5. Obtain empirical Bayes predictions of the random effects based on the REML estimates.
6. Plot the empirical Bayes predictions for the years against time.
7. Fit the model using fixed effects for years instead of random effects. Explain how and why the estimated coefficient of log fuel price changes compared with the model that treats year as random.

9.3 U.S. production data

The data are from Munnell (1990) and were described in exercise 8.3, where we considered a three-level model for state productivity.

1. Fit the three-level model described in exercise 8.3 using `xtmixed` with the `reml` option.
2. Add a random effect for year to this model. Choose the model specification in `xtmixed` that minimizes the number of random effects at a level. Fit the model by REML.
3. Compare the models using a likelihood-ratio test.
4. Interpret the variance-component estimates for the model in step 2.

9.4 Video-ratings data

The Vancouver Sedative Recovery Scale (VSRS) was developed to measure recovery from sedation following pediatric open heart surgery. Macnab et al. (1994) report a study where 16 ICU staff were trained by videotape instruction to use the VSRS. To determine whether videotaped instruction produced adequate skill, an interobserver reliability study was carried out.

In a balanced incomplete design, 16 staff each rated a different subset of 16 videotaped case examples using the VSRS so that each case was rated by six raters and each rater rated six cases. Two experts also rated all cases. The data, shown in table 9.4, were also analyzed by Dunn (2004).

Table 9.4: Rating data for 16 cases in incomplete block design

Raters	Cases or Videos															
	1	2	3	4	5	6	7	8	9	10	11	12	13	14	15	16
1	.	.	.	6	.	.	.	0	18	.	12	.	14	.	.	19
2	.	.	.	5	.	4	.	.	13	13	.	0	.	.	16	.
3	.	.	20	.	13	1	.	0	15	.	.	.	.	15	.	.
4	9	.	.	.	.	.	22	.	19	18	.	.	14	14	.	.
5	11	0	21	.	.	.	.	.	19	.	10	.	.	.	19	.
6	.	1	.	.	13	.	22	.	16	.	.	0	.	.	.	22
7	12	0	.	0	.	1	22	0	.	.	.	.	.	.	.	.
8	.	1	19	5	14	.	.	.	.	15	.	.	14	.	.	.
9	.	.	.	9	16	.	15	.	.	.	10	.	.	12	19	.
10	10	.	17	7	.	.	.	.	.	.	.	0	.	13	.	20
11	.	0	.	.	.	.	.	0	.	.	.	0	13	16	22	.
12	.	.	20	.	.	.	20	0	.	12	10	0	.	.	.	.
13	6	.	.	.	15	.	.	0	.	17	.	.	.	.	19	22
14	8	.	.	.	16	1	.	.	.	.	11	0	15	.	.	.
15	.	.	21	.	.	1	18	.	.	.	.	.	16	.	21	22
16	.	0	.	.	.	1	.	.	.	18	15	.	.	10	.	21
17	10	0	21	8	18	4	16	0	18	16	10	0	16	12	19	20
18	10	0	21	5	19	7	18	0	16	15	12	0	14	14	20	21

Source: Macnab et al. (1994)

The long version of these data in `videos.dta` contains the following variables:

- `rater`: rater identifier (j)
- `video`: video (case example) identifier (i)
- `y`: VSRS score
- `novice`: dummy variable for rater being a novice (not an expert)

In generalizability theory, the design of this study can be viewed as a one-facet crossed design. Here the *objects of the measurement* are the cases shown in the videos, and these are crossed with raters, who can be thought of as *conditions* of a *facet* of the measurement.

We wish to generalize from this particular set of raters to a *universe* of raters, treating raters as a random sample of all *admissible* raters. Denoting cases or videos as i and raters as j, the model can be written as

$$y_{ij} = \beta + \zeta_{1i} + \zeta_{2j} + \epsilon_{ij}, \quad \psi_1 \equiv \text{Var}(\zeta_{1i}), \psi_2 \equiv \text{Var}(\zeta_{2j}), \theta \equiv \text{Var}(\epsilon_{ij})$$

Here ϵ_{ij} represents the sum of the case by rater interaction and any other sources of error. A good book on generalizability theory is Shavelson and Webb (1991).

1. Fit this model for the nonexperts using REML.
2. Interchange `rater` and `video` in your command for step 1 and compare the estimates.
3. If the intention is to use the VSRS score from one rater for an *absolute decision*, not just for the purpose of ranking cases rated by the same rater, the generalizability coefficient is defined as

$$\phi = \frac{\psi_1}{\psi_1 + \psi_2 + \theta}$$

   where $\psi_2 + \theta$ represents the measurement error variance for absolute decisions. Obtain an estimate of this coefficient by plugging in the estimated variance components from step 1 or 2.
4. If the intention is to use the mean score from N_r raters for an absolute decision, the measurement error variance can be divided by N_r and the generalizability coefficient is defined as

$$\phi = \frac{\psi_1}{\psi_1 + (\psi_2 + \theta)/N_r}$$

   Estimate this generalizability coefficient for $N_r = 6$.

9.5 Neighborhood-effects data ⸢Solutions⸣

We now consider the data from Garner and Raudenbush (1991), Raudenbush and Bryk (2002), and Raudenbush et al. (2004) previously analyzed in exercises 2.4 and 3.1. In exercise 2.4, we considered two separate models, one with a random intercept for schools and the other with a random intercept for neighborhoods. Here we will include both random intercepts in the same model.

The variables in `neighborhood.dta` are

- Student level
  - `attain`: a measure of educational attainment

  - p7vrq: primary seven (year 7 of primary school) verbal reasoning quotients
  - p7read: primary seven reading test scores
  - dadocc: father's occupation scaled on the Hope–Goldthorpe scale in conjunction with the Registrar General's social-class index (Willms 1986)
  - dadunemp: dummy variable for father being unemployed (1: unemployed; 0: not unemployed)
  - daded: dummy variable for father's schooling being past the age of 15
  - momed: dummy variable for mother's schooling being past the age of 15
  - male: dummy variable for student being male
- Neighborhood level
  - neighid: neighborhood identifier
  - deprive: social deprivation score, derived from poverty concentration, health, and housing stock of local community
- School level
  - schid: school identifier

1. Fit a model for student educational attainment without covariates but with random intercepts of neighborhood and school by ML.
2. Include a random interaction between neighborhood and school, and use a likelihood-ratio test to decide whether the interaction should be retained (use a 5% level of significance).
3. Include the neighborhood-level covariate deprive. Discuss both the estimated coefficient of deprive and the changes in the estimated standard deviations of the random effects due to including this covariate.
4. Remove the neighborhood-by-school random interaction (which is no longer significant at the 5% level) and include all student-level covariates. Interpret the estimated coefficients and the change in the estimated standard deviations.
5. For the final model, estimate the residual intraclass correlations due to being in
   a. the same neighborhood but not the same school
   b. the same school but not the same neighborhood
   c. both the same neighborhood and the same school
6. ❖ Use the supclust command to see if estimation can be simplified by defining a virtual level-3 identifier.

9.6 Nitrogen data

Littell et al. (2006) describe an experiment to evaluate the yield of two varieties of crop at five levels of nitrogen fertilization. The five levels of nitrogen were applied to 15 relatively large whole plots that formed a 3×5 grid. Because of substantial north–south and east–west gradients, the levels of nitrogen were applied in an incomplete Latin-square design, with each nitrogen level occurring in each row, but not in each column, as shown in table 9.5.

Table 9.5: Latin-square design for nitrogen fertilization experiment

	Col 1	Col 2	Col 3	Col 4	Col 5
Row 1	Nit 1	Nit 2	Nit 5	Nit 4	Nit 3
Row 2	Nit 2	Nit 1	Nit 3	Nit 5	Nit 4
Row 3	Nit 3	Nit 4	Nit 1	Nit 2	Nit 5

Each whole plot was split into two subplots to which the two crops were randomly assigned.

Letting i denote the subplots, j the rows, and k the columns in which the whole plots are arranged, a crossed random-effects model for crop yield y_{ijk} can be written as

$$y_{ijk} = \beta_1 + \zeta_{1j} + \zeta_{2k} + \zeta_{3jk} + \epsilon_{ijk}$$

where ζ_{1j} is the random effect of row, ζ_{2k} is the random effect of column, ζ_{3jk} is the random interaction between row and column, and ϵ_{ijk} is a residual error term. Note that the number of clusters is arguably too small for estimating variance components.

The variables in `nitrogen.dta` are

- `row`: row identifier (j)
- `col`: column identifier (k)
- `N`: nitrogen level (1, 2, 3, 4, 5)
- `G`: genotype or crop variety (1, 2)
- `y`: yield of the crop (y_{ijk})

1. Fit the model given above using REML. Interpret the estimated variance components.
2. Fit a model with the same random part as in step 1 but also including fixed effects of nitrogen level (treated as unordered), genotype, and their interaction.
3. Perform a Wald test for the nitrogen by genotype interaction. Omit the interaction terms if this test is not significant at the 5% level.
4. Interpret the estimated regression coefficients.

9.7 Olympic skating data

In the 1932 Lake Placid Winter Olympics, seven figure skating pairs were judged by seven judges using two different criteria (program and performance). The ratings are provided by Gelman and Hill (2007) and are shown in table 9.6, where the countries of origin of judges and pairs are also given as abbreviations for France, the United States, Hungary, Canada, Norway, Austria, Finland, and the United Kingdom.

Table 9.6: Ratings of seven skating pairs by seven judges using two criteria (program and performance) in the 1932 Winter Olympics

	Judge													
	1 (Hun)		2 (Nor)		3 (Aus)		4 (Fin)		5 (Fra)		6 (UK)		7 (US)	
Pair														
1 (Fra)	5.6	5.6	5.5	5.5	5.8	5.8	5.3	4.7	5.6	5.7	5.2	5.3	5.7	5.4
2 (US)	5.5	5.5	5.2	5.7	5.8	5.6	5.8	5.4	5.6	5.5	5.1	5.3	5.8	5.7
3 (Hun)	6.0	6.0	5.3	5.5	5.8	5.7	5.0	4.9	5.4	5.5	5.1	5.2	5.3	5.7
4 (Hun)	5.6	5.6	5.3	5.3	5.8	5.8	4.4	4.8	4.5	4.5	5.0	5.0	5.1	5.5
5 (Can)	5.4	4.8	4.5	4.8	5.8	5.5	4.0	4.4	5.5	4.6	4.8	4.8	5.5	5.2
6 (Can)	5.2	4.8	5.1	5.6	5.3	5.0	5.4	4.7	4.5	4.0	4.5	4.6	5.0	5.2
7 (US)	4.8	4.3	4.0	4.6	4.7	4.5	4.0	4.0	3.7	3.6	4.0	4.0	4.8	4.8

The data in `olympics.dta` contain one row for each combination of skating pair and judge with separate variables for the two criteria. The variables are

- `pair`: skating pair
- `judge`: judge
- `pcountry`: country of origin of pair
- `jcountry`: country of origin of judge
- `program`: rating for program (criterion 1)
- `performance`: rating for performance (criterion 2)

1. Write down a linear model for the program rating of judge j for pair k that includes a fixed overall intercept and random intercepts for judges and pairs. Interpret the random intercepts in the context of this example.

2. Fit the model from step 1 using REML.

3. Test whether both random intercepts are needed by comparing the model from step 1 with the two nested models having only one random intercept (using a 5% level of significance). Note, however, that the number of clusters is really too small to rely on these asymptotic tests.

4. Extend the model by constructing a dummy variable for judge and pair coming from the same country and including this as a covariate. Interpret the estimated regression coefficient and comment on its magnitude.

5. ❖ Consider joint models for program and performance ratings.

   a. Reshape the data to long form, stacking the ratings using the two criteria into one response variable, and produce a dummy variable for the criterion being performance.
   b. Fit the same model as in step 4 to the responses using both criteria.
   c. The model above makes these strong assumptions (among others):
      - The mean rating is the same for the two criteria after controlling for whether the pair and the judge are from the same country

- The ratings for the same skating pair from the same judge using two different criteria are conditionally independent given the covariate and the random intercepts for judges and pairs

Fit an extended model that relaxes these assumptions.

9.8 Smoking and birthweight data

We now consider the data from Abrevaya (2006) that were used in chapter 3. There we ignored the fact that mothers are nested in U.S. states of residence crossed with the states in which they were born. Here we consider the effects of the mother's state of residence and state of birth.

The variables in `smoking.dta` that we will use here are

- `momid`: mother identifier
- `birwt`: birthweight (in grams)
- `stateres`: mother's state of residence
- `mplbir`: mother's place (state) of birth
- `male`: dummy variable for child being male

1. Fit a model for birthweight with random effects for state of residence and state of birth and a fixed effect of `male`.
2. Fit the same model as in step 1 but also include a random interaction between state of residence and state of birth. Compare this model with the model from step 1 using a likelihood-ratio test. Use a 5% level of significance to decide which model to retain.
3. Fit the model from step 2 but with an additional random intercept for mothers. Mothers are nested within the combinations of state of residence and state of birth (mothers who moved between births could not be matched and were not included in the data). Compare this model with the model from step 2 using a likelihood-ratio test.
4. Interpret the parameter estimates for the model chosen in step 3.

9.9 Cigarette-consumption data

The dataset for this exercise concerns cigarette consumption from 1963 to 1992 in 48 U.S. states and is from Baltagi, Griffin, and Xiong (2000). In exercise 6.4, this long panel dataset was described and analyzed using the `xtpcse` and `xtgls` commands to accommodate cross-sectional dependence between states. Here we will consider a two-way error-components model instead.

1. The consumer price index is a weighted average of prices of a basket of consumer goods and services. In this dataset, `CPI` is 100 times the consumer price index in a given year divided by the consumer price index in 1983. Convert `price`, `pimin`, and `NDI` to 1983 U.S. dollars using the consumer price index. Such real prices and real income are adjusted for inflation over years. In the analyses below, you will use the natural logarithms of real prices and real disposable income.

2. Fit a model to the logarithm of the number of packs of cigarettes sold per person of smoking age with a random effect of state, a random effect of year, and the following covariates: the logarithm of real cigarette price, the logarithm of real disposable income, the logarithm of the real minimum price in adjoining states (a proxy for casual smuggling across states), and year (treated as continuous).

3. Calculate the estimated longitudinal within-state intraclass correlation and the estimated cross-sectional within-year intraclass correlation.

4. Extend the model by allowing the level-1 residuals to follow an AR(1) process over time.

5. Use a likelihood-ratio test to decide (at the 5% level) whether the AR(1) process is needed.

6. Can the random part of the model selected in step 5 be simplified?

7. ❖ Derive an expression for the longitudinal within-state correlation as a function of the time lag between occasions. What are the corresponding estimated lag-1 and lag-29 correlations?

9.10 ❖ STAR data

In this exercise, we consider multiple membership models for the STAR study used in exercises 8.6 and 8.7 and documented in Finn et al. (2007). We will analyze the dataset star_mm.dta, which was derived from star1.dta. The dataset contains only the second-grade data for the subset of children who were in the study from kindergarten through second grade; remained in the same school; have teacher information for kindergarten, first, and second grade; and have a reading score in second grade.

The variables in star_mm.dta we will analyze here are

- schid school identifier
- reading: total score for Stanford achievement test (SAT) in reading
- wttr1–wttr30: weight variables for multiple membership model
- gr0wttr1–gr0wttr8 weight variables for kindergarten teacher
- gr1wttr1–gr1wttr12 weight variables for grade 1 teacher
- gr2wttr1–gr2wttr11 weight variables for grade 2 teacher

By second grade, the children have been taught reading for three years. It is possible that reading achievement depends on which teachers the child has had, in addition to which school the child is in. Each child has been taught by at most three teachers (and it turns out in this dataset, each teacher teaches only one grade so each child has been taught by three different teachers). We will let $j(ikg)$ denote the teacher j who taught child i in school k in grade g ($g = 0, 1, 2$ for kindergarten, first grade, second grade, respectively).

Then a multiple membership model can be written as

$$y_{ik} = \beta + \sum_{g=0}^{2} \frac{1}{3} \zeta_{j(ikg)}^{(2)} + \zeta_k^{(3)} + \epsilon_{ik}, \qquad (9.4)$$

where $\frac{1}{3}\zeta_{j(ikg)}^{(2)}$ is the contribution of teacher $j(ikg)$ to the total effect of teachers on student ik's reading attainment, $\zeta_k^{(3)}$ is the effect of school k on reading, and ϵ_{ik} is a child-level residual. Each child is a member of at most three different teachers and each year of teaching contributes $1/3$ to the overall teacher effect. The teacher random intercepts have variance $\psi^{(2)}$, and the school random intercepts have variance $\psi^{(3)}$. All random intercepts are mutually uncorrelated and uncorrelated across schools and teachers.

In order to fit the model in xtmixed, we label the teachers within each school from 1 to n_k, where in this dataset $n_k \leq 30$ in all schools. We define 30 school-level random coefficients $u_{aik}^{(3)}$, $a = 1, \ldots, 30$, for teachers, along with 30 weight variables w_{aik} taking the value $1/3$ for child ik if the corresponding teacher ever taught that child and zero otherwise. The model can then be written as

$$y_{ik} = \beta + \sum_{a=1}^{30} w_{aik} u_{aik}^{(3)} + \zeta_k^{(3)} + \epsilon_{ik}$$

1. The required weight variables w_{aik} are called wttr1 to wttr30. What should be the sum of these variables for each child? Verify that this is correct. Also list the data to make sure you understand the weight variables.

2. Fit the model in xtmixed. Keep in mind that the random coefficients $u_{aik}^{(3)}$ are uncorrelated with identity covariance structure and that you can specify two random parts for schools.

3. Interpret the estimates.

4. The model makes the strong assumption that the kindergarten, first-grade, and second-grade teachers all contribute equally to the child's reading in second grade. However, the teachers encountered the children at different stages of development, and the effect of previous teachers may fade over time. The weight variables gr0wttr1–gr0wttr8, gr1wttr1–gr1wttr12, and gr2wttr1–gr2wttr11 are analogous to the weight variables used in step 1 except that they are specific to the kindergarten, grade 1, and grade 2 teachers, respectively (in this dataset, no teacher taught more than one grade, with grade identified by the third character of the variable names). There were at most 8 different kindergarten teachers, 12 different first-grade teachers, and 11 different second-grade teachers per school. For child ik, gr0wttr1 takes the value $1/3$ if the child was taught by the first kindergarten teacher of school k and zero otherwise, and analogously for the other teachers and grades.

a. Confirm that each set of variables takes the value $1/3$ once for each child.

b. Fit a model identical to the multiple membership model in (9.4) except that $\zeta_{j(ikg)}^{(2)}$ has grade-specific variance $\psi_g^{(2)}$. This can be achieved by using eight random coefficients of gr0wttr1–gr0wttr8 for kindergarten, and similarly for first and second grade, again keeping in mind that you can specify separate random parts to obtain different variances.

c. Interpret the estimates.

d. Compare the model with the model fit in step 2 using a likelihood-ratio test.

e. Can the selected model be simplified?

9.11 Different kinds of intraclass correlations

In this exercise, you will derive some of the intraclass correlations given in section 9.7.2.

1. Derive $\rho(\text{secondary})$.

2. Derive $\rho(\text{secondary}, \text{primary})$.

3. Derive $\rho(\text{primary}|\text{secondary})$.

A Useful Stata commands

Here we list commands and special options that are useful for handling multilevel and longitudinal data. These commands and options have all been illustrated in the book and are listed in the subject index.

Here we assume that the response variable is y, a covariate is x, and the cluster identifier is cluster. Sometimes we will assume that the data have three levels with repeated observations on subjects (identifier subject) nested in clusters and time variable year. We will not assume that subject takes on unique values across clusters, that is, subjects may be numbered from 1 within each cluster.

by — Repeat Stata command on subsets of the data

- Create a unit identifier, counting from 1 within each cluster (in ascending order of x) in two-level data:
  by cluster (x), sort: generate *varname* = _n
- As above, but counting from 1 within each subject in three-level data:
  by cluster subject (x), sort: generate *varname* = _n
- Create a lagged variable within the subjects:
  by cluster subject (year), sort: ///
  generate *varname2* = *varname1*[_n-1]
  See also Stata's time-series operator L, under help tsvarlist

egen — Extensions to generate

- Form cluster means:
  egen *varname2* = mean(*varname1*), by(cluster)
- Form cluster sizes:
  egen *varname2* = count(*varname1*), by(cluster)
- Construct variable taking the value 1 for one unit per cluster:
  egen *varname* = tag(cluster)
- Construct cluster identifier consisting of consecutive integers:
  egen *varname* = group(cluster)
- Construct unique subject identifier in three-level data:
  egen *varname* = group(cluster subject)

graph twoway — Two-way graphs

- Plot cluster-specific regression lines (spaghetti plot): After sorting by cluster and x, use twoway line with the connect(ascending) or connect(L) option (see also xtline command)

- Make a trellis graph of two-way plots for the individual clusters: Use the `twoway` command with the `by(cluster)` option; `by(cluster, compact, cols(5))` produces a compact trellis with graphs arranged in five columns

`merge` — Merge datasets

- Combine individual-level data `lev1.dta` (level-1 variables) with cluster-level data `lev2.dta` (level-2 variables). First sort both files by `cluster`
    - Read `lev2.dta` and use command
      `merge 1:m cluster using lev1.dta`
    - Read `lev1.dta` and use command
      `merge m:1 cluster using lev2.dta`

`reshape` — Convert data from wide to long form and vice versa

- Convert data from wide form (one line per cluster; variables `y1 y2 y3` are responses for units 1, 2, 3) to long form (multiple lines per cluster, with responses in different rows of one variable `y`), and new variable `occ` taking the values 1, 2, 3 to identify the values that were in the variables `y1 y2 y3`:
  `reshape long y, i(cluster) j(occ)`
- Convert data from long form to wide form:
  `reshape wide y, i(cluster) j(occ)`
- If variable names are `bygog f1gog f2gog` for base year, follow-up 1, and follow-up 2, respectively, use
  `reshape long @gog, i(cluster) j(occ) string`
  Here `@` is the position of the string variables that indicate the panel wave, and the `string` option is necessary when the panel wave labels are not numeric

`statsby` — Collect statistics for a command across a by-list

- Syntax: `statsby` *exp_list*, `by(`*varname*`):` *command*
- Save intercepts and slopes of least-squares regression fit to each cluster:
  `statsby b[_cons] b[x], by(cluster): regress y x, ///`
  `saving(estimates)`

`xtdescribe` — Describe missingness pattern of xt data

- Find missing-data pattern for balanced longitudinal data (assume every value of `year` that occurs for anyone should have occurred for everyone and `subject` takes a unique value for each subject):
  `xtset subject year`
  `xtdescribe if y<.`

`xtsum` — Summarize xt data

- Obtain within- and between-cluster standard deviations:
  `xtset cluster`
  `xtsum y x`

References

Abrevaya, J. 2006. Estimating the effect of smoking on birth outcomes using a matched panel data approach. *Journal of Applied Econometrics* 21: 489–519.

Acock, A. C. 2010. *A Gentle Introduction to Stata.* 3rd ed. College Station, TX: Stata Press.

Adams, M. M., H. G. Wilson, D. L. Casto, C. J. Berg, J. M. McDermott, J. A. Gaudino, and B. J. McCarthy. 1997. Constructing reproductive histories by linking vital records. *American Journal of Epidemiology* 145: 339–348.

Agresti, A., and B. Finlay. 2007. *Statistical Methods for the Social Sciences.* 4th ed. Englewood Cliffs, NJ: Prentice Hall.

Allison, P. D. 1995. *Survival Analysis Using SAS: A Practical Guide.* Cary, NC: SAS Institute.

———. 2005. *Fixed Effects Regression Methods for Longitudinal Data Using SAS.* Cary, NC: SAS Institute.

Amemiya, T., and T. E. MaCurdy. 1986. Instrumental-variable estimation of an error-components model. *Econometrica* 54: 869–880.

Anderson, T. W., and C. Hsiao. 1981. Estimation of dynamic models with error components. *Journal of the American Statistical Association* 76: 598–606.

———. 1982. Formulation and estimation of dynamic models using panel data. *Journal of Econometrics* 18: 47–82.

Arellano, M., and S. Bond. 1991. Some test of specification for panel data: Monte Carlo evidence and an application to employment equations. *Review of Economic Studies* 58: 277–297.

Baltagi, B. H. 2008. *Econometrics of Panel Data.* 4th ed. Chichester, UK: Wiley.

Baltagi, B. H., J. M. Griffin, and W. Xiong. 2000. To pool or not to pool: Homogeneous versus heterogeneous estimators applied to cigarette demand. *Review of Economics and Statistics* 82: 117–126.

Baltagi, B. H., S. H. Song, and B. C. Jung. 2001. The unbalanced nested error component regression model. *Journal of Econometrics* 101: 357–381.

Balzer, W., N. Boudreau, P. Hutchinson, A. M. Ryan, T. Thorsteinson, J. Sullivan, R. Yonker, and D. Snavely. 1996. Critical modeling principles when testing for gender equity in faculty salary. *Research in Higher Education* 37: 633–658.

Bandini, L. G., A. Must, J. L. Spadano, and W. H. Dietz. 2002. Relation of body composition, parental overweight, pubertal stage, and race-ethnicity to energy expenditure among premenarcheal girls. *American Journal of Clinical Nutrition* 76: 1040–1047.

Battese, G. E., R. M. Harter, and W. A. Fuller. 1988. An error-components model for prediction of county crop areas using survey and satellite data. *Journal of the American Statistical Association* 83: 28–36.

Beck, N., and J. N. Katz. 1995. What to do (and not to do) with time-series cross-section data. *American Political Science Review* 89: 634–647.

Begg, M. D., and M. K. Parides. 2003. Separation of individual-level and cluster-level covariate effects in regression analysis of correlated data. *Statistics in Medicine* 22: 2591–2602.

Bingenheimer, J. B., and S. W. Raudenbush. 2004. Statistical and substantive inferences in public health: Issues in the application of multilevel models. *Annual Review of Public Health* 25: 53–77.

Bland, J. M., and D. G. Altman. 1986. Statistical methods for assessing agreement between two methods of clinical measurement. *Lancet* 327: 307–310.

Bliese, P. 2009. Multilevel Modeling in R (2.3): A brief introduction to R, the multilevel package and the nlme package. http://cran.r-project.org/doc/contrib/Bliese_Multilevel.pdf.

Bliese, P. D., and R. R. Halverson. 1996. Individual and nomothetic models of job stress: An examination of work hours, cohesion, and well-being. *Journal of Applied Social Psychology* 26: 1171–1189.

Bollen, K. A., and P. J. Curran. 2006. *Latent Curve Models: A Structural Equation Perspective*. Hoboken, NJ: Wiley.

Boot, J. C. G., and G. M. de Wit. 1960. Investment demand: An empirical contribution to the aggregation problem. *International Economic Review* 1: 3–30.

Borenstein, M., L. V. Hedges, J. P. T. Higgins, and H. R. Rothstein. 2009. *Introduction to Meta-Analysis*. Chichester, UK: Wiley.

Boudreau, N., J. Sullivan, W. Balzer, A. M. Ryan, R. Yonker, T. Thorsteinson, and P. Hutchinson. 1997. Should faculty rank be included as a predictor variable in studies of gender equity in university faculty salaries? *Research in Higher Education* 38: 297–312.

Broota, K. D. 1989. *Experimental Design in Behavioural Research*. New Delhi: New Age International.

Brown, H., and R. I. Prescott. 2006. *Applied Mixed Models in Medicine.* 2nd ed. Chichester, UK: Wiley.

Browne, W. J., H. Goldstein, and J. Rasbash. 2001. Multiple membership multiple classification (MMMC) models. *Statistical Modelling* 1: 103–124.

Bryk, A. S., and S. W. Raudenbush. 1987. Application of hierarchical linear models to assessing change. *Psychological Bulletin* 101: 147–158.

Cameron, A. C., and P. K. Trivedi. 2005. *Microeconometrics: Methods and Applications.* Cambridge: Cambridge University Press.

Caudill, S. B., J. M. Ford, and D. L. Kaserman. 1995. Certificate-of-need regulation and the diffusion of innovations: A random coefficient model. *Journal of Applied Econometrics* 10: 73–78.

Cornwell, C., and P. Rupert. 1988. Efficient estimation with panel data: An empirical comparison of instrumental variables estimators. *Journal of Applied Econometrics* 3: 149–155.

Crowder, M. J., and D. J. Hand. 1990. *Analysis of Repeated Measures.* London: Chapman & Hall/CRC.

Curran, P. J., E. Stice, and L. Chassin. 1997. The relation between adolescent alcohol use and peer alcohol use: A longitudinal random coefficients model. *Journal of Consulting and Clinical Psychology* 65: 130–140.

Davis, C. S. 2002. *Statistical Methods for the Analysis of Repeated Measurements.* New York: Springer.

De Boeck, P., and M. Wilson. ed. 2004. *Explanatory Item Response Models: A Generalized Linear and Nonlinear Approach.* New York: Springer.

DeMaris, A. 2004. *Regression with Social Data: Modeling Continuous and Limited Response Variables.* Hoboken, NJ: Wiley.

Demidenko, E. 2004. *Mixed Models: Theory and Applications.* New York: Wiley.

Dempster, A. P., C. M. Patel, M. R. Selwyn, and A. J. Roth. 1984. Statistical and computational aspects of mixed model analysis. *Journal of the Royal Statistical Society, Series C* 33: 203–214.

Diez Roux, A. V. 2002. A glossary for multilevel analysis. *Journal of Epidemiology and Community Health* 56: 588–594.

Diggle, P. J., P. J. Heagerty, K.-Y. Liang, and S. L. Zeger. 2002. *Analysis of Longitudinal Data.* 2nd ed. Oxford: Oxford University Press.

Dohoo, I. R., W. Martin, and H. Stryhn. 2010. *Veterinary Epidemiologic Research.* 2nd ed. Charlottetown, Canada: VER Inc.

Dohoo, I. R., E. Tillard, H. Stryhn, and B. Faye. 2001. The use of multilevel models to evaluate sources of variation in reproductive performance in dairy cattle in Reunion Island. *Preventive Veterinary Medicine* 50: 127–144.

Duncan, C., K. Jones, and G. Moon. 1998. Context, composition and heterogeneity: Using multilevel models in health research. *Social Science & Medicine* 46: 97–117.

Dunn, G. 1992. Design and analysis of reliability studies. *Statistical Methods in Medical Research* 1: 123–157.

————. 2004. *Statistical Evaluation of Measurement Errors: Design and Analysis of Reliability Studies*. London: Arnold.

Dupuy, H. J. 1978. Self-representations of general psychological well-being of American adults. Paper presented at the American Public Health Association Meeting, Los Angeles.

Ebbes, P., U. Böckenholt, and M. Wedel. 2004. Regressor and random-effect dependencies in multilevel models. *Statistica Neerlandica* 58: 161–178.

Everitt, B. S. 1995. The analysis of repeated measures: A practical review with examples. *Statistician* 44: 113–135.

Finn, J. D., J. Boyd-Zaharias, R. M. Fish, and S. B. Gerber. 2007. *Project STAR and Beyond: Database User's Guide*. Lebanon, TN: HEROS.

Fitzmaurice, G. M. 1998. Regression models for discrete longitudinal data. In *Statistical Analysis of Medical Data: New Developments*, ed. B. S. Everitt and G. Dunn, 175–201. London: Arnold.

Fitzmaurice, G. M., N. M. Laird, and J. H. Ware. 2011. *Applied Longitudinal Analysis*. 2nd ed. Hoboken, NJ: Wiley.

Fox, J. 1997. *Applied Regression Analysis, Linear Models, and Related Methods*. Thousand Oaks, CA: Sage.

Frees, E. W. 2004. *Longitudinal and Panel Data: Analysis and Applications in the Social Sciences*. Cambridge: Cambridge University Press.

Frets, G. P. 1921. Heredity of headform in man. *Genetica* 3: 193–400.

Garner, C. L., and S. W. Raudenbush. 1991. Neighborhood effects on educational attainment: A multilevel analysis. *Sociology of Education* 64: 251–262.

Garrett, G. 1998. *Partisan Politics in the Global Economy*. Cambridge: Cambridge University Press.

Gelman, A., and J. Hill. 2007. *Data Analysis Using Regression and Multilevel/Hierarchical Models*. Cambridge: Cambridge University Press.

Goldberg, D. P. 1972. *The Detection of Psychiatric Illness by Questionnaire.* Oxford: Oxford University Press.

Goldstein, H. 1987. Multilevel covariance component models. *Biometrika* 74: 430–431.

———. 1991. Multilevel modelling of survey data. *Statistician* 40: 235–244.

———. 2011. *Multilevel Statistical Models.* 4th ed. Chichester, UK: Wiley.

Goldstein, H., P. Huiqi, T. Rath, and N. Hill. 2000. *The Use of Value Added Information in Judging School Performance.* London: Institute of Education.

Goldstein, H., J. Rasbash, M. Yang, G. Woodhouse, H. Pan, D. L. Nuttall, and S. Thomas. 1993. A multilevel analysis of school examination results. *Oxford Review of Education* 19: 425–433.

Greene, W. H. 2012. *Econometric Analysis.* 7th ed. Upper Saddle River, NJ: Prentice Hall.

Greenland, S., J. J. Schlesselman, and M. H. Criqui. 1986. The fallacy of employing standardized regression coefficients and correlations as measures of effect. *American Journal of Epidemiology* 123: 203–208.

Gregoire, A. J. P., R. Kumar, B. S. Everitt, A. F. Henderson, and J. W. W. Studd. 1996. Transdermal oestrogen for the treatment of severe postnatal depression. *Lancet* 347: 930–933.

Griliches, Z., and J. A. Hausman. 1986. Errors in variables in panel data. *Journal of Econometrics* 31: 93–118.

Grunfeld, Y. 1958. The determinants of corporate investment. PhD diss., University of Chicago.

Gutierrez, R. G. 2011. xtmixed_corr: Stata module to compute model-implied intracluster correlations after xtmixed. Boston College Department of Economics, Statistical Software Components S457297. http://ideas.repec.org/c/boc/bocode/s457297.html.

Hall, S., R. I. Prescott, R. J. Hallman, S. Dixon, R. E. Harvey, and S. G. Ball. 1991. A comparative study of Carvedilol, slow-release Nifedipine, and Atenolol in the management of essential hypertension. *Journal of Pharmacology* 18: S35–S38.

Hand, D. J., F. Daly, A. D. Lunn, K. J. McConway, and E. Ostrowski. 1994. *A Handbook of Small Data Sets.* London: Chapman & Hall.

Hausman, J. A. 1978. Specification tests in econometrics. *Econometrica* 46: 1251–1271.

Hausman, J. A., and W. E. Taylor. 1981. Panel data and unobservable individual effects. *Econometrica* 49: 1377–1398.

Hayes, R. J., and L. H. Moulton. 2009. *Cluster Randomised Trials.* Boca Raton, FL: Chapman & Hall/CRC.

Hedeker, D., and R. D. Gibbons. 2006. *Longitudinal Data Analysis*. Hoboken, NJ: Wiley.

Hedeker, D., R. D. Gibbons, M. du Toit, and Y. Cheng. 2008. *SuperMix: Mixed Effects Models*. Lincolnwood, IL: Scientific Software International.

Heeringa, S. G., B. T. West, and P. A. Berglund. 2010. *Applied Survey Data Analysis*. Boca Raton, FL: Chapman & Hall/CRC.

Hill, H. C., B. Rowan, and D. L. Ball. 2005. Effects of teachers' mathematical knowledge for teaching on student achievement. *American Educational Research Journal* 42: 371–406.

Hox, J. J. 2010. *Multilevel Analysis: Techniques and Applications*. 2nd ed. New York: Routledge.

Hsiao, C. 2003. *Analysis of Panel Data*. 2nd ed. Cambridge: Cambridge University Press.

Johnson, V. E., and J. H. Albert. 1999. *Ordinal Data Modeling*. New York: Springer.

Kohler, U., and F. Kreuter. 2009. *Data Analysis Using Stata*. 2nd ed. College Station, TX: Stata Press.

Kontopantelis, E., and D. Reeves. 2010. metaan: Random-effects meta-analysis. *Stata Journal* 10: 395–407.

Kreft, I., and J. de Leeuw. 1998. *Introducing Multilevel Modeling*. London: Sage.

Krueger, A. B. 1999. Experimental estimates of education production functions. *Quarterly Journal of Economics* 114: 497–532.

Lawson, A. B., W. J. Browne, and C. L. Vidal Rodeiro. 2003. *Disease Mapping with WinBUGS and MLwiN*. New York: Wiley.

Lillard, L. A., and C. W. A. Panis, ed. 2003. *aML User's Guide and Reference Manual*. Los Angeles, CA: EconWare.

Littell, R. C., G. A. Milliken, W. W. Stroup, R. D. Wolfinger, and O. Schabenberger. 2006. *SAS for Mixed Models*. 2nd ed. Cary, NC: SAS Institute.

Lloyd, T., M. B. Andon, N. Rollings, J. K. Martel, J. R. Landis, L. M. Demers, D. F. Eggli, K. Kieselhorst, and H. E. Kulin. 1993. Calcium supplementation and bone mineral density in adolescent girls. *Journal of the American Medical Association* 270: 841–844.

MacDonald, A. M. 1996. An epidemiological and quantitative genetic study of obsessionality. PhD diss., Institute of Psychiatry, University of London.

Macnab, A. J., M. Levine, N. Glick, N. Phillips, L. Susak, and M. Elliott. 1994. The Vancouver sedative recovery scale for children: Validation and reliability of scoring based on videotaped instruction. *Canadian Journal of Anesthesia* 41: 913–918.

Magnus, P., H. K. Gjessing, A. Skrondal, and R. Skjærven. 2001. Paternal contribution to birth weight. *Journal of Epidemiology and Community Health* 55: 873–877.

Morgan, S. L., and C. Winship. 2007. *Counterfactuals and Causal Inference: Methods and Principles for Social Research*. Cambridge: Cambridge University Press.

Mosteller, F. 1995. The Tennessee study of class size in the early school grades. *Future of Children* 5: 113–127.

Mundlak, Y. 1978. On the pooling of time series and cross section data. *Econometrica* 46: 69–85.

Munnell, A. H. 1990. Why has productivity growth declined? Productivity and public investment. *New England Economic Review* January/February: 3–22.

Naumova, E. N., A. Must, and N. M. Laird. 2001. Tutorial in Biostatistics: Evaluating the impact of 'critical periods' in longitudinal studies of growth using piecewise mixed effects models. *International Journal of Epidemiology* 30: 1332–1341.

Neuhaus, J. M., and J. D. Kalbfleisch. 1998. Between- and within-cluster covariate effects in the analysis of clustered data. *Biometrics* 54: 638–645.

Nuttall, D. L., H. Goldstein, R. Prosser, and J. Rasbash. 1989. Differential school effectiveness. *International Journal of Educational Research* 13: 769–776.

O'Connell, A. A., and D. B. McCoach, ed. 2008. *Multilevel Modeling of Educational Data*. Charlotte, NC: Information Age Publishing.

Palta, M., and C. Seplaki. 2002. Causes, problems and benefits of different between and within effects in the analysis of clustered data. *Health Services and Outcomes Research Methodology* 3: 177–193.

Pan, W. 2002. A note on the use of marginal likelihood and conditional likelihood in analyzing clustered data. *American Statistician* 56: 171–174.

Papke, L. E. 1994. Tax policy and urban development: Evidence from the Indiana enterprise zone program. *Journal of Public Economics* 54: 37–49.

Parks, R. W. 1967. Efficient estimation of a system of regression equations when disturbances are both serially and contemporaneously correlated. *Journal of the American Statistical Association* 62: 500–509.

Paterson, L. 1991. Socio-economic status and educational attainment: A multi-dimensional and multi-level study. *Evaluation & Research in Education* 5: 97–121.

Phillips, S. M., L. G. Bandini, D. V. Compton, E. N. Naumova, and A. Must. 2003. A longitudinal comparison of body composition by total body water and bioelectrical impedance in adolescent girls. *Journal of Nutrition* 133: 1419–1425.

Potthoff, R. F., and S. N. Roy. 1964. A generalized multivariate analysis of variance model useful especially for growth curve problems. *Biometrika* 51: 313–326.

Prosser, R., J. Rasbash, and H. Goldstein. 1991. *ML3 Software for 3-level Analysis: User's Guide for V. 2*. London: Institute of Education, University of London.

Rabe-Hesketh, S., and B. S. Everitt. 2007. *A Handbook of Statistical Analyses Using Stata*. 4th ed. Boca Raton, FL: Chapman & Hall/CRC.

Rabe-Hesketh, S., A. Skrondal, and H. K. Gjessing. 2008. Biometrical modeling of twin and family data using standard mixed model software. *Biometrics* 64: 280–288.

Rao, J. N. K. 2003. *Small Area Estimation*. Hoboken, NJ: Wiley.

Rasbash, J. 2005. Cross-classified and multiple membership models. In *Encyclopedia of Statistics in Behavioral Science*, ed. B. S. Everitt and D. Howell, 441–450. London: Wiley.

Rasbash, J., and W. J. Browne. 2001. Modelling non-hierarchical structures. In *Multilevel Modelling of Health Statistics*, ed. A. H. Leyland and H. Goldstein, 93–105. Chichester, UK: Wiley.

Rasbash, J., F. A. Steele, W. J. Browne, and H. Goldstein. 2009. *A User's Guide to MLwiN Version 2.10*. Bristol: Centre for Multilevel Modelling, University of Bristol. http://www.bristol.ac.uk/cmm/software/mlwin/download/manual-print.pdf.

Raudenbush, S. W. 1984. Magnitude of teacher expectancy effects on pupil IQ as a function of the credibility of expectancy induction: A synthesis of findings from 18 experiments. *Journal of Educational Psychology* 76: 85–97.

———. 1989. The analysis of longitudinal, multilevel data. *International Journal of Educational Research* 13: 721–740.

———. 1993. A crossed random effects model for unbalanced data with applications in cross-sectional and longitudinal research. *Journal of Educational Statistics* 18: 321–349.

Raudenbush, S. W., and A. S. Bryk. 2002. *Hierarchical Linear Models: Applications and Data Analysis Methods*. 2nd ed. Thousand Oaks, CA: Sage.

Raudenbush, S. W., A. S. Bryk, Y. F. Cheong, and R. Congdon. 2004. *HLM 6: Hierarchical Linear and Nonlinear Modeling*. Lincolnwood, IL: Scientific Software International.

Sham, P. 1998. *Statistics in Human Genetics*. London: Wiley.

Shavelson, R. J., and N. M. Webb. 1991. *Generalizability Theory: A Primer.* Newbury Park, CA: Sage.

Singer, J. D., and J. B. Willett. 2003. *Applied Longitudinal Data Analysis: Modeling Change and Event Occurrence.* Oxford: Oxford University Press.

———. 2005. Growth curve modeling. In *Encyclopedia of Statistics in Behavioral Science,* ed. B. S. Everitt and D. Howell, 772–779. London: Wiley.

Skrondal, A., and S. Rabe-Hesketh. 2004a. *Generalized Latent Variable Modeling: Multilevel, Longitudinal, and Structural Equation Models.* Boca Raton, FL: Chapman & Hall/CRC.

———. 2004b. *Generalized Latent Variable Modeling: Multilevel, Longitudinal, and Structural Equation Models.* Boca Raton, FL: Chapman & Hall/CRC.

———. 2009. Prediction in multilevel generalized linear models. *Journal of the Royal Statistical Society, Series A* 172: 659–687.

Skrondal, A., and S. Rabe-Hesketh, ed. 2010. *Multilevel Modelling, Vol. I—Linear Multilevel Models: Model Formulation and Interpretation.* London: Sage.

Snijders, T. A. B. 2004. Multilevel analysis. In Vol. 2 of *The SAGE Encyclopedia of Social Science Research Methods,* ed. M. S. Lewis-Beck, A. E. Bryman, and T. F. Liao, 673–677. London: Sage.

———. 2005. Power and sample size in multilevel linear models. In *Encyclopedia of Statistics in Behavioral Science,* ed. B. S. Everitt and D. R. Howell, 1570–1573. London: Wiley.

Snijders, T. A. B., and R. J. Bosker. 2012. *Multilevel Analysis: An Introduction to Basic and Advanced Multilevel Modeling.* 2nd ed. London: Sage.

Spiegelhalter, D., A. Thomas, N. Best, and W. Gilks. 1996a. *BUGS 0.5 Examples, Volume 1.* Cambridge: MRC Biostatistics Unit.

———. 1996b. *BUGS 0.5 Examples, Volume 2.* Cambridge: MRC Biostatistics Unit.

StataCorp. 2011. *Stata Longitudinal-Data/Panel-Data Reference Manual, Release 12.* College Station, TX: Stata Press.

Steenbergen, M. R., and B. S. Jones. 2002. Modeling multilevel data structures. *American Journal of Political Science* 46: 218–237.

Stock, J. H., and M. W. Watson. 2011. *Introduction to Econometrics.* 3rd ed. Englewood Cliffs, NJ: Prentice Hall.

Streiner, D. L., and G. R. Norman. 2008. *Health Measurement Scales: A Practical Guide to Their Development and Use.* 4th ed. Oxford: Oxford University Press.

Swaminathan, H., and H. J. Rogers. 2008. Estimation procedures for hierarchical linear models. In *Multilevel Modeling of Educational Data*, ed. A. A. O'Connell and D. B. McCoach, 469–520. Charlotte, NC: Information Age Publishing.

Therneau, T. M., and P. M. Grambsch. 2000. *Modeling Survival Data: Extending the Cox Model*. New York: Springer.

Train, K. E. 2009. *Discrete Choice Methods with Simulation*. 2nd ed. Cambridge: Cambridge University Press.

Vella, F., and M. Verbeek. 1998. Whose wages do unions raise? A dynamic model of unionism and wage rate determination for young men. *Journal of Applied Econometrics* 13: 163–183.

Verbeke, G., and G. Molenberghs. 2000. *Linear Mixed Models for Longitudinal Data*. New York: Springer.

―――. 2003. The use of score tests for inference on variance components. *Biometrics* 59: 254–262.

Vermunt, J. K., and J. Magidson. 2005. *Technical Guide for Latent GOLD 4.0: Basic and Advanced*. Belmont, MA: Statistical Innovations.

Vittinghoff, E., S. C. Shiboski, D. V. Glidden, and C. E. McCulloch. 2005. *Regression Methods in Biostatistics: Linear, Logistic, Survival, and Repeated Measures Models*. New York: Springer.

Vonesh, E. F., and V. M. Chinchilli. 1997. *Linear and Nonlinear Models for the Analysis of Repeated Measurements*. New York: Marcel Dekker.

Weiss, R. E. 2005. *Modeling Longitudinal Data*. New York: Springer.

West, B. T., K. B. Welch, and A. T. Galecki. 2007. *Linear Mixed Models: A Practical Guide Using Statistical Software*. Boca Raton, FL: Chapman & Hall/CRC.

Whaley, S. E., M. Sigman, C. Neumann, N. Bwibo, D. Guthrie, R. E. Weiss, S. Alber, and S. P. Murphy. 2003. The impact of dietary intervention on the cognitive development of Kenyan school children. *Journal of Nutrition* 133: 3965S–3971S.

Wight, D., G. M. Raab, M. Henderson, C. Abraham, K. Buston, G. Hart, and S. Scott. 2002. Limits of teacher delivered sex education: Interim behavioural outcomes from randomised trial. *British Medical Journal* 324: 1430–1433.

Willett, J. B., J. D. Singer, and N. C. Martin. 1998. The design and analysis of longitudinal studies of development and psychopathology in context: Statistical models and methodological recommendations. *Development and Psychopathology* 10: 395–426.

Willms, J. D. 1986. Social class segregation and its relationship to pupils' examination results in Scotland. *American Sociological Review* 51: 224–241.

Wooldridge, J. M. 2009. *Introductory Econometrics: A Modern Approach.* 4th ed. Cincinnati, OH: South-Western.

———. 2010. *Econometric Analysis of Cross Section and Panel Data.* 2nd ed. Cambridge, MA: MIT Press.

Ziliak, J. P. 1997. Efficient estimation with panel data when instruments are predetermined: An empirical comparison of moment-condition estimators. *Journal of Business and Economic Statistics* 15: 419–431.

Author index

Subject index